D0937278

GLOBAL CLINICAL TRIALS

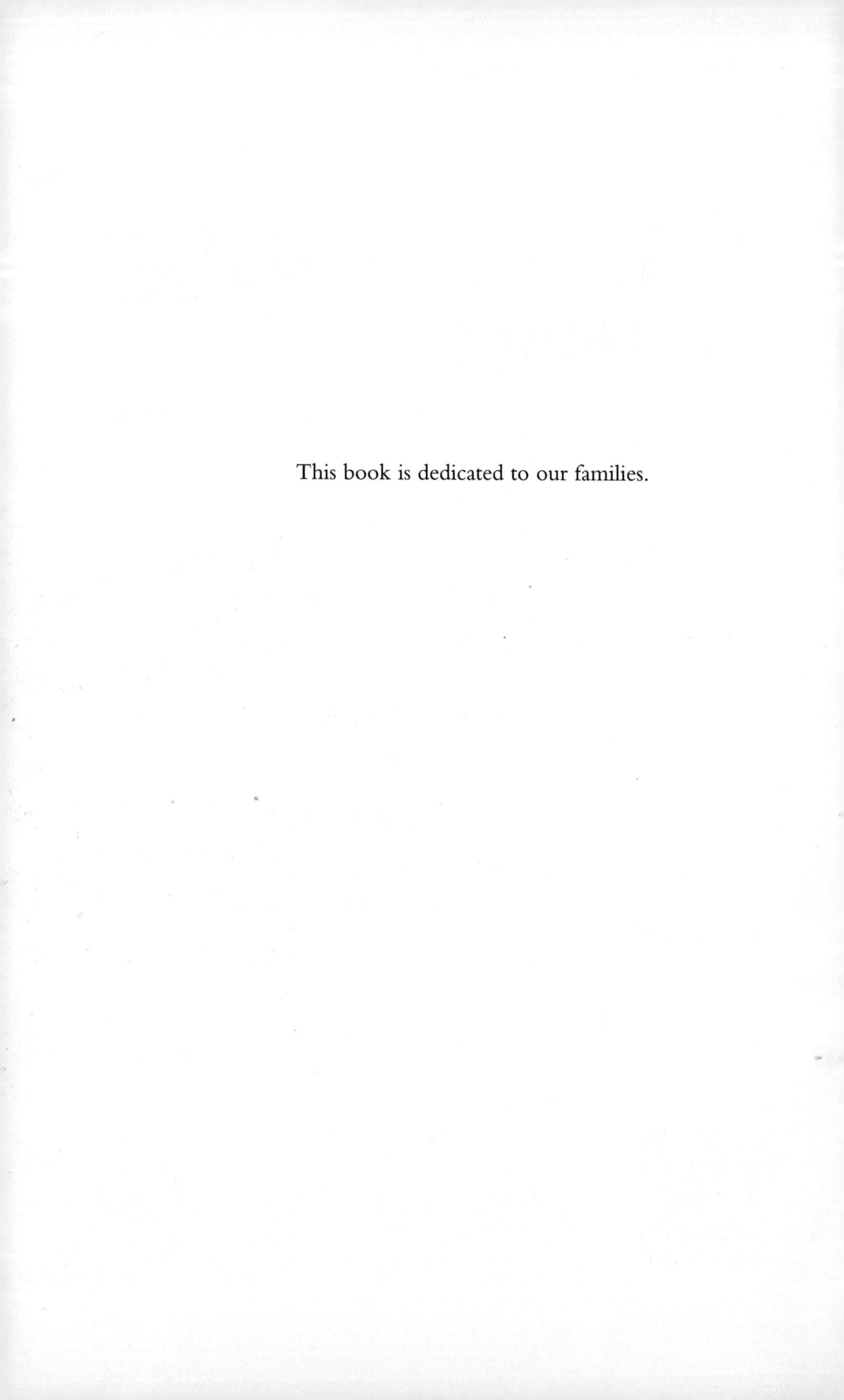

This book is dedicated to our families.

GLOBAL CLINICAL TRIALS

Effective Implementation and Management

RICHARD CHIN, M.D.
CEO, Institute for OneWorld Health,
South San Francisco, CA, USA.

MENGHIS BAIRU, M.D.
Executive Vice President & General Manager,
Elan Biopharmaceuticals,
South San Francisco, CA, USA

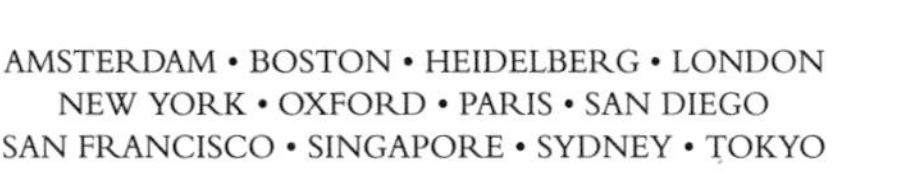
AMSTERDAM • BOSTON • HEIDELBERG • LONDON
NEW YORK • OXFORD • PARIS • SAN DIEGO
SAN FRANCISCO • SINGAPORE • SYDNEY • TOKYO

Academic Press is an imprint of Elsevier

Academic Press is an imprint of Elsevier
32 Jamestown Road, London NW1 7BY, UK
225 Wyman Street, Waltham, MA 02451, USA
525 B Street, Suite 1800, San Diego, CA 92101-4495, USA

First edition 2011

British Library Cataloguing-in-Publication Data
A catalogue record for this book is available from the British Library

Library of Congress Cataloging-in-Publication Data
A catalog record for this book is available from the Library of Congress

ISBN : 978-0-12-381537-8

For information on all Academic Press publications
visit our website at www.elsevierdirect.com

Typeset by TNQ Books and Journals Pvt. Ltd.

Printed and bound in United States of America

11 12 13 14 10 9 8 7 6 5 4 3 2

CONTENTS

ABOUT THE EDITORS

Richard Chin, M.D. is the CEO of Institute for OneWorld Health, the first U.S. nonprofit pharmaceutical company. OneWorld Health, mostly funded by the Bill and Melinda Gates Foundation, develops affordable drugs for neglected patients in the developing world. Dr. Chin has extensive expertise in drug development, including over 45 INDs and 10 drug registrations. Some of the drugs he has overseen include Rituxan, Lucentis, Tysabri, TNKase, Raptiva, Xolair, Cathflo, Prialt, Protropin, Nutropin, Pulmozyme, Azactam, Maxipime, and Bapineuzuma, among others. His previous roles include CEO of a NASDAQ-listed company, Senior Vice President of Global Development at Elan, and Head of Clinical Research for the Biotherapeutics Unit at Genentech. He was named by Businessweek in 2006 as one of the youngest 99 public company CEOs in the United States. Dr. Chin earned an M.D. from Harvard and the equivalent of a J.D. from Oxford, where he studied as a Rhodes Scholar. His previous textbook, Principles and Practice of Clinical Trial Medicine was also published by Elsevier. Dr. Chin serves as Associate Professor at UCSF School of Medicine and was previously on the adjunct faculty at Stanford University School of Medicine. Dr. Chin may be contacted at richardchin@clinicaltrialist.com.

Menghis Bairu, physician, scientist and Sr executive business professional—has more than two decades of international experience in the pharmaceutical/healthcare industry and in working to leverage that experience in the non-profit global health arena. His healthcare expertise spans research and development, commercial, managed care, public health and access sectors of the biopharmaceutical industry, with particular emphasis on the clinical development of important, new therapies. He is currently Executive Vice President and General Manager for Elan Biopharmaceuticals, an internationally renowned neuroscience/biotech company. He was previously the Head of Global Development and Chief Medical Officer for Elan where, on a global basis, Dr. Bairu was responsible for Clinical Development, Biometrics, Regulatory, CMC, Quality/Compliance, Safety and Risk Management, Clinical Operations, and Medical Affairs.

Prior to joining Elan, Dr. Bairu worked for more than five years at Genentech, a period wherein his clinical, commercial and managed care experience and responsibilities increased exponentially, culminating in his medical marketing role with Rituxan, Genentech's first multibillion dollar product. Dr. Bairu worked as Medical Consulting Director for Fremont Health Corporation/Industrial Indemnity before joining Genentech. He received his undergraduate degree in Business Administration from Instituto VII Tecnico Commerciale in Milan, Italy, attended John Hopkins University (Public Health) and the University of Milan, faculty of medicine & surgery in Milan-Italy where he received his Medical Degree.

Complementing the corporate and philanthropic elements of healthcare advances has always been a focus for Dr. Bairu. In 1999, he co-founded International Medical Foundation, Inc., a 501(c) 3 foundation whose mission is to acquire medical equipment and materials for donation to non-profit institutions and government health care entities in developing and emerging countries.

FOREWORD

Drugs, biologics and medical devices were once local affairs although it wasn't long before these products were shipped from country to country. The consumer protections provided in laws passed in many countries starting around 1900 focused on products in the marketplace which were dangerous or misrepresented and if there were proactive interactions between government regulators and manufacturers it involved inspections and bench testing of vaccines and serum therapies. In 1962 in the United States, in response to the harm caused by thalidomide, sweeping regulatory changes were introduced, the most important being a section of the law which required that prior to marketing a new drug efficacy must be determined by 'adequate and well controlled studies' in humans. While human drug clinical trials had their beginnings in trials of drugs for tuberculosis, high blood pressure and diabetes, the new law made such studies mandatory for all drugs in the US, except those on the market since before 1938. The law also added post-marketing safety reporting requirements, and over the years, along with clinical trials became an international endeavor.

The United States, Japan and many other countries continued to have a strong preference for clinical trials done within their own borders. But as international harmonization efforts and changes in local regulations evolved, medical product development became an international enterprise. Concerns arose that the ethnic and genetic differences between populations would make reliance on foreign trials uncertain. But, in fact, the real challenge is not the genetic differences in drug metabolism, but the wide differences in health care delivery and the practice of medicine from country to country. There are variations in the local practices with respect to informed consent, differences of opinion on the ethical use of placeboes. Logistical challenges persist for the regulators, who have the authority of law at home, but are visiting officials not necessarily able to read or speak the local language when they inspect a foreign trial or manufacturing site. The management of a trial with oversight by multiple in-country regulators can be challenging. But the challenges notwithstanding, the manufacturing, the preclinical investigations, the clinical trials for registration, the post-marketing study requirements, and the safety surveillance is now a global effort, and a better one for it.

This first textbook on global clinical trials that explores the principles and practice of global clinical trials is important and welcome. Just as a multi-center trial differs from a single site trial, global clinical trials differ from a domestic one in many ways, including regulations, ethical issues, and logistics. It is important for those conducting international trials to fully understand the country and region specific differences and requirements. Without a good understanding, many issues including logistic challenges, scientific validity, and privacy violations can ensue. In this book, the authors drawn from a broad range of industry and academia address the critical issues often seen in global clinical trials.

The scope is comprehensive, covering countries from virtually every continent. In each chapter, the authors examine the country and region specific considerations for conducting trials. This book is a landmark publication, and it is likely to stand as the standard reference book for global clinical trials for many years to come.

David Feigel,
former Head of CDRH at FDA

CONTRIBUTORS

Sebastian Antonelli MD
Medical and Scientific Services, Quintiles Latin America, Buenos Aires, Argentina

Jennifer Aquino
Customer and Project Management, Cenduit, Raleigh, NC, USA

Irina Baeumer MD PhD
Medical and Safety Services, ICON Clinical Research, Germany

Sue Bailey
Global Sales Operations, Quintiles Sub-Saharan Africa, Irene, South Africa

Menghis Bairu MD
Elan Biopharmaceuticals, South San Francisco, CA, USA

Alin Balalau MD
ICON Clinical Research, Romania

Mary Bareilles MD
ICON Clinical Research, Japan

Havanakwavo Chikoto PhD
Triclinium Clinical Trial Project Management, Sandown, South Africa

Edwin Chia
Project Services, Quintiles Asia, Singapore

Richard Chin MD
Institute for OneWorld Health, South San Francisco, CA, USA

Mateusz Chrząszcz MD
Site Start-Up, Quintiles, Warsaw, Poland

Marianne Coetzee MSc Pharm
Triclinium Clinical Trial Project Management, Sandown, South Africa

Gillian Corken
Quintiles Sub-Saharan Africa, Irene, South Africa

Romillie E. Cruz MD
Medical and Safety Services, ICON Clinical Research, Singapore

Victoria Datsenko MD PhD
ICON Clinical Research, Ukraine

Jogin Desai MD
Cenduit, Raleigh, NC, USA

James (Dachao) Fan MD
Medical and Safety Services, ICON Clinical Research, Singapore

Carlos Fernando de Oliveira MD PhD
Latin America Pharmacovigilance, PPD, Morrisville, NC, USA

Janos Filakovszky MD
EEME, Quintiles, Vienna, Austria

Suzanne Gagnon MD
Medical, Safety and Scientific Services, ICON Clinical Research, USA

Edda Gomez-Panzani MD
Clinical Sciences – Neuroendocrinology, Ipsen US, Brisbane, CA, USA

Antonio Guimaraes MA
Human Resources Latin America, PPD, Wilmington, NC, USA

Rajesh Jain MBBS, CCRA, PGDGM
ECCRO, Mumbai, India

Jay Johnson BS MA
Corporate Communication, Quintiles, Durham, NC, USA

Lynn Katsoulis PhD
Triclinium Clinical Trial Project Management, Sandown, South Africa

Masanobu Kimura
ICON Clinical Research, Japan

Pavel Lebeslé
ICON Clinical Research, Czech Republic

Barbara Lilienfeld
International Corporate Events Network, Johannesburg, South Africa

Monica Lizano MD MBA
Site Management, Quintiles Latin America, San José, Costa Rica

Anna Paula Más RPh
Regulatory Affairs Country Management Latin America, PPD, São Paulo, Brazil

Mohua Maulik PhD
Clinical Trials Registry – India, National Institute of Medical Statistics, Indian Council of Medical Research, New Delhi, India

Vladimir Misik PhD
EEME, Quintiles, Vienna, Austria

Hiroshi Naito
ICON Clinical Research, Japan

Alan Ong MBA
Quintiles Global Central Laboratories Asia, Singapore

Mariano Parma MBA
Strategic Development Latin America, PPD, Buenos Aires, Argentina

Viola-Marie Raubenheimer
Quintiles Africa, Pretoria, South Africa

Svetlana Riekstina MD
ICON Clinical Research, Latvia and Estonia

Mira Serhal BS, MPH
American University of Beirut, Beirut, Lebanon

S.D. Seth MBBS, MD
Clinical Trials Registry – India, National Institute of Medical Statistics, Indian Council of Medical Research, New Delhi, India

Surinder Singh MD
Central Drugs Standard Control Organization, Ministry of Health and Family Welfare, Government of India, New Delhi, India

Tracy Southwood Dip Pharm
Triclinium Clinical Trial Project Management, Sandown, South Africa

Margaret Ann Snowden MPH
Biostatistics and Data Management, Aeras, Rockville, MD, USA

Monika Stepniewska DVM
ICON Clinical Research, Poland

Victor Strugo BPharm Hons
Triclinium Clinical Trial Project Management, Sandown, South Africa

Veronica Suarez
Quintiles Latin America, Buenos Aires, Argentina

Krisztina Szabo MD
ICON Clinical Research, Hungary

J. Rick Turner PhD
Cardiac Safety Services, Quintiles, Raleigh, NC, USA

Henrietta Ukwu MD FACP FRAPS
Global Regulatory Affairs, PPD, Blue Bell, PA, USA

Sorika van Niekerk MS
Clinical Operations, Quintiles Sub-Saharan Africa, Irene, South Africa

Nermeen Varawalla MD PhD MBA
ECCRO, London, UK

Daniel Vazquez
Quintiles Latin America, Buenos Aires, Argentina

Johan Venter BPharm, MSc Pharm, PhD
Business Development, Quintiles Sub-Saharan Africa, Irene, South Africa

Elizabeth Villeponteaux MA MSLS ELS
Global Medical Writing, PPD, Morrisville, NC, USA

Greg Voinov MD
Clinical Operations, France

Hiromi Wakita PhD
ICON Clinical Research, Japan

Janice B. Wilson PhD
Wilson Quality & Compliance Consulting, Antioch, CA, USA

Yaw Asare-Aboagye
Clinical Trial Services, United Therapeutics, Durnham, NC, USA

Jenny Zhang MD MHA
Tigermed Consulting, Shanghai, PR China

Yuriy Zuykov MD
ICON Clinical Research, Russia

SECTION 1

Overview

Background

Richard Chin
Institute for OneWorld Health, South San Francisco, CA, USA

Contents

1.1. INTRODUCTION

Not since almost a century and a half ago, when the modern economy was created by a confluence of new inventions – telegraph, railroad, refined oil, and Bessemer steel – have we seen the wholesale transformation of entire industries that we are seeing today. Propelled by the Internet, microprocessors, mobile phones, and standardized shipping containers, the world has over the last decade or so been fast-forwarding into globalization with a rapidity and scale that is hardly imaginable.

As with other industries such as information technology and traditional manufacturing, drug development has become highly globalized. Globalization brings both significant benefits and potential difficulties. Done correctly, global clinical trials lower the cost of drugs for everyone, distribute modern science and medicine broadly, and enhance the health of patients around the world. However, the endeavor of drug development is complicated and risky, and it is extremely important to conduct trials

Global Clinical Trials
ISBN 978-0-12-381537-8, Doi:10.1016/B978-0-12-381537-8.10001-9

correctly. The goal of this book is to help physicians and clinical trial professionals everywhere to achieve that objective.

1.2. GROWTH OF GLOBAL CLINICAL TRIALS

The number of international trials has been growing rapidly. According to a report from the Office of the Inspector General (OIG), as of 2008, 80 percent of marketing applications for drugs and biologicals approved by the US Food and Drug Administration (FDA) contained data from US clinical trials conducted outside the USA [1]. Table 1.1 shows the breakdown from the report [1]. According to the same source, over half of the sites and subjects in these applications were from non-US sources.

Table 1.1 Food and Drug Administration (FDA) marketing approval for drugs and biologicals containing clinical data approved in fiscal year 2008

Marketing applications	Drugs	Biologicals	Drugs and biologicals
Applications with only US data	15	9	24
Applications with non-US and US data	82	5	87
Applications with only non-US data	9	1	10
Totals	106	15	121

These numbers are based on 121 applications with sufficient information to determine whether the data were non-US or US.
Source: OIG analysis of FDA marketing applications approved in fiscal year 2008.

As can be seen in Table 1.2, 57 percent of patients in new drug applications (NDAs) and 87 percent of patients in biologicals license applications (BLAs) were from non-US sites. The largest proportion of non-US patients came from Western Europe, as of 2008 approvals (Figure 1.1) [1]. Figure 1.1

Table 1.2 Number and percentage of non-US subjects and sites from clinical trials supporting drug- and biological-marketing applications approved in fiscal year 2008

	Drugs	Biologicals	Drugs and biologicals
No. of non-US and US subjects	92.859	206,842	299,701
No. of non-US subjects	52.820	179,712	232,532
Percentage of non-US subjects	56.9%	86.9%	77.6%
No. of non-US and US trial sites	11,227	717	11,944
No. of non-US trial sites	6,129	356	6,485
Percentage of non-US trial sites	54.6%	49.7%	54.3%

These numbers are based on data from 193 clinical trials with complete subject and site information.
Source: OIG analysis of FDA marketing applications approved in fiscal year 2008.

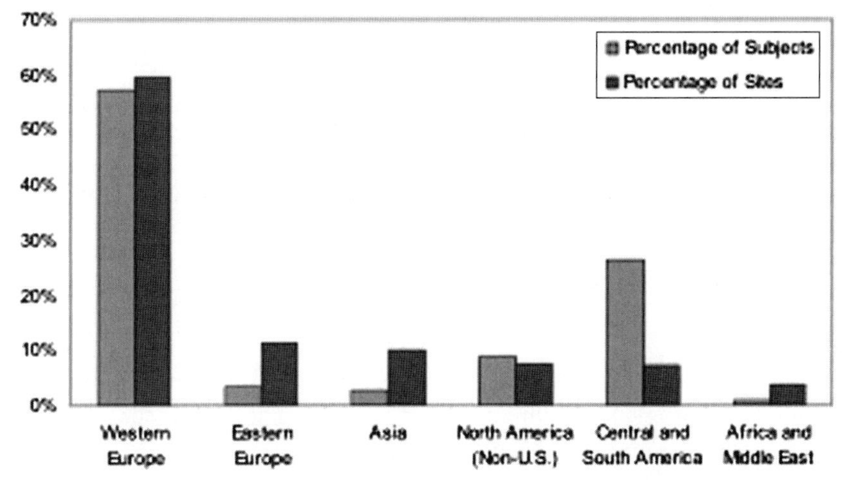

Figure 1.1 Percentage of non-US clinical trial subjects and sites by region for FDA marketing applications approved in fiscal year 2008. These numbers are based on data from 193 clinical trials with complete subject and site information. *Source:* OIG analysis of FDA marketing applications approved in fiscal year 2008.

is based on patients in approved applications, and therefore reflects the proportion of patients enrolled into trials over the past 10 years or so.

The proportion of non-US patients is growing rapidly (Figure 1.2) [1]. This trend is not confined to lower quality or less prominent clinical trials. An analysis by Glickman and colleagues recently found that between 1995

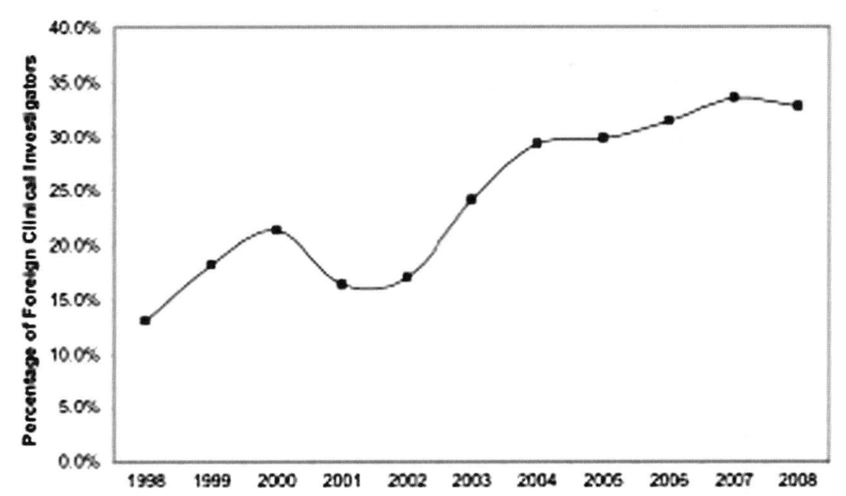

Figure 1.2 Trend in non-US clinical investigators as a percentage of all clinical investigators identified in investigational new drug (IND) applications from 1998 to 2008.

and 2005, the number of countries represented in a sample of studies published in the New England Journal of Medicine doubled from 33/150 to 70/150 [2].

These figures are largely consistent with the numbers calculated by Tufts Center for Drug Development, which estimates that in 2007, 57 percent of the patients in clinical trials came from the USA, 14 percent from Western Europe, and 29 percent from rest of the world.

1.3. DRIVERS OF GLOBALIZATION

The trend toward globalization has been fueled by several factors. The rapidly increasing cost of clinical trials – increasing at 20 percent per year – has been driving much of the effort. Conducting clinical trials in developing countries can easily reduce costs by 50 percent or more, and in some cases can reduce costs by up to 90 percent.

More importantly, recruitment of patients tends to be much faster in developing countries. In many cases, recruitment can be increased by 100 percent, and in some cases by 500 percent or more. In an industry such as pharmaceutical development, speed of enrollment can be a very strong driver for globalization.

In addition, in some diseases such as oncology and rheumatology, it can be difficult to recruit patients who are naïve to therapy into clinical trials in developed countries. In the USA and Europe, for example, most patients with severe rheumatoid arthritis may already have been exposed to biologicals such tumor necrosis factor (TNF) inhibitors. It can be almost impossible to enroll patients who are TNF inhibitor naïve into clinical trials in those countries. Developing countries can offer a larger pool of treatment-naïve patients.

In other cases, the diseases of interest only exist in those countries, or the prevalence is much higher in those countries. Certain forms of hepatitis are one example. Global health has become an important area of drug development, largely owing to the investment by the Bill and Melinda Gates Foundation. For diseases such as malaria and visceral leishmaniasis, trials must be conducted in developing countries because that is where the diseases exist.

More recently, quality has also become a driver in some instances. Many countries that were previously being used in clinical trials because of cost advantages have now evolved. Rather than being lower cost, lower quality regions, they have become more expensive and also higher quality. In many

Eastern European countries, for example, the clinical trial data from top clinical trial sites are often more reliable than in the USA.

1.4. ISSUES OF GLOBALIZATION

Of course, with globalization can come a variety of potential issues.

One of the first issues, which will be discussed in Chapter 2, is potential ethical issues. An ethical problem may arise if a drug is being tested in a population where people may never stand to benefit from their voluntary acts of charity, namely in enrolling in clinical trials. For example, because of this issue, the government of Thailand insisted that human immunodeficiency virus (HIV) vaccine studies in Thailand could only proceed if the vaccine would be made available to Thai patients upon successful development. Another ethical issue is the difficulty in obtaining true informed consent in some countries and populations. This will be discussed further.

A second issue is regulatory harmonization, or lack thereof.

Prior to the International Conference on Harmonisation (ICH), which brought the Global Cooperation Group (GCP) regulations in line across the USA, Europe, and Japan, the varied regulatory requirements for each country made international trials very onerous. The ICH was convened to harmonize regulations in those three regions, and involved the European Commission, European Federation of Pharmaceutical Industries Associations, Japanese Ministry of Heath and Welfare, Japanese Pharmaceutical Manufacturers' Association, Centers for Drug Evaluation and Research, Biological Evaluations and Research, and Pharmaceutical Research and Manufacturers of America. After ICH, the developed countries largely harmonized their regulatory requirements and international trials became a lot more practicable. Today, most countries' implementation of ICH is reasonably similar and the vast majority of regulations are the same. For example, reporting requirements for suspected unexpected serious adverse reactions (SUSARs) are fairly standard.

However, even in the USA and European Union (EU), some regulations have started to diverge. For example, the USA no longer follows the Declaration of Helsinki, because the declaration considers placebo-controlled trials to be unethical in cases where an active drug is available, and the USA disagrees. New privacy regulations, especially in the EU, financial disclosure requirements in the USA, and various other non-ICH requirements are also making the clinical trial landscape more complicated than before.

In developing countries, implementation of ICH guidelines may be varied, and there may be additional requirements. In many countries, for example, Phase I studies are much more difficult to perform than in developed countries.

A third potential issue is logistics. Getting a drug in and out of the countries, through customs, can sometimes be difficult. Getting blood and tissue samples across customs can be even harder.

1.5. CONTROL AND INSPECTIONS OF FOREIGN SITES

In general, most regulatory authorities do not extend jurisdiction over conduct of studies in other countries. That is, the Spanish regulatory agency will not care that patients are being dosed in a clinical trial in Japan. However, almost all regulatory authorities will require sponsors to keep them updated of new safety events that occur anywhere in the world. A few agencies, such as the FDA, will require investigators to follow US laws if they are conducting studies under US investigational new drug (IND) applications. Of course, it is up to the sponsor to decide whether to place foreign studies under the US IND or keep them separate.

Historically, FDA has inspected very few sites. The FDA inspected only 1 percent of non-US sites as of 2008 [1]. Other countries have had a varied record with respect to audits of foreign sites. With the recent heparin contamination issue, however, inspections of foreign sites are likely to increase. The FDA recently set up an office in China, for example. It is also engaged in working more with other countries' regulatory agencies to coordinate inspections.

1.6. ETHNIC AND GENETIC DIFFERENCES

Apart from the issues outlined above, international trials in general can raise the issue of ethnic, genetic, and other differences, as variability in these factors may make extrapolation of results from one geographical region to another problematic.

As a general rule, these differences have not historically resulted in changes to the risk/benefit ratio. There are very few drugs that have a favorable risk benefit in one patient's population and not in another. There are, however, some drugs where the dosing needs to be adjusted in different populations; for example, tissue plasminogen activator has a substantially lower dose in Japan than in the USA.

The differences between populations and geographies can be divided into intrinsic and extrinsic differences [3]. Extrinsic differences include environmental differences, such as climate and weather; cultural differences, such as languages and diet; medical treatment differences, such as treatment patterns and diagnosis; and population differences, such as prevalence of smoking. Intrinsic factors include genetic factors, such as polymorphisms in drug metabolism and genetic diseases; and physiological differences, such as size and cardiovascular function. While E5 classifies diseases as an intrinsic factor, in many cases it can be an extrinsic factor, such as prevalence of tuberculosis in the general population.

Whenever international trials are conducted, and even when an international trial is conducted, differences in these factors can play a role in the interpretability of the clinical trial. With a broader array of countries now being involved in clinical trials, these factors can sometimes be magnified.

1.7. USABILITY OF INTERNATIONAL CLINICAL DATA IN REGULATORY FILINGS

In global clinical trials, the goal is usually to file the data in another country for registration, typically in the USA and EU. Fortunately, ICH E5 (Ethnic factors in the acceptability of foreign clinical data) has been adopted by many countries, and helps significantly in standardizing the expectations of the regulators [3].

In the USA, 21 CFR 314.106 (The acceptance of foreign data in a new drug application) is the overarching regulation that governs what foreign data are acceptable. The code makes it clear that data from US patients are not needed if three criteria are met. First, the foreign data must be relevant and applicable to the US population. Second, investigators who are competent must have performed the studies. Third, the FDA must have confidence or have the ability to validate the verity of the data [4].

In the rest of the regulations, the FDA makes it clear that studies that are solely conducted outside the USA do no need to be performed under an IND. However, they must be conducted under all local regulations and ICH.

In the EU, the rules are largely similar, and foreign data are usually acceptable.

However, although regulations allow the approval of drugs with 100 percent data from patients outside the country, in many cases, regulatory authorities strongly prefer at least some of the data to be based on patients from within the country.

The principles outlined in ICH E5 are clear. Where possible, it is desirable to reduce duplication of data and it is beneficial to use foreign data where possible. The approval requirements for the drug remain the same, but the goal is to fit the foreign data into the framework of existing regulatory requirements.

To that end, the types of studies and data required by the local registration requirements must be fulfilled. For example, if a QTc study is required, that must be performed. Once a clinical package is assembled that meets the normal regulatory requirements, that package can be considered to be complete.

A complete package would contain [3]:

- clinical efficacy and safety data, and dose response, performed under GCP and utilizing approvable endpoints and appropriate controls; the medical diagnosis and definitions must be applicable in the new region
- pharmacokinetics/pharmacodynamics/dose response/safety/efficacy in the population studied
- pharmacokinetics and where possible pharmacodynamics/dose response in the new region.

Once a "complete" package has been assembled, there may or may not be a question as to whether the population in the studies is similar enough to the target population that the data can be extrapolated. If there is doubt, bridging studies should be done to demonstrate the appropriate extrapolation.

In order to decide whether the data can be extrapolated, the drug's sensitivity to ethnic factors needs to be determined. With some drugs in some populations, there will be little ethnic impact on safety and efficacy, while in others, there may be substantial effects. The degree of potential concern depends on both the drug and the population.

Drugs that are metabolized by enzymes that are heterogeneous across populations, such as G6PD, are likely to be more problematic. Drugs that are passively excreted without undergoing metabolism are less likely to be ethnically sensitive, as are drugs with a wide therapeutic margin or a flat dose–response curve. In some cases, it is known how ethnically sensitive drugs in the same class have been, and that can help to determine the likelihood that the new drugs will be ethnically sensitive.

The following lists summarize some of the factors that may make a drug more or less likely to be ethnically sensitive.

Properties of a compound making it less likely to be sensitive to ethnic factors [3]:

- linear pharmacokinetics (pharmacokinetics)
- a flat pharmacodynamic (effect–concentration) curve for both efficacy and safety in the range of the recommended dosage and dose regimen (this may mean that the medicine is well tolerated)
- a wide therapeutic dose range (again, possibly an indicator of good tolerability)
- minimal metabolism or metabolism distributed among multiple pathways
- high bioavailability, thus less susceptibility to dietary absorption effects
- low potential for protein binding
- little potential for drug–drug, drug–diet, and drug–disease interactions
- non-systemic mode of action
- little potential for inappropriate use.

Properties of a compound making it more likely to be sensitive to ethnic factors [3]:

- non-linear pharmacokinetics
- a steep pharmacodynamic curve for both efficacy and safety (a small change in dose results in a large change in effect) in the range of the recommended dosage and dose regimen
- a narrow therapeutic dose range
- highly metabolized, especially through a single pathway, thereby increasing the potential for drug–drug interaction
- metabolism by enzymes known to show genetic polymorphism
- administration as a prodrug, with the potential for ethnically variable enzymatic conversion
- high intersubject variation in bioavailability
- low bioavailability, thus more susceptible to dietary absorption effects
- high likelihood of use in a setting of multiple comedications
- high likelihood for inappropriate use, e.g. analgesics and tranquilizers.

Depending on the likelihood of ethnic sensitivity, more or fewer data may be required on the comparative pharmacokinetics, pharmacodynamics, and clinical relationship to dose. At a minimum, though, basic absorption, distribution, metabolism, and excretion (ADME) data are required. Some drug–drug interaction and food–drug interaction data may be required as well, particularly if certain drugs and food differ significantly between two populations.

In many cases, there is sufficient evidence about the pharmacokinetics or pharmacodynamics endpoint and their relationships to clinical data that demonstrating equivalence at the pharmacokinetics or pharmacodynamics level may be sufficient. Occasionally, there may be a need for additional clinical data.

1.8. WORLD HEALTH ORGANIZATION PREQUALIFICATION PROCESS

The World Health Organization (WHO) is not a regulatory agency, but for many countries that cannot afford a robust regulatory system or have not yet developed one, the WHO is a very important source of validation for drugs and vaccines. Because there are many drugs and vaccines that are inadequately tested, counterfeit, adulterated, or otherwise inappropriate for treatment of patients, many countries rely on the WHO's prequalification system to qualify the drugs and vaccines they purchase.

In addition to drugs and vaccines, the WHO qualifies or prequalifies contraceptives, diagnostics, and quality control laboratories.

1.8.1. History of Prequalification

In 2001, the WHO set up a drug prequalification system "to facilitate access to medicines that meet unified standards of quality, safety and efficacy for HIV/AIDS, malaria and tuberculosis", as well as for some additional diseases [5]. Through the process, the WHO determines whether a product "meets the requirements recommended by the WHO and are manufactured in compliance with current good manufacturing practices" [6]. The original intent was to help agencies such as UNAIDS, UNICEF, UNFPA, and the World Bank determine which drugs were of high and dependable quality. However, the system has now evolved to a point where multiple agencies, countries, and other groups rely on the certification process to ensure that the drugs they purchase are of high quality.

As of 2010, 42 proprietary and 61 generic medicines had undergone the prequalification process successfully. Many of the drugs are for diseases with an extremely high burden of illness in developing countries, such as drugs for HIV, malaria, and tuberculosis.

The stated criteria for assessment of the drugs are "the same as those used by the European Agency for the Evaluation of Medicinal Products (EMEA) and the USA Food and Drug Administration (FDA). In other words, the prequalification assessment team evaluates the required data, including in vivo bioequivalence tests carried out by the manufacturers" [5]. However, in practice, the ability of the WHO to conduct the assessment is limited by the resources and personnel available, which are substantially less than those available to the FDA and EMEA.

Although the WHO prequalification process is intended to be akin to harmonization systems such as the Pharmaceutical Inspection Convention and Pharmaceutical Inspection Co-operation Scheme (PIC/S) and ICH, in practice it has evolved into a pseudoregulatory approval system for those countries without a robust regulatory system.

1.8.2. Prequalification Process for Drugs

The prequalification process (PP) starts when a medicine is added to the prequalification program. Inclusion in the program is usually predicated on the medicine being on the WHO Essential Medicines list. Also, normally, for a drug to be considered for the program, the expectation is that a funding source is available to purchase it.

The first step in the process is to prepare the criteria. So, as a first step, a draft "general procedure for product prequalification" is prepared by the WHO, and the procedure is posted on the WHO website. It is also reviewed by experts. The procedure is then revised and submitted to the WHO Expert Committee for Specifications for Pharmaceutical Preparations. Once approved, the procedure is processed by the WHO Executive Board and World Health Assembly. Finally, it is published in Technical Report Series (TRS) and on the WHO website.

Once the procedure has been approved, the individual drugs can be prequalified. As a first step in the process, an invitation for expression of interest (EOI) is published by the WHO. The interested manufacturers then submit their dossiers. The dossiers are screened and assessed. This is followed by a site inspection. A report is prepared on findings and recommendations and the evaluation results are then published. There is continued sampling and periodic re-evaluation and reinspection. Finally, delisting removes the drug from the list. Throughout the process, a good deal of transparency exists; the updates on the process of the drug are often publicly available.

In general, the EOI is open to all interested manufacturers, although in some cases of high need, the WHO may specifically invite parties to submit a product.

The dossier is similar to a regulatory submission, in that it must include information about the product efficacy, safety, and manufacturing in the format specified by the WHO, along with an appropriate cover letter and product samples. For generic products, also known as multisource products, the focus of assessment is often on the manufacturing and product quality. The dossier must include:

- details of the product
- marketing authorization status
- active pharmaceutical ingredient (API) information:
  - properties of the API(s)
  - sites of manufacture
  - route of synthesis
  - specifications
  - stability testing
- finished product information:
  - formulation
  - sites of manufacture
  - manufacturing procedure
  - specifications for excipients
  - specifications for the finished pharmaceutical product (FPP)
  - container/closure system(s) and other packaging
  - stability testing
- product information:
  - summary of product characteristics
  - package leaflet
  - labeling
- summaries of:
  - quality
  - biopharmaceutics (pharmacokinetics/pharmacodynamics, interchangeability) [6].

For innovator medicines, which are defined as "Medicines that are approved or registered by countries with a stringent regulatory agency (EU, USA, Japan, and others)", and where patents are still in force and generic equivalents are not legally available, the process is somewhat different in that the focus is on review of existing approvals. The dossier must include:

- assessment report(s) issued by the respective national medicines regulatory authorities
- WHO-type batch certificate from the manufacturer
- WHO-type certificate of a pharmaceutical product, issued by one of the national medicines regulatory authorities of the ICH region and/or associated countries, together with the approved summary of product characteristics [6].

The submission guidelines can be found at the following websites: Guideline on submission of documentation for prequalification of multi-source (generic) finished pharmaceutical products (FPPs) used in the treatment of HIV/AIDS, malaria and tuberculosis at http://healthtech.who.int/pq/info_applicants/Guidelines/GuideGenericSubmitDocFPPs_08_2005_WoAnnexes.pdf and Guide on submission of documentation for prequalification of innovator finished pharmaceutical products (FPPs) used in the treatment of HIV/AIDS, malaria and tuberculosis and approved by drug regulatory authorities (DRAs) in the International Conference on Harmonization (ICH) region and associated countries, including the EU, Japan and USA at: http://healthtech.who.int/pq/info_applicants/Guidelines/GuideSubmittingDocFPPs_DRA _ICH_08_2005.pdf

After the dossiers have been screened for completeness, an assessment team is assembled. The WHO often relies on a team of experts drawn from national regulatory authorities. The team conducts review of the data, conducts site inspections, and compiles a report. The WHO then will determine whether the drug should be prequalified and the results will be communicated to the sponsor. The WHO will also publish the results as a WHO Public Assessment Report.

Following the process, the manufacturers need to keep the WHO informed of any changes or variance to the previous manufacturing process, and re-evaluations are performed at least at five-year intervals.

1.8.3. Prequalification Process for Vaccines

The prequalification for vaccines is older than the one for drugs. It was established in 1989 and is now fairly advanced. The purpose of the assessment is "to verify that the vaccines: (a) meet the specifications of the relevant UN agency; and (b) are produced and overseen in accordance with the principles and specifications recommended by the WHO, for good manufacturing practice (GMP), and for good clinical practice (GCP)" [7]. Countries and international agencies within the United Nations (UN) as well as outside such as Global Alliance for Vaccines and Immunization rely on the assessments by the WHO through this process as they decide where to procure their vaccines.

The WHO relies heavily on the assessment performed by the National Regulatory Authority (NRA) in the country of manufacture for the vaccine. It is important that the regulatory authority is strong and is able to assess the vaccine properly. The WHO invests in strengthening the regulatory system where necessary. In addition, the WHO performs quality assurance random sample checking to monitor the quality of vaccines and performs postmarketing surveillance to investigate adverse events.

The first step, therefore, in obtaining prequalification for vaccines, is to apply to the NRA for the country of manufacture. The sponsor must allow the authority to share data with the WHO. The NRA must have undergone an assessment by the WHO and been found to meet the requirements of the WHO. Once the marketing authorization from the NRA has been approved (except in special cases such as emergencies), the sponsor and the WHO should hold a pre-evaluation meeting.

A product summary file is then submitted, with the following content [7]:

- Chapter 1: General information
- Chapter 2: Personnel
- Chapter 3: Premises and equipment
- Chapter 4: Vaccine composition
- Chapter 5: Production
- Chapter 6: Quality control
- Chapter 7: Stability
- Chapter 8: Clinical experience
- Chapter 9: Production and distribution data
- Chapter 10: Update of Regulatory Authority actions relevant to the product.

The Common Technical Document format can be used instead, so long as appropriate cross-references to the WHO format are included. The packaging and labeling must meet the specific WHO requirements as outlined in the WHO International Guideline on Packaging and Shipping of Vaccine.

The WHO's target is to complete the initial review of the application within three months. If the application is adequate, the WHO will test samples of the vaccine and perform site visits.

Once the evaluation is complete, the WHO informs the sponsor, the NRA, and the UN agencies.

In emergency cases, a fast-track process expedites the process.

After the prequalification, reassessments are performed, usually every two years. The reassessment usually requires submission of information detailing changes to the product or manufacturing, retesting of samples, and a site visit, among other actions.

There is, in addition, continual random testing of vaccine samples, as well as monitoring of complaints about the vaccines from the field.

REFERENCES

[1] Levinson DR. Challenges to FDA's Ability to Monitor and Inspect Foreign Clinical Trials. Department of Health and Human Services 2010. Office of Inspector General.
[2] Glickman SW, McHutchinson JG, Peterson ED, Cairns CB, Harrington RA, Califf RM, Schulman KA. Ethical and scientific implications of the globalization of clinical research. New England Journal of Medicine 2009;360(8):816–23.
[3] International Conference on Harmonisation of Technical Requirements for Registration of Pharmaceuticals for Human Use (1998). Ethnic factors in the acceptability of foreign clinical data, E5(R1).
[4] US 21 CFR 314.106. The acceptance of foreign data in a new drug application.
[5] World Health Organization. The WHO Prequalification Project. http://www.who.int/mediacentre/factsheets/fs278/en/index.html (accessed July 15, 2010).
[6] World Health Organization. Procedure for prequalification of pharmaceutical products. http://www.who.int/vaccines-documents/DocsPDF06/812.pdf (accessed August 1, 2010).
[7] World Health Organization. Procedure for assessing the acceptability, in principle, of vaccines for purchase by United Nations agencies. http://who.int/vaccines-documents/DocsPDF06/812.pdf (accessed August 1, 2010).

CHAPTER 2

Bioethical Considerations in Global Clinical Trials

Menghis Bairu
Elan Biopharmaceuticals, South San Francisco, CA, USA

Contents

2.1. INTRODUCTION

Globalization is now a powerful driver in biomedical research, creating myriad benefits and opportunities, yet fraught with ethical challenges. By some estimates, more than 30 percent of all clinical trials are underway outside the USA and Western Europe, a number predicted to double in the next decade [1].

The majority of these studies will be conducted in lesser developed countries (LDCs) already struggling under the burden of poverty, illiteracy, and disease. The absence of clearly defined, judicious and universally accepted standards of ethical conduct for biomedical research has triggered confusion and discord among sponsors, researchers, and host countries, often at the expense of the patients we are sworn to protect. These controversies resonate far beyond the research community.

In 1964, the World Medical Association published the Declaration of Helsinki – "Ethical Principles for Medical Research Involving Human Subjects" [2]. Grounded in the 1947 Nuremburg Code and amended eight times over the years, the Declaration of Helsinki is the cornerstone of ethical human research standards.

Guidelines promulgated by the World Health Organization's Council for International Organizations of Medical Science (CIOMS) and the 1996

Global Clinical Trials
ISBN 978-0-12-381537-8, Doi:10.1016/B978-0-12-381537-8.10002-0

International Conference on Harmonisation of Technical Requirements for Registration of Pharmaceuticals for Human Use Good Clinical Practice (ICH-GCP) supplement and refine Declaration of Helsinki principles [3,4]. Adapted or codified by governmental agencies, advocacy groups, non-governmental organizations (NGOs), ethicists, and academic institutions, these standards provide guideposts for research in the developing world. In the best case, they foster the globalization of clinical research.

Nonetheless, a growing gap exists between ethical responsibility and real-world behaviors and constraints, which threatens the mutual benefits inherent in globalizing medical research. Critics, for example, contend that the Helsinki guidelines are either lax or too rigid, non-binding or unenforceable. Rather than providing oversight, institutional review boards (IRBs) function as rubber stamps for corporate interests or host nations desirous of the healthcare resources promised by clinical trials. All too often, directives to publish protocols and negative trial outcomes are ignored or overlooked.

The extent of this lack of oversight, and its potential consequences, was brought alarmingly to life in a report issued by the US Department of Health and Human Services' Office of the Inspector General (OIG) in June 2010. In fiscal year 2008, according to the OIG study, fully 80 percent of the applications for drugs and biologicals approved for US use utilized data from overseas clinical trials. More than 50 percent of clinical trial participants and sites were located outside the USA. And yet the US Food and Drug Administration (FDA), charged with ensuring the safety of participants and verifying the accuracy of trial data, was able to inspect only 0.7 percent of these sites, less than half the number of domestic trial locations reviewed. Further, the veracity of such data and their applicability, for example, to the US population, is raising alarms among researchers about the ethical and scientific implications of such research.

In effect, the unprecedented "offshoring" of medical research is proving as susceptible to ethical lapses and corrupting influences as the financial services industry. Will this gap become unbridgeable as pharmaceutical companies and other sponsors, facing ever-mounting research costs and competitive pressures, outsource ever greater percentages of clinical trials? Are the international standards of conduct now in place failing to provide an ethical safety net? This is the critical question we face.

Several seemingly ill-considered, poorly administered, or allegedly risky research trials have triggered charges of negligence and ethical relativism not only in the robust debate of the medical journals, but also in international headlines and halls of justice.

In 1997, the Centers for Disease Control and Prevention (CDC) and National Institutes of Health (NIH) published a study of randomized, placebo-controlled trials to test the effectiveness of short-course zidovudine (AZT) treatment in preventing perinatal transmission of human immunodeficiency virus (HIV). (An intensive zidovudine regimen had already been proven effective and adapted in developed countries.) The trials were conducted in sub-Saharan Africa and Thailand [5].

The use of placebos with a proven treatment available triggered an ethical firestorm. Hundreds of infants "needlessly contracted HIV infection", charged Drs Peter Laurie and Sidney Wolfe in the New England Journal of Medicine (NEJM) [6]. Comparing the trials to the Tuskegee syphilis study, Marcia Angell, NEJM editor-in-chief, condemned the "widespread exploitation of vulnerable Third World populations for research programs that could not be carried out in the sponsoring country" [7].

Lost in the clamor were findings demonstrating that the short-course (less costly) zidovudine regimen had shown a 51 percent reduction in perinatal HIV transmission. The use of placebo controls, CDC officials argued, had provided "the most rapid, accurate, and reliable answer to the question of the value of the intervention being studied compared to the local standard of care".

In the developing world, "local standard of care" is an ethical slippery slope. "It is an unfortunate fact", the CDC noted, "that the current standard of perinatal care for the HIV-infected pregnant women in the sites of the studies does not include any HIV prophylactic intervention at all" [8].

2.2. ETHICAL STANDARDS AND INTERNATIONAL LAW

A recent US Supreme Court opinion (Pfizer Inc. v. Abdullahi) suggests that the ethical guidelines set forth by Helsinki et al. are rapidly attaining the status of international human rights law, statutes that can be enforced and remedies imposed in the world's courts [9].

US top court rejects Pfizer Nigeria lawsuit appeal

By James Vicini

WASHINGTON, June 29, 2010 (Reuters) –The Supreme Court declined to hear an appeal by Pfizer Inc. of a ruling that reinstated US lawsuits by Nigerian families who said the drug maker tested an experimental antibiotic on their children without getting adequate consent. The justices, without comment, let stand a ruling by a US appeals court in New York that allowed the lawsuits involving alleged harm caused by the drug, Trovan, to go forward...[10]

The case, a class-action suit, is linked to a devastating 1996 outbreak of meningococcal meningitis in the state of Kano, Nigeria; specifically, Pfizer's alleged failure during a two-week trial of the experimental antibiotic, Trovan, to obtain written informed consent from the families of children involved in the trial or witness statements attesting to verbal consent.

More than 200 children were enrolled in the study: half were given Trovan; the rest treated with a low dosage of the FDA-approved antibiotic Rocephin. Ultimately, 11 subjects died and dozens more were left blind, deaf, paralyzed, or brain-damaged. The lawsuit further alleges that Pfizer's researchers did not advise the subjects' families of the availability of a proven drug regimen, or make clear that Trovan – later pulled off the market – was an experimental therapy. Pfizer originally argued that no international norm exists requiring physicians to obtain informed consent for the use of experimental drugs. That defense was later dropped after a series of setbacks in Nigerian courts.

A Washington Post story [11] on the trials triggered a worldwide outcry with its opening descriptions of the slow death of a 10-year-old girl known only as Subject 6587-0069, while Pfizer researchers at the Kano Infectious Disease Hospital allegedly stood by, unwilling to break their protocol.

The challenge before us is often framed as a simple question: "Is it appropriate to apply the precise set of ethical standards and procedures adapted in the developed world to clinical trials conducted in developing countries?"

Thus far, the answer has proved elusive.

2.3. CODES OF CONDUCT

Opportunities for researchers conducting trials in the developing world are manifold: cost savings, enhanced subject availability, shortened time-lines, eased regulatory constraints regarding compliance, documentation, training, expanded markets, the potential for new treatments or cures. As are the ethical obligations. The Declaration of Helsinki posits, "the promise of direct, tangible, and significant benefit to the host population", a goal often overlooked.

Despite the mass overseas migration of resources and talent, the "10/90 gap" persists: less than 10 percent of world healthcare expenditures are devoted to malaria, tuberculosis, diarrhea, and other debilitating illnesses that account for 90 percent of the global disease burden. Research essentially follows the money focusing on degenerative diseases – arthritis, obesity, cardiovascular disease – illnesses of the developed world. Only 1 percent of the drugs developed in the past 25 years have targeted tropical diseases, an ethical doublestandard that recalls the inequities of the colonial era.

2.3.1. Responsibility to Patients

The health of my patient shall be my first consideration [12].

World Medical Association Declaration of Geneva (1948)

What follows is a compilation of the principles, challenges and responsibilities inherent in human biomedical research; a list of best practices drawn from the Declaration of Helsinki and other standards. It is incomplete and open-ended, given ever-evolving real-world conditions and circumstances, yet consistent in its regard for both rigorous science and the well-being of research subjects. In effect, it is a set of parameters within which clinical trials can proceed effectively and ethically in the developing world.

- **Research protocols**. Each trial must be framed in a research protocol incorporating a statement of ethical considerations that includes information regarding funding, sponsorship, institutional affiliations, potential conflicts of interest, incentives, and provisions for treatment or compensation to subjects harmed as a consequence of participation. The protocol must state the subjects' right to poststudy interventions, including compassionate usage, identified as beneficial.
- **Responsibility to minimize risk to patients**. Implicit in every proposed trial is a careful assessment of the predictable risks and potential burdens to the individuals and communities involved. Clinical trials may not proceed without the confidence that risks can be satisfactorily managed. (An exception may exist in instances where potential benefits clearly outweigh inherent risk.) The efficacy of proposed trial therapies must be weighed against the best proven intervention. When risk outweighs potential benefits or conclusive evidence of positive results is obtained, a trial must be halted.
- **Responsibility to inform the patient and to obtain consent**. Ethical research is inseparable from subject rights. The physician/researcher's obligation is to ensure that a potential subject fully understands the purpose, potential risks and benefits of a proposed trial, a working definition of informed consent. Informed consent must be freely given, preferably in writing. Participants must be made aware of their right to refuse to participate, and the right to withdraw at any time.

Where appropriate, family members, local healthcare professionals, and community leaders may be consulted. However, no competent individual may be enrolled in a research trial without prior agreement.

- **Responsibility to special populations**. Some research populations – those who cannot give or refuse consent for themselves and those vulnerable to coercion or undue influence – must be afforded special protections; the physician/researcher must make every effort to determine whether a potential subject is in a dependent relationship or under duress. A dependent relationship may also exist between a potential subject and a researcher. In such situations, a qualified individual independent of this relationship should be empowered to determine informed consent.
- **Non-written consent/translations**. It is the researcher's responsibility to ensure that proposed subjects understand all pertinent information about a proposed trial, including potential outcomes. This information must be provided in the subject's native language or precisely translated. If written informed consent cannot be obtained, non-written consent must be documented, witnessed, and verified.
- **Patients unable to consent**. For incompetent, unconscious or incapacitated individuals, informed consent must be obtained from a legally authorized representative. If a potential subject deemed incompetent is able to give consent, the physician/researcher must procure his or her consent in addition to that of a legal representative. Involvement of incompetent or incapacitated individuals in clinical trials can be justified only if their physical or mental condition is a necessary characteristic of the research population. Such individuals must be excluded from research that: can be undertaken with competent persons; entails significant risk or burden; has no likelihood of benefit for them, except when the research may promote the health of the population they represent.
- **Responsibility not to induce**. A body of evidence in the literature emerging from India, China and other LDCs documents numerous instances of sponsors compromising the ethical responsibility "not to induce"; in one example, offering "illiterate blue-collar workers more per month to participate in studies than they earn at their jobs and …providing medication that is worth more than their annual salary" [13]. In effect, sponsors were compelling subjects to take risks they normally would not take. Such practices, though not technically illegal, run counter to ethical proscriptions. Money and other inducements

constitute coercion, particularly among impoverished populations, distorting judgment and negating the principle of informed consent.

- **Ethical differences in different countries**. Controversies over research trials in LDCs are often rooted in the conscious or unconscious fallacy that ethical standards differ from region to region, reflecting local beliefs and customs. All ethical guidelines mandate that research subjects are entitled to minimum guarantees that are transnational and non-negotiable.
- **Research in non-industrial countries**. Clearly, the failure to abide by ethical standards during clinical trials has a much greater impact in the developing world where subjects often lack the social safety net provided by legal protections and other guarantees in the developed world. Ethical standards include but are not limited to the following:
  - Every study must have a valid scientific design and be performed by qualified persons.
  - A study must be responsive to the host country's health needs and strike a reasonable balance between predictable risks and foreseeable benefits.
  - Subjects' interests should not be subordinated to those of science and society.
  - Informed consent must be obtained without duress or coercion.
  - Research must hold the promise of direct, tangible, and significant benefit to the host country population, if not to the subjects themselves.
- **Responsibility to maintain privacy/confidentiality**. Every subject's right to privacy must be respected. The physician/researcher's responsibility is to ensure that subjects' personal information is kept confidential. Guidelines on privacy/confidentiality in clinical research are set forth in the US Health Insurance Portability and Accountability Act's "Standards for Privacy of Individually Identifiable Health Information" [14].
- **Responsibility to provide continued therapy.** Subjects participating in a clinical study must be informed of its outcome. They are entitled to share any resultant benefits. If interventions do not exist or have been proved ineffective at the conclusion of a research trial, the researcher may attempt an unproven intervention if it is the considered judgment that it offers hope of saving life, re-establishing health, or alleviating suffering. Such an intervention requires informed consent from the subject or a legally authorized representative.

- **Responsibility to compensate for injuries**. The Declaration of Helsinki mandates that research protocols include "provisions for treating and/or compensating subjects who are harmed as a consequence of participation in a research study". In addition, these subjects should be compensated for any resultant loss of income.
- **Institutional review board (IRB)/ethics committee (EC)**
  - **Background**. In 1975, the Declaration of Helsinki was amended to include the principle of oversight of clinical trials by "Independent Committees", a function now assignable to IRBs and ethics committees. In the developing world, ethical oversight often lags behind developed world standards and assumptions. India and China have become vast proving grounds for pharmaceutical testing, yet lack both the staff and healthcare infrastructure to host effective and ethical trials. Where they exist, critics say, on-site ethics committees often lack expertise and experience, or function as rubber stamps for sponsors or health ministries eager to import care for their populaces.
  - In one recent study, 44 percent of nearly 700 researchers conducting trials in LDCs reported that their protocols had not been reviewed by local ethics committees or government ministries [15].
  - According to a 2005 survey, only 150 of India's approximately 14,000 general hospitals are adequately equipped to host clinical trials. Of these, only half have IRBs, many without the expertise to evaluate protocols [13].
  - Of published clinical trials conducted in China in 2004, 90 percent proceeded without a prior ethical review of protocols. Eighteen percent satisfied the requirements for obtaining subjects' informed consent [16].
  - In 1991, the US Department of Health and Human Services issued a set of guidelines ("Federal Policy for the Protection of Human Subjects"), referred to as the "Common Rule", incorporating standards for IRB membership, function, operations, review, and record-keeping in trials funded or regulated by federal agencies and the FDA.
  - **IRB/EC requirements**. A previously appointed IRB/EC is required to review a clinical trial's protocols and provide comment, guidance, and approval. Funding mechanisms, sponsors, affiliations, potential conflicts, and incentives are subject to IRB review. IRBs and ethics committees function independently of investigators,

sponsors, and the host nation. Board oversight includes, but is not limited to:

- compliance with local laws and regulations
- assurance that the justification for a trial is clearly articulated and defensible, particularly where the population is vulnerable to exploitation
- ongoing monitoring of trial information.

- **Centralized IRBs**. The sheer number, size, and complexity of research trials in the developing world create inordinate demands on sponsors, researchers, and host countries. Centralized IRBs – independently staffed and accredited oversight committees – can help to enforce best practices and standards, share non-proprietary information, and reduce redundancy. One model for the approach is the National Cancer Institute's Central Institutional Review Board Initiative [17].

2.3.2. Scientific Responsibility

- **Publication**. The Declaration of Helsinki obligates authors and publishers to publish well-documented and accurate accounts of clinical trials – including negative data – and to reject "reports of experimentation not in accordance" with its principles. The International Committee of Medical Journal Editors has issued a set of standards covering trial design, access to data, control over the publication of results, and other issues. As noted elsewhere, information regarding funding, institutional affiliations, and potential conflicts of interest must be made public.
- **Equipoise**. On the face of it, the equipoise requirement for research protocols/trials in the developing world seems irreconcilable with the therapeutic obligation to provide "optimal medical care" to a patient. Real-world challenges – cost, delivery, access – to optimal medical care make a clear-cut determination of equipoise (genuine doubt in the mind of a researcher about the relative merits of a set of medical interventions) and other ethical guidelines exceedingly difficult to assess. A deeper consideration of equipoise underlies and shades the issue of ethics and clinical research in the developing world, be it cosmetics testing in India or dementia trials in sub-Saharan Africa. How can we rigorously advance scientific knowledge and provide respectful and compassionate care to the billions of human beings who daily endure the burdens of disease, malnutrition, and poverty? To the mother of a malnourished infant dying of malaria in the Sudan, a guideline is another unkept promise.

Dare we stop or go forward?

I envision a two-tier model. The gap (the burden of disease, poor nutrition, illiteracy, political instability, inadequate healthcare, etc.) between the developed world and the economically advancing BRIC nations (Brazil, Russia, India, China) is narrowing, making the implementation of ICH and other ethical guidelines practicable. In sub-Saharan Africa and elsewhere, it looms as a chasm. Sponsors, research organizations, host nations, and IRBs must take into account "the highest practically obtainable standards" when framing risk, benefit, and other parameters in LDCs. They must ask what I believe is a critical ethical question: Do we have any intention of marketing an approved drug in the country in which its trial is conducted? If not, are we not simply buying patients?

Only when we have answered this question can we decide whether or not to move forward.

The continuing involvement of LDCs in clinical trials far outweighs the negatives generated by a handful of ill-advised or poorly administered protocols; negatives, it must be noted, that have grievously impacted both sponsors and subjects. Providing local populations with potential early access to innovative treatments, as well as the ancillary healthcare benefits such trials generate, is an inherent good, as is the responsibility to ground local physicians and researchers in the principles and guidelines of ethical research. Differences in local standards of care and other potential pitfalls can be bridged.

2.4. GOOD CLINICAL PRACTICE

The operating principles of the World Health Organization's Initiative for Vaccine Research(IVR) argue for "capacity building" in GCP and bioethics in the developing world. At the same time, they warn of the extent of the challenges ahead:

> *All vaccine trials should meet Good Clinical Practice (GCP), and ethical and applicable regulatory requirements. However, most institutes and clinical sites located in developing countries with a high prevalence of diseases such as malaria, tuberculosis or AIDS,* have no experience in GCP, *nor the capacity to conduct good quality trials or ethical reviews. Capacity building for research institutes in developing countries to conduct such trials is therefore a priority [emphasis added] [18].*

The understandable impulse to exclude, whether because of bad clinical experiences or unfortunate colonial histories, should be resisted. I would argue that well thought-out, ethically grounded protocols can be implemented anywhere, and for the benefit of all.

REFERENCES

[1] Lamberti MJ, Space S, Gammbrill S. Going global. Applied Clinical Trials 2004;13: 84–92.
[2] World Medical Association. Declaration of Helsinki: Ethical Principles for Medical Research Involving Human Subjects, http://www.wma.net/en/30publications/10policies/b3/index.html; 1964. (accessed July 30, 2010).
[3] Council for International Organizations of Medical Sciences (CIOMS). International Ethical Guidelines for Biomedical Research Involving Human Subjects, http://www.cioms.ch/publications/layout_guide2002.pdf (accessed July 31, 2010).
[4] International Conference on Harmonisation. Good Clinical Practice Guidelines, http://www.ich.org/LOB/media/MEDIA482.pdf
[5] Centers for Disease Control. Update on CDC Collaborative research studies on perinatal HIV prevention in the developing world: preliminary results find short-course AZT effective, http://www.cdc.gov/nchstp/od/Perinatal; 1998.
[6] Lurie P, Wolfe SM. Unethical trials of interventions to reduce perinatal transmission of the human immunodeficiency virus in developing countries. New England Journal of Medicine 1997;337:853–5.
[7] Angell M. The ethics of clinical research in the third world. New England Journal of Medicine 1997;337:847–9.
[8] Varmus H, Satcher D. Ethical complexities of conducting research in developing countries. New England Journal of Medicine 1997;337:1003–5.
[9] Pfizer Inc. v. Rabi Abdullahi, et al. Supreme Court 130 S. Ct. 534, http://caselaw.findlaw.com/us-2ndcircuit/1442878.html; 2010.
[10] Vicini, J. US top court rejects Pfizer Nigeria lawsuit appeal. Reuters (June 29), 2010.
[11] Bagenda D, Muske-Mudido P. A look at ... ethics and AIDS. Washington Post (September 28), Section C, 3, 1997.
[12] World Medical Association. Physician's Oath. Declaration of Geneva.http://www.cirp.org/library/ethics/geneva/; 1948.
[13] Nundy S, Gulhati CM. A new colonialism? Conducting clinical trials in India. New England Journal of Medicine 2005;352:1633–6.
[14] US Health Insurance Portability and Accountability Act. Standards for Privacy of Individually Identifiable Health Information. US Department of Health and Human Services; 1996.
[15] Glickman SW, McHutchison JG, Peterson ED, Cairns CB, Harrington RA, et al. Ethical and scientific implications of the globalization of clinical research. New England Journal of Medicine 2009;360:2793.
[16] Zhang D, Yin P, Freemantle N, Jordan R, Zhong N, Cheng KK. An assessment of the quality of randomized controlled trials conducted in China. Trials 2008;9:22.
[17] National Cancer Institute. Central Institutional Review Board Initiative, http://www.ncicirb.org/CIRB_Handbook.pdf; 2009. (accessed August 2010).
[18] World Health Organization, Initiative for Vaccine Research Capacity Building in Good Clinical Practice and Bioethics, http://www.who.int/vaccine_research/capacity_strengthening/en/ (accessed October 2010).

CHAPTER 3

United States Regulations

Janice B. Wilson* and Edda Gomez-Panzani**
*Wilson Quality & Compliance Consulting, Antioch, CA, USA
**Clinical Sciences – Neuroendocrinology, Tercica Inc., Ipsen Group, Brisbane, CA, USA

Contents

3.1. OVERVIEW OF THE FOOD AND DRUG ADMINISTRATION IN THE USA

3.1.1. History

In the early eighteenth century, there was no regulation anywhere in the world concerning food, drugs, or medical activities. Anybody could manufacture, sell and import/export foods, beverages, and drugs without concern for their origin, the purity of their ingredients, or the effects on those who used them.

In the USA, one of the first attempts to regulate foods was an Act passed in Massachusetts on March 8, 1785. This Act stated:

> *Whereas some evilly disposed persons, from motives of avarice and filthy lucre, have been induced to sell diseased, corrupted, contagious, or unwholesome provisions, to the great nuisance of public health and peace: Be it therefore*

Global Clinical Trials
ISBN 978-0-12-381537-8, Doi:10.1016/B978-0-12-381537-8.10003-2

> *enacted … that if any person shall sell any such diseased, corrupted, contagious or unwholesome provisions, whether for meat or drink, knowing the same without making it known to the buyer … he shall be punished by fine, imprisonment, standing in the pillory, and binding to the good behaviors, or one or more of these punishments, to be inflicted according to the degree and aggravation of the offence [1].*

A few years later, in the mid-nineteenth century, US Congress enacted the first law on drugs. This was the 1848 Drug Importation Act and it was aimed at preventing the importation of adulterated drugs from other countries.

The origin of the Food and Drug Administration (FDA) dates to 1862 when a single chemist, Dr Charles M. Wetherill, was appointed to the US Department of Agriculture, giving birth to the Division of Chemistry; the scope of the division of chemistry was purely scientific.

In 1883, Dr Harvey Wiley was appointed chief chemist of the US Bureau of Chemistry. Dr Wiley, with the help of what he called his "poison squad" (consisting of five volunteers who acted as guinea-pigs agreeing to consume dubious food additives to determine how they impacted their health), researched and tested different foods and found them to be adulterated with chemical preservatives. Although completely unethical by today's standards, this practice allowed Dr Wiley to document the ill-effects of the additives and to start campaigning for a national food and drug law.

In part because of the initial efforts by Dr Wiley, the first statute regulating drugs was passed in the form of the 1906 Food and Drugs Act. This Act prohibited interstate transport of misbranded or adulterated foods, drinks, and drugs and stated that the food or drug label could not be false or misleading [2]. This Act added, for the first time, regulatory functions to the Division of Chemistry which ultimately became the Food and Drug Administration in 1930.

Although soon after the FDA recognized that The Food and Drugs Act had become obsolete, no action was taken until five years later when the 1937 sulfanilamide tragedy, in which a drug used to treat infections was diluted with the solvent diethylene glycol (antifreeze) and marketed as Elixir Sulfanilamide, caused the death of 107 people, mostly children. This disaster drove Congress to act and, in 1938, The Federal Food, Drug, and Cosmetic Act was passed. This law required, for the first time, preapproval of any new chemical entity or combination whereby the manufacturer had to prove the compound was safe before being authorized to market it. It also required all drugs to be appropriately labeled to ensure they were safely used and prohibited any false claim being made. However, it is interesting that, at that time, only the *safety* of the drugs was a concern. It was not until 1962 that the regulations were amended (Kefauver-Harris Drug Amendments) to state that the manufacturers had to prove not only the safety of the drugs but their *efficacy* before they were allowed to market them.

In 1981, mostly prompted by the 1979 Belmont Report (see subsection 3.1.2), the FDA amended the regulations for the protection of human subjects. Eight years later, in 1987, the regulations were once again revised to allow patients with serious diseases, for which no alternative treatments are available, access to investigational drugs.

3.1.2. Regulations and Ethical Medical Research

Records of experimentation on human subjects date back several centuries; however, it was not until fairly recently that the rights of the human participants took a prominent place in the forefront of human clinical research. Probably the first well-known case that came to light and became the focus of public scrutiny was "The doctors' trial" (1946–1947). This was the first of 12 war crimes trials held by the International Military Tribunal in Nuremberg, Germany, at the end of World War II. The defendants, mostly physicians, were accused of having been involved in sadistic human experiments. The Nuremberg Code is a set of 10 research principles and it was the first document of its kind to ensure the rights of the human participants in clinical research. It was created in 1947 as a result of this trial [3].

Almost two decades later, in 1964, the World Medical Association, an international organization of physicians, developed the Declaration of Helsinki (Helsinki, Finland) [4]. The Declaration of Helsinki is based on the Nuremberg Code and a revised version of the Hippocratic Oath (the Declaration of Geneva; 1948). Since its inception, it has undergone six amendments, the last one in 2008, and two clarifications. Interestingly, although the USA was the primary sponsor of the Nuremberg Code on which it was based, no US representative signed the Declaration of Helsinki. Even under these circumstances, the principles in the Declaration of Helsinki to ensure the protection and the safety of the participants are still relevant, especially in respect of the responsibilities of the clinical investigators [3].

Following the Declaration of Helsinki, in 1968, the Medicines Act [5] was passed in the UK to prevent the harmful and illegal use of medicines (prescription-only medicines prescribed by a physician, pharmacy medicines that do not need a physician's prescription but have to be dispensed by a pharmacist, and over-the-counter drugs). This Act controlled the licensing, manufacture, supply, storage, dispensing, promotion, prescribing, sale, use, and advertising of medicinal substances, and clinical trials.

In 1974, Congress passed the National Research Act, which for the first time mandated the establishment of institutional review boards (IRBs) and the need for them to review all federally funded human research. In the same year, the National Commission for the Protection of Human Subjects of Biomedical and Behavioral Research was created. Until then, the protection of human subjects participating in research was voluntary. In 1979, the Commission published "The Belmont Report: Ethical Principles and Guidelines for the Protection of Human Subjects of Research". This report was never formally adopted by Congress; however, to date it is considered by some as the USA's contribution to the international guidelines for protecting human subjects in research [6].

Concerns regarding the treatment of human subjects participating in clinical studies, the quality of the data, and the interpretation of results led to a collection of regulations and guidelines that defines the responsibilities of all the parties (e.g. sponsor, investigators, monitors, IRBs) involved in clinical research and is consistent with the Declaration of Helsinki; this collection is known as Good Clinical Practice (GCP).

The definition of GCP is: "An international ethical and scientific quality standard for designing, conducting, recording and reporting trials that involve the participation of human subjects and ensures that the rights, safety, and well-being of the trial subjects are protected and that the data reported are credible" [7].

Legal requirements for the conduct of clinical research involving human participants are at present mostly based on the GCP guidelines and guaranteed by the need to obtain IRB approval.

GCP was established in the USA in 1978 and has been evolving ever since. Around the same time GCP was instituted, the European Community increased its focus on harmonization of regulatory requirements for pharmaceuticals and talks on this topic began with the USA and Japan. During the World Health Organization (WHO) International Conference of Drug Regulatory Authorities (ICDRA) in Paris in 1989, an action plan began to develop. In the following year, regulatory agencies and industry representatives from Europe, Japan, and the USA began planning an International Conference on Harmonisation (ICH). The International Conference on Harmonisation of Technical Requirements for Registration of Pharmaceuticals for Human Use was the result of efforts mainly from the USA, Europe, and Japan (plus Australia, Canada, the Nordic countries, and the WHO) and was adopted in 1997 in all three countries. In Japan it was adopted as a law, and in Europe and the USA as a guideline (in the USA in the federal register).

As a result of this harmonization, and to ensure that only good-quality, safe, and effective drugs are developed, GCP has most recently been redefined as the "International Conference on Harmonisation (ICH) Harmonised Tripartite Guideline for Good Clinical Practice E6" and was adopted as FDA guidance in May 1997.

All clinical trials involving human participants in the USA, regardless of the development phase, must be conducted following GCP.

3.1.3. Preclinical and Chemistry, Manufacturing, and Control

Before studies in humans can begin, studies in animals (preclinical studies) must be completed. The purpose of these studies is to prove that the compound being studied is safe for use in humans. The chemistry of the drug must also be characterized.

The first step is to determine the chemical, physical, and biological characteristics of the compound. Data on the manufacturing of the compound, as well as information to prove that the compound will remain stable (stability tests) throughout the time it will be used in the preclinical studies, are required before administering it to animals.

Once the compound has been characterized and deemed stable, preclinical studies can begin. The type of studies and their design can vary depending on the compound but the purpose is to characterize the toxic effects of a compound on different organs and biological processes and their relationship to exposure. Simply stated, the majority of preclinical studies pertain to the toxicology and safety of the molecule.

The minimum required preclinical studies include single and repeated dose toxicity studies, pharmacokinetic and pharmacodynamic studies, genotoxicity, carcinogenicity, and reproductive studies.

Repeated dose toxicity studies include multiple dose administrations done over a period similar to the indication and exposure expected during the clinical studies in humans (they rarely exceed 12 months). These studies also must be conducted in two mammalian species (one rodent and one non-rodent). Frequently, additional preclinical studies are performed while a drug is simultaneously undergoing studies in humans.

Carcinogenicity studies are necessary for any compound that is intended for use for six months (either continuous six months or for intermittent periods that add to a total six-month or more exposure). These studies entail daily administration of the compound, are usually conducted in rodents, and last approximately the life of the animal (approximately two years).

Reproductive toxicity studies assess the effects of the compound on fertility, reproduction, and fetal toxicity; they involve repeated administration of the compound to the animals before, during, and after a gestational period. These studies must be completed before submitting a New Drug Application (NDA) to the FDA.

All the studies must be conducted under the ICH good laboratory practice (GLP) guidelines [8] (the toxicology equivalent of GCPs).

3.1.4. Regulatory Agencies Involved in the Regulation of Drugs and Devices

The FDA is an agency within the Department of Health and Human Services and consists of centers and offices that regulate various aspects of foods, drugs, and cosmetics. The primary centers and officers that are involved with the development and registration of a new drug are listed in Table 3.1, along with a brief description provided by the FDA [1].

Table 3.1 Regulatory agencies involved in the regulation of drugs and devices

Office of the Commissioner (OC)	The OC provides centralized agency-wide program direction and management services to support effective administration and FDA's consumer protection efforts within its regulatory framework and to put available resources to the most efficient use
Center for Biologics Evaluation and Research (CBER)	The CBER is the center within the FDA that regulates biological products for human use under applicable federal laws, including the Public Health Service Act and the Federal Food, Drug and Cosmetic Act
Center for Drug Evaluation and Research (CDER)	The CDER is the center within the FDA that regulates all drugs (over-the-counter and prescription) and encompasses more than just medicines. Items such as fluoride toothpaste, antiperspirants, and medicated shampoos are considered drugs
Center for Devices and Radiological Health (CDRH)	The CDRH is the center within the FDA that is responsible for regulating firms that manufacture, repackage, relabel, and/or import medical devices sold in the USA. In addition, CDRH regulates radiation-emitting electronic products (medical and non-medical) such as lasers, X-ray systems, ultrasound equipment, microwave ovens, and color televisions

3.2. THE FDA'S LEGAL AUTHORITY

The federal laws concerning the FDA are part of the Food, Drug and Cosmetic Act, and are codified in Title 21, Chapter 9, of the US Code. These laws give authority to the FDA to oversee and regulate the safety of food, drugs, and cosmetics. The laws were passed by Congress in 1938 as a direct fallout of the infamous sulfanilamide tragedy that resulted in the death of approximate 100 people, many of them children. The codes have been amended many times since then, with the most recent amendments relating to bioterrorism preparedness.

3.2.1. Regulations and Guidance Related to Product Registration

3.2.1.1. Code of Federal Regulations – Title 21 – Food and Drugs

The Code of Federal Regulations (CFR) is the codification of the general and permanent rules published in the Federal Register by the executive departments and agencies of the federal government. Title 21 of the CFR is reserved for rules of the FDA. Each title (or volume) of the CFR is revised once each calendar year. A revised Title 21 is issued on approximately April 1 of each year and is usually available on the FDA website several months later. The parts of Title 21 that are applicable to developing drugs/clinical trial studies or new drug applications are mostly found in the parts listed in Table 3.2.

In addition to the Title 21 regulations, the FDA expects drug manufacturers to comply with the requirements set forth by the ICH regulations; in particular, those that are referenced for clinical studies or new drug development – the efficacy series (Table 3.3).

Table 3.2 Parts of Code of Federal Regulations (CFR) Title 21 applicable to drugs and clinical trials

CFR Title 21 Part	Title
11	Electronic Records; Electronic Signatures
50	Protection of Human Subjects
54	Financial Disclosure by Clinical Investigators
58	Good Laboratory Practice for Non-Clinical Laboratory Studies
56	Institutional Review Boards
312	Investigational New Drug Application
314	New Drug Approval

Table 3.3 International Conference on Harmonisation (ICH) regulations: efficacy series

ICH guidance document	Title
E1, E2A–E	Drug Safety Planning and Reporting
E3	Clinical Trial Study Reports
E6	Good Clinical Practices
E8	General Considerations for Clinical Trials

3.3. REQUIREMENTS FOR APPROVAL

3.3.1. Registration Process

For both biologicals and drug products in the USA, the pathway to registration starts with the investigational new drug (IND) application. The laws codified in Title 21 of the CFR require that a drug be the subject of an approved marketing application before it is transported or distributed across state lines. However, a marketing application cannot be granted until a sponsor demonstrates that the drug is suitable for marketing, and this requires testing or clinical trial studies/investigations. Since all sponsors want to be able to market their drug throughout the USA, they must seek an exemption from that legal requirement of having an approved marketing application in order to ship the new drug to investigators in various states. The IND application is the means through which the sponsor gets an exemption to ship its IND across state lines.

Different types of drugs or medical devices must adhere to the requirements of an IND as well as the marketing application process for the new drug or device. Overall, it is a two-step process that initially requires the submission of the application to investigate the new drug, biological, or device and finally the application to market the new drug, biological, or device. The details of the requirements are codified in 21 CFR sections as listed below.

For biological products:

- IND – Investigational New Drug Application (21 CFR 312)
- BLA – Biologics License Application (21 CFR 600–680).

For drugs:

- IND – Investigational New Drug Application (21 CFR 312)
- NDA – New Drug Application (21 CFR 314).

For medical devices:

- 510(k) – (21 CFR 807)

- IDE – Investigational Device Exemption (21 CFR 812)
- PMA – Pre-Market Application (21 CFR 814).

3.3.2. Types of Investigational New Drug Application

Before initiating any clinical study with an IND, the sponsor must submit an investigational new drug application (INDA) to the Center for Drug Evaluation and Research (CDER). CDER is the division of the FDA that makes sure that all prescription and over-the-counter marketed drugs are safe and effective. There are three types of INDA:

- **The Investigator INDA** is submitted by a physician who both initiates and conducts a clinical investigation. The physician can submit an INDA for a new unapproved drug or for an approved drug for a new indication or on a new patient population.
- **The Emergency Use INDA** is sought when it is beneficial to use an experimental drug in an emergency situation that does not allow time for submission of an INDA in accordance with 21 CFR, Section 312.23 or 312.34, or when patients might not meet the criteria for an existing protocol.
- **The Treatment INDA** is submitted for experimental drugs showing promise in clinical testing for serious or immediately life-threatening conditions while the final clinical work is conducted and the FDA review takes place.

The INDA should include the general investigational plan, all the available protocols for any study that is planned to be conducted in humans, the investigator brochure and all information on the investigational drug available at the time. In general, the INDA comprises the following (21 CFR 312.23):

- preclinical data: animal pharmacology and toxicology studies that support the safety of the drug for its use in humans
- any previous experience with the investigational drug in humans: usually studies conducted in other countries
- chemistry, manufacturing, and control (CMC) information such as nomenclature, structure and general drug substance properties, characterization, manufacturing process, control of drug substance, reference standards of materials, information on the container closures systems, and stability under different conditions/lengths of time enough to support the expected conditions at which the drug will be exposed.

As new protocols are developed, they must be submitted to the FDA under the same INDA. The objective of submitting an INDA before any clinical study can be conducted in humans is to allow the FDA to assure the safety and the rights of the participants as well as the quality of the data being collected. Once the INDA has been submitted to the FDA, the sponsor is allowed to ship the IND to the clinical sites/investigators as soon as authorization is received from the FDA, or 30 days after the FDA receives the INDA unless the FDA notifies the sponsor that the clinical research is subject to a clinical hold.

If the FDA were to identify significant deficiencies (e.g. human participants would be exposed to an unreasonable and significant risk of illness or injury) that required the clinical research plan to be put on hold, subjects may not be given the investigational drug until the deficiencies are corrected and the FDA gives authorization to proceed [9,10].

The INDA must be updated on an annual basis. This update consists of any new information about the drug and the results of studies that were completed since the last submission or available information on ongoing studies such as the updated enrollment numbers, adverse event information, and an overview of the status. If there are any changes or updates to the clinical plan that was originally submitted, they should also be provided to the Agency at this time.

If there are any significant changes to the IND (e.g. a compound that was intended solely for subcutaneous administration can now be administered orally), the clinical studies [e.g. a dose–response study where after a data and safety monitoring board (DSMB) assessment it is determined that the highest dose is not safe, therefore requiring a change to the study design] or at any time the sponsor deems it necessary, an amendment to the INDA may be filed.

There are certain instances when a compound is found to have additional indications than that for which it was originally registered. In this case, the compound is submitted to another division of the FDA and receives another IND number. An example of this would be a compound that treats a hormonal syndrome, for which an INDA was submitted to the Endocrinology Division, is found to also have antiproliferative effects and now will be submitted to the Oncology Division.

If a study is conducted under a US IND, then sites outside the USA are also subject to FDA regulations, including filing of the forms (e.g. 1572) required by the FDA. It may be advisable and logistically simpler to split the study into US and non-US arms and separate the protocol into two sister studies that are operationally separate but will be analyzed together as a single study.

3.3.2.1. Drugs and Biologicals

The INDA or BLA must contain information that demonstrates that the drug is reasonably safe for testing in humans and that it can be manufactured in a consistent manner, and contain clinical trial protocols designed to show the safety and efficacy of the drug to humans. The qualifications of the investigators are also included in the INDA/BLA to ensure the FDA that those conducting the study are qualified to do so.

Human safety is demonstrated with preclinical data obtained during animal pharmacology and toxicology studies. Safety for humans can also include information related to previous use of the drug in other countries.

The CMC section of the INDA is where the manufacturing information pertaining to the drug maker (manufacturer), drug composition and stability, and the controls used for manufacturing the drug substance and the drug product are listed. The CMC section is extremely important because it is here that the sponsor assures the FDA that the drug can be adequately and consistently produced in the quantity and of the quality needed.

The section detailing the protocols and investigator information must be detailed enough to demonstrate that the initial trials will not expose human subjects to unnecessary risks and that the investigators who will oversee the studies are qualified to conduct the studies. It is also in this section where the sponsor commits:

- to obtain informed consent from the research subjects
- to obtain review of the study by an IRB
- to use an independent safety committee that will provide ongoing safety reviews.

Once the INDA has been submitted, there is a required waiting period of 30 calendar days before any clinical trials can be initiated, which allows the FDA the opportunity to review the IND for safety and risk to research subjects.

The regulations in Table 3.4 apply to the IND application process.

Table 3.4 Regulations relating to the investigational new drug (IND) application process

21 CFR Part 312	Investigational New Drug Application
21 CFR Part 314	INDA and NDA Applications for FDA Approval to Market a New Drug (New Drug Approval)
21 CFR Part 316	Orphan Drugs
21 CFR Part 58	Good Lab Practice for Non-Clinical Laboratory [Animal] Studies
21 CFR Part 50	Protection of Human Subjects
21 CFR Part 56	Institutional Review Boards
21 CFR Part 201	Drug Labeling
21 CFR Part 54	Financial Disclosure by Clinical Investigators

3.3.2.2. Medical Devices

As with drugs or biologicals, an exemption is required to allow sponsors to ship unapproved medical devices across state lines. For medical devices, the investigational device exemption (IDE) allows the investigational device to be used in a clinical study and, like the IND, provides a means to collect the safety and effectiveness data that are required to support a Premarket Approval (PMA) application or a Premarket Notification [510(k)] submission to the FDA. All clinical evaluations of investigational devices, unless exempt, must have an approved IDE before the study is initiated.

Clinical evaluation of devices that have not been cleared for marketing requires:

- an IDE approved by an IRB; if the study involves a significant risk device, the IDE must also be approved by the FDA
- informed consent from all patients
- labeling for investigational use only
- monitoring of the study
- required records and reports.

The approved IDE permits a device to be shipped lawfully only for the purpose of conducting investigations without complying with other requirements of the Food, Drug, and Cosmetic Act that apply to devices in commercial distribution.

For IDEs, the regulations that apply in addition to 21 CFR 50, 54, and 56, are sections 812, Investigational Device Exemptions, and 820 Subpart C, Design Controls of the Quality System Regulation.

3.4. USE OF FOREIGN CLINICAL DATA BY THE FDA

For a number of years the FDA has accepted clinical studies conducted outside the USA in support of safety and efficacy claims for drugs, biological products, and medical devices as long as the studies were conducted under an INDA or IDE and were governed by informed consent and IRB requirements.

Under 21 CFR 312.120(c)(1), FDA will accept a foreign clinical study involving a drug or biological product not conducted under an IND only if the study conforms to whichever of the following provides greater protection of the human subjects:

- the ethical principles contained in the 1989 version of the Declaration of Helsinki

- the laws and regulations of the country in which the research was conducted.

The same held true for medical devices; however, the requirements are codified under 21 CFR 814.15(a) and (b).

In April 2008, the FDA published a regulatory change ending the need for clinical trials conducted outside the USA to comply with the Declaration of Helsinki [4]. The ruling in which the FDA revised the regulations in 21 CFR Part 312 in summary to accept foreign studies conducted in accordance with GCP became effective on October 27, 2008. Again, the GCP regulations to which the FDA is deferring are those developed through the ICH, in particular, E6. There is concern that some of the requirements of the Declaration of Helsinki will be lost, most notably the protection of underprivileged subjects relative to the use of placebos. However, in addition to ICH E6, the FDA defines GCP as including the requirement of review and approval by an independent ethics committee (IEC) before initiation of the study and during the ongoing study, as well as obtaining and documenting freely given informed consent of study subjects.

In September 2009, the FDA and the European Medicines Agency (EMA) entered into the agreement known as the EMA-FDA GCP Initiative [11]. The initiative is a confidentiality arrangement in effect through 2010 that allows the sharing of:

- advance drafts of legislation and/or regulatory guidance documents
- information related to the authorization and supervision of medicinal products
- information, including inspection reports, about GCP inspections for specific products.

The reference GCP standard for the inspections is again ICH E6: the Clinical Trial Directive 2001/20/EC for the EMA, and Title 21 CFR Parts 11, 50, 54, 56, 58, 312, and 314 for the FDA.

In June 2010, the US Office of the Inspector General (OIG) issued the results of a study entitled "Challenges to the FDA's Ability to Monitor and Inspect Foreign Clinical Trials" [12]. There were two primary objectives of the study:

- to determine the extent to which sponsors submitted data from foreign clinical trials to support drug and biological marketing applications approved by the FDA in 2008
- to determine the extent to which the FDA monitors and inspects foreign clinical trials that support marketing applications.

The primary findings of the study were that:

- Eighty percent of approved marketing applications contained data from foreign clinical trials. A large proportion of clinical trial subjects and investigator sites (over 50%) were located outside the USA. Central and South America had the highest number of subjects per site, while, Western Europe had the most sites and subjects.
- The FDA inspected clinical investigators at only 1.2 percent of the clinical trial sites for the all applications approved in fiscal year 2008. As for sites, the FDA inspected 1.9 percent of domestic sites and 0.7 percent of foreign sites.
- Challenges to conducting foreign inspections and data limitations inhibit the FDA's ability to monitor foreign clinical trials. Sponsors are increasingly conducting early phase studies outside the USA without INDs since, as stated previously, this is allowed under 21 CFR 312. Without an IND, sponsors do not have to notify the FDA of clinical trials before they start and this may mean that the FDA is totally unaware of a clinical trial until after the data have been submitted for review.

To remediate the findings, the OIG made recommendations that were accepted in total or in part by the FDA in its response that was also published along with the study report in June 2010. These recommendations were that:

- The FDA should require standardized electronic clinical trial data and create an internal database. The OIG found that the FDA was unable to locate some data that had been submitted on paper. An electronic submission would ensure that reviewers had all information from sponsors at the time of data review for a submission. Standardized data would also aid in the selection of clinical trial sites to inspect and in meeting review timelines. The internal database would allow the FDA to identify trends and conduct analyses to determine where trials are being conducted and identify areas of risk, both of which could lead to focusing oversight in regions or at sites where trends may suggest that below-par subject health and safety exists.
- The FDA should monitor trends in foreign clinical trials not conducted under INDs and, if necessary, take steps to encourage sponsors to file INDs. An IND allows the FDA to review the study protocol before subjects are enrolled and it gives more opportunity to ensure data integrity through real-time inspections by the agency. Upon review, the FDA can request modifications to the protocol or prevent the trial from starting if concerns warrant such actions. In foreign trials, even though

they are conducted under ICH E6, the FDA loses or never has this level of oversight as it does in the USA. By monitoring trends, the FDA may be able to determine if the rights, safety, and well-being of subjects have been compromised. While this will be discovered after the fact, the information can be leveraged when trying to convince sponsors to file an IND for all proposed drug and biological development products.

- The FDA should continue to explore ways to expand its oversight of foreign clinical trials by:
  - Continuing to develop inspectional agreements with foreign regulatory bodies such as the one reached with the EMA. This would allow the FDA to maximize its resources allocated to inspecting foreign clinical trial sites.
  - Inspecting clinical trials in more countries. The OIG emphasized that by doing so the FDA would communicate the importance of complying with US regulations. It specifically recommended that the FDA target trials in countries that it has not yet inspected or in those countries where GCP has been recently adopted.
  - Looking to new models of oversight such as a quality risk management approach. This approach would allow the FDA to focus on identifying and analyzing risk factors unique to each investigational drug or biological.

The FDA agreed to all three recommendations and its response indicated that it has ongoing efforts or is developing new procedures to address each recommendation. In particular, the FDA cited the site selection tool that was being piloted at the time of the report as a means toward the standardization of clinical trial data. The FDA also committed to trending data to determine whether there is a difference in data integrity and human subject protection between domestic and foreign clinical trial sites. With respect to expanding its oversight of foreign clinical trials, if the agreement with the EMA proves to be successful, it would indeed leverage this partnership with other regulatory bodies.

These recommendations and the FDA's response mean that sponsors can expect more FDA oversight of foreign clinical trials.

REFERENCES

[1] US Food and Drug Administration. www.fda.gov/aboutFDA/Centersoffices/default.htm

[2] The "Wiley Act", Public Law No. 59–384. http://www.fda.gov/regulatoryinformation/legislation/ucm148690.htm

[3] British Medical Journal 313, 1448. http://www.cirp.org/library/ethics/nuremberg/ (December 7, 1996)
[4] http://www.historycentral.com/Today/HelsinkiAccords.html
[5] http://www.legislation.gov.uk/ukpga/1968/67
[6] Office of Human Subjects Research, National Institutes of Health. Regulations and Ethical Guidelines. http://ohsr.od.nih.gov/guidelines/belmont.html
[7] ICH Guidelines for Good Clinical Practices (E6-R1).
[8] FDA – 21 CFR Part 58 – Good Laboratory Practices.
[9] Code of Federal Regulations. Title 21 – Food and Drugs. Subpart B – Investigational New Drug Application. 312.20 Requirement for an IND. (2009)
[10] Code of Federal Regulations. Title 21 – Food and Drugs. Subpart C – Administrative Actions. 312.40 General requirements for use of an investigational new drug in a clinical investigation. (2009)
[11] bhttp://www.fda.gov/InternationalPrograms/FDABeyondOurBordersForeignOffices/EuropeanUnion/EuropeanUnion/EuropeanCommission/ucm189508.htm
[12] OEI-01-08-0 0510 Challenges to the FDA's ability to monitor and inspect foreign clinical trials.

CHAPTER 4

European Union Regulations

Janice B. Wilson
Wilson Quality & Compliance Consulting, Antioch, CA, USA

Contents

4.1. OVERVIEW OF THE EUROPEAN UNION

The European Union (EU) is composed of 27 member states: Belgium, Denmark, France, Germany, Greece, Ireland, Italy, Luxembourg, the Netherlands, Portugal, Spain, United Kingdom, Finland, Sweden, Austria, Cyprus, Estonia, Latvia, Lithuania, Malta, Poland, Slovakia, Slovenia, Czech Republic, Hungary, Romania, and Bulgaria.

Committed to regional integration, the EU was established by the Treaty of Maastricht in 1993 upon the foundations of the European Communities. The actual birth of the EU began with the establishment of the European Coal and Steel Community (ECSC) shortly after World War II. It existed from 1945 to 1957. The founding members of the community were Belgium, France, Italy, Luxembourg, the Netherlands, and West Germany.

In 1957, the six countries signed the Treaties of Rome, which extended the cooperation of the ECSC into other areas, creating the European Economic Community (EEC). Two other separate communities were

Global Clinical Trials
ISBN 978-0-12-381537-8, Doi:10.1016/B978-0-12-381537-8.10004-4

established as well, the Customs Union and the European Atomic Energy Community.

In 1965, an agreement was reached that led to the signing of the Merger Treaty in Brussels, which came into force in 1967 with the formation of a single set of institutions for the three communities collectively referred to as the European Community (EC).

The enlargement of the EC began in 1973 with the inclusion of Denmark, Ireland, and the United Kingdom. Greece joined in 1981 and Spain and Portugal in 1986. After the fall of the Iron Curtain in 1990, the former East Germany became a part of the community. Other member states continued to join, Austria, Sweden, and Finland in 1995, with the largest number joining in 2004 with the inclusion of Malta, Cyprus, Slovenia, Estonia, Latvia, Lithuania, Poland, the Czech Republic, Slovakia, and Hungary. Romania and Bulgaria joined in 2007. Today, there are four official candidates for inclusion, Croatia, Iceland, Macedonia, and Turkey. To join the EU, a country must meet the Copenhagen criteria, defined at the 1993 Copenhagen European Council.

Iceland, Liechtenstein, Switzerland, and Norway are four Western European countries that are not part of the EU but have partly committed to the EU's regulations.

The EU has four decision-making institutions: the Council of Ministers, the European Commission, the European Parliament, and the European Court of Justice. The Council of Ministers has the final decision in legal matters and is presided over for a period of six months by each of the 27 member states in turn. The Council of Ministers is located in Brussels.

For additional information regarding the economics and regulations of the EU, please refer to the EU's website, the source for much of the information in this section of this chapter [1].

4.2. OVERVIEW OF THE EUROPEAN MEDICINES AGENCY

The European's agency for the evaluation of medicinal products is today known as the European Medicines Agency (EMA). From its establishment in 1995 to 2004, the EMA was known as the European Agency for the Evaluation of Medicinal Products (EMEA).

The EMA, located in London, operates as a decentralized body (as opposed to a regulatory authority) of the EU.

An independent Management Board whose members are appointed to act in the public interest governs the EMA. The 35-member Board sets the Agency's budget, approves the annual work program and is responsible for ensuring that the Agency works effectively and cooperates successfully with partner organizations across the EU and beyond. The members do not represent any government, organization, or sector. All members are required to make an annual declaration of any direct or indirect interests they have in the pharmaceutical industry. The 35 members of the Management Board consist of:

- one representative of each of the 27 member states
- two representatives of the European Commission
- two representatives of the European Parliament
- two representatives of patients' organizations
- one representative of doctors' organizations
- one representative of veterinarians' organizations.

In addition to the members, the Management Board has one observer from each of the three European Economic Area–European Free Trade Association (EEA-EFTA) states, Iceland, Liechtenstein, and Norway.

4.2.1. Authorization and Regulation of Drugs

Medicines can be authorized in the EU by using either the centralized authorization procedure or national authorization procedures.

4.2.1.1. Centralized Authorization Procedure

The EMA is responsible for the centralized procedure (also known as the Community authorization procedure) for human and veterinary medicines. This procedure results in a single marketing authorization (called a community marketing authorization) that is valid across the EU, as well as in the EEA/EFTA states (Iceland, Liechtenstein, and Norway). The centralized procedure is compulsory for human medicines that are:

- derived from biotechnology processes, such as genetic engineering
- advanced-therapy medicines, such as gene therapy, somatic cell therapy, or tissue-engineered medicines
- intended for the treatment of HIV/AIDS, cancer, diabetes, neurodegenerative disorders, or autoimmune diseases and other immune dysfunctions
- officially designated "orphan medicines" (medicines used for rare diseases).

For medicines that do not fall within these categories (the "mandatory scope"), companies have the option of submitting an application for a centralized marketing authorization to the EMA, as long as the medicine concerned is a significant therapeutic, scientific, or technical innovation, or if its authorization would be in the interest of public health.

Applications through the centralized procedure are submitted directly to the EMA. Evaluation by the Agency's relevant scientific committee takes up to 210 days, at the end of which the committee adopts an opinion on whether the medicine should be marketed or not. This opinion is then transmitted to the European Commission, which has the ultimate authority for granting marketing authorizations in the EU.

Once a community marketing authorization has been granted, the marketing-authorization holder can begin to make the medicine available to patients and healthcare professionals in all EU countries.

4.2.1.2. National Authorization Procedures

Each EU member state has its own procedures for the authorization, within its own territory, of medicines that fall outside the scope of the centralized procedure.

There are also two possible routes available to companies for the authorization of such medicines in several countries simultaneously:

- **Decentralized procedure**. Using the decentralized procedure, companies may apply for simultaneous authorization in more than one EU country of medicines that have not yet been authorized in any EU country and that do not fall within the mandatory scope of the centralized procedure.
- **Mutual-recognition procedure**. In the mutual-recognition procedure, a medicine is first authorized in one EU member state, in accordance with the national procedures of that country. Following this, further marketing authorizations can be sought from other EU countries in a procedure whereby the countries concerned agree to recognize the validity of the original, national marketing authorization.

In some cases, disputes arising in the decentralized or mutual-recognition procedures can be referred to the EMA for arbitration as part of a referral procedure. Otherwise, the EMA does not get involved in the authorization of medicines via the national authorization procedure.

For additional information, the reader is referred to the EMA website, the source of much of the background information presented in this section of this chapter [2].

4.3. EMA COMMITTEES

As with the various branches/divisions of the US Food and Drug Administration (FDA), there are six scientific committees in the EMA that carry out scientific evaluation on applications from pharmaceutical companies. The committees and a brief description of their responsibilities are listed in Table 4.1. These committees normally meet on a monthly basis and are comprised of members nominated by the member states.

4.3.1. EMA Authority

The Committee for Medicinal Products for Human Use (CHMP) has issued a large number of guidelines related to good clinical practice (GCP). These guidelines are not legally binding, but have the same regulatory status as the FDA's guidance documents.

The standards that are included in EU regulations and many of the CHMP's guidelines are developed by one of four European organizations listed in Table 4.2. The CHMP has the authorization to modify, but generally adopts the standards as they are issued. These organizations, in turn, on a volunteer basis may incorporate all or parts of the standards developed by the International Organization for Standardization (ISO) and the International Electrotechnical Commission (IEC) into the standards they develop.

4.4. REGULATIONS AND GUIDANCE RELATED TO PRODUCT AUTHORIZATION

While the EMA (CHMP) guidelines for GCP are not legally binding, to get a product approved for marketing in the EU to which the guidelines have not been applied to an investigational article of that product, companies must explain in a document called an Expert Report why they chose not to apply the guidelines. This formal document must be submitted along with the marketing application.

For products eligible for or requiring centralized approval, a company submits an application for a marketing authorization to the EMA. A single evaluation is carried out through the CHMP. If the relevant committee concludes that quality, safety, and efficacy of the medicinal product are sufficiently proven, it adopts a positive opinion. This is sent to the European Commission to be transformed into a marketing authorization valid for the whole of the EU. The EMA's Committee on Orphan Medicinal Products (COMP) administers the granting of orphan drug status.

Table 4.1 European Medicines Agency (EMA) committees

Committee	**Primary responsibility**
Committee for Medicinal Products for Human Use (CHMP)	The CHMP was established by Regulation (EC) No. 726/2004. It replaced the former Committee for Proprietary Medicinal Products (CPMP). It is responsible for preparing the EMA's opinions on all questions concerning medicines for human use
Committee for Medicinal Products for Veterinary Use (CVMP)	The CVMP was established by Regulation (EC) No. 726/2004. It is responsible for preparing the EMA's opinions on all questions concerning veterinary medicinal products
Committee for Orphan Medicinal Products (COMP)	The COMP is responsible for reviewing applications from persons or companies seeking "orphan medicinal product designation" for products they intend to develop for the diagnosis, prevention or treatment of life-threatening or very serious conditions that affect not more than five in 10,000 persons in the European Union
Committee on Herbal Medicinal Products (HMPC)	The HMPC was established in September 2004, replacing the CPMP Working Party on Herbal Medicinal Products. The Committee was established in accordance with Regulation (EC) No. 726/2004 and Directive 2004/24/EC, which introduced a simplified registration procedure for traditional herbal medicinal products in European Union member states
Paediatric Committee (PDCO)	The main responsibility of the PDCO is to assess the content of pediatric investigation plans and adopt opinions on them in accordance with Regulation (EC) 1901/2006 as amended. This includes the assessment of applications for a full or partial waiver and assessment of applications for deferrals
Committee for Advanced Therapies (CAT)	The CAT was established in accordance with Regulation (EC) No 1394/2007 on advanced-therapy medicinal products (ATMPs). It is a multidisciplinary committee, gathering together some of the best available experts in Europe to assess the quality, safety, and efficacy of ATMPs, and to follow scientific developments in the field

Table 4.2 European standards organizations

CEN	European Committee for Standardization
CENELEC	European Committee for Electrotechnical Standardization
EOTA	European Organization for Technical Approvals: provides technical assessments of construction products to determine their fitness for intended uses
ETSI	European Telecommunications Standards Institute

Since July 2008, all new applications for the marketing authorization of new pharmaceutical products have had either to include data from pediatric studies [previously agreed with the Paediatric Committee (PDCO)] or to demonstrate that a waiver or a deferral of these studies has been obtained by the PDCO. From January 2009, this obligation was extended to most variations (new therapeutic indications) of already authorized products.

The majority of existing medicines throughout the EU member states remain authorized nationally, but the majority of genuinely novel medicines are authorized through the EMA.

The EMA has a staff of about 500, and decentralizes its scientific assessment of medicines by working through a network of about 3500 experts throughout the EU. The EMA draws on the resources of National Competent Authorities (NCAs) of EU member states.

The CHMP is obliged by the regulation to reach decisions within 210 days, though the clock is stopped if it is necessary to ask the applicant for clarification or further supporting data. Taking into consideration the allowance of time for applicants to respond to the 210 day questions, this compares well with the average of 500 days taken by the US FDA [3].

4.5. EMA GOOD CLINICAL PRACTICES

The EMA recognizes that GCP is an international ethical and scientific quality standard for designing, recording, and reporting trials that involve the participation of human subjects [1].

As with the FDA, the EMA considers compliance with this standard as assurance to the public that the rights, safety, and well-being of trial subjects are protected, consistent with the principles that have their origin in the Declaration of Helsinki, and that the clinical trial data are credible.

Requirements for the conduct of clinical trials in the EU, including GCP and good manufacturing practice (GMP) and inspections of these, have been implemented in the Clinical Trial Directive (Directive 2001/20/EC) and the GCP Directive (Directive 2005/28/EC).

Because member states may authorize products via the national authorization procedures, information concerning the activities in member states should be reviewed at the Heads of Medicines Agencies website [4]. The Heads of Medicines Agencies is a network of the heads of the national competent authorities whose organizations are responsible for the regulation of medicinal products for human and veterinary use in the EEA.

4.5.1. European Union Harmonization

As one of many parts of the EU's process of harmonizing the regulations and standards of the member states, the EU has issued guidelines and guidance documents. These can be found on http://ec.europa.eu/enterprise/pharmaceuticals/eudralex.

Work involving harmonization and coordination of GCP-related activity at EU level is carried out in the EMA's Inspections Sector. The Inspections Sector is involved in coordinating GCP inspections for the centralized procedure. Volume 10 of the Rules Governing Medicinal Products in the EU (the Eudralex), Clinical Trials Guidelines, brings together information on clinical trial authorization, safety monitoring, GCP inspections, and GCP and GMP requirements for clinical trials in the EEA.

The number of individual guidelines in the six chapters of this volume are too numerous to list in this book. However, this is where you will find all the guidelines and directives related to the application process, monitoring and pharmacovigilance, quality requirements for the investigational medicinal product (IMP), inspections to prepare for, legislation related to GCP and other documents, regulations, and guidance such as International Conference on Harmonisation (ICH) E6.

4.5.2. International Clinical Trials

As with many of the new drug submissions in the USA containing clinical trials conducted outside the USA, a large percentage of marketing authorization applications come from countries outside the EU in whole or in part. The EU adopted the ICH-GCP guideline in July 1996. ICH has developed unified standards for Europe, the USA, and Japan. As listed in the previous chapter, the relevant ICH guidelines are ICH E1, E2A-E, E3, E6, and E8. Additional information on ICH can be found on the ICH website [5].

The EU requires that clinical trials included in marketing authorization applications in the EEA be conducted in accordance with GCP (Directive

2001/83/EC Annex I, as amended by Directive 2003/63/EC) and the ethical standards of Directive 2001/20/EC as initially published on May 1, 2001. All the member states at that time had until May 1, 2003 to implement, prepare and publish the EU Directive into the national law and then another year before the effectiveness date of May 1, 2004. Most countries met the deadline, but some, especially the most recent member states, were late. As of January 2008, the EU Directive had been implemented in all 27 member states. The three EEA countries (Norway, Iceland, and Lichtenstein) comply with the EU Directive as well.

The EU Directive (2001/20/EC) applies to all clinical trials on investigational medicinal products, Phase II–IV, except for non-interventional trials. On October 8, 2003, 2003/94/EC including the GMP rules (Annex 13) was published. This was followed by 2005/28/EC on April 8, 2005: GCP, Trial Master File, and IMP. From 2004 to 2007, the latest editions of guidance documents were issued, to support the 2001 Directive.

Again, even though each member state has implemented the EU Directive into its national law, as each country has its own interpretation and some have additional requests, it is not sufficient to follow the EU Directive in case of a clinical trial in a member state: the local law should always be reviewed and followed.

Additional information on GCP requirements in the EU can be found on the websites of international organizations such as the Council for International Organizations of Medical Science (CIOMS) and the World Medical Association (WMA), responsible for the Declaration of Helsinki.

4.5.3. Clinical Trials Conducted in Developing Countries

The EMA has published a strategy paper on acceptance of clinical trials conducted in developing countries, for evaluation in marketing authorization applications, the EMEA strategy paper: "Acceptance of clinical trials conducted in developing countries, for evaluation in marketing authorization applications" [6]. The strategy is now being developed by an EMA working group, which includes GCP inspectors, assessors, and representatives from patient groups.

As in the USA's OIG June 2010 report [7], the EMA has determined that the number of clinical trials conducted, and of patients recruited into clinical trials, in countries outside the "traditional" Western European and North American research areas has been increasing for a number of years. Through its tracking of the geographical origin of patients in pivotal trials submitted through its centralized marketing authorization applications (MAA) process, it determined that approximately one-quarter of patients recruited in pivotal trials submitted between 2005 and 2008 were recruited in countries in Latin America, Asia, the Commonwealth of Independent States, and Africa. The EMA's findings are delineated in the report "Clinical trials submitted in marketing authorization applications to the EMA: overview of patient recruitment and the geographical location of investigator sites" (updated with data from marketing authorization applications submitted in 2009) [8], published in November 2010, a mere five months after the OIG's report.

Thus, as with the OIG's recommendations to the FDA (accepted by the FDA), the EMA's strategy includes but is not limited to:

- increasing the number of inspections in these geographical areas during the clinical trials as well as review of submissions
- more transparency during the planning and development of clinical trials
- increased training and other practical applications related to the EMA requirements.

As part of the Agency's overall strategy on the acceptance of clinical trials conducted in third countries, EMA published on May 28, 2010 a draft, "Reflection paper on ethical and GCP aspects of clinical trials conducted in developing countries for evaluation in marketing authorization applications for medicines for human use submitted to the EMA" [9]. The document was under public consultation until September 30, 2010, and highlights the need for cooperation between international regulatory authorities. The paper proposes a series of measures to ensure a robust framework for the oversight and conduct of clinical trials, no matter where in the world investigators' sites are located and patients are recruited.

4.5.4. EMA-FDA GCP Initiative

The EMA and the US FDA launched a joint initiative to collaborate on international GCP inspection activities in September 2009. This initiative comes under the scope of the confidentiality arrangements between the European Commission, the EMA, and the FDA. The key objectives of the initiative listed below included the sharing of information on inspection planning, policy, and outcomes, and the conduct of collaborative inspections.

The key objectives of the EMA-FDA GCP initiative, which began with an 18-month pilot phase on September 1, 2009, are:

- to conduct periodic information exchanges on GCP-related information in order to streamline sharing of GCP inspection planning information, and to communicate effectively and in a timely manner on inspection outcomes

- to conduct collaborative GCP inspections by sharing information, experience, and inspection procedures, cooperating in the conduct of inspections and sharing best-practice knowledge
- to share information on interpretation of GCP, by keeping each regulatory agency informed of GCP-related legislation, regulatory guidance and related documents, and to identify and act together to benefit the clinical research process.

More detailed information about the EMA-FDA GCP initiative can be found in the document "EMEA-FDA GCP Initiative-Terms of engagement and procedures for participating authorities" which is found on the EMA website as Doc Ref. EMEA/INS/GCP/538414/2008 FDA Mod. 20 August 2009.

4.6. REQUIREMENTS FOR MARKETING AUTHORIZATION APPLICATIONS

The requirements for the EU's MAAs are very similar to those for investigational new drug (IND) applications in the USA. However, in addition to consideration of the fact that ICH E6 was also adopted by the USA and Japan, in May 2010, the EMA published document EMA/712397/2009, "Reflection paper on ethical and GCP aspects of clinical trials of medicinal products for human use conducted in third countries and submitted in marketing authorization applications to the EMA", in an effort to provide guidance to foreign sponsors wishing to submit MAAs in the EU. In this reference document, the EMA clearly indicates that clinical trials conducted in third countries (non-EU member state and North America) and used in MAAs must be conducted on the basis of principles equivalent to the ethical principles and principles of good clinical practice applied to clinical trials in the EU. The primary principles specifically read like those of the Belmont Report, Helsinki Declaration, and ICH E6; during medical research, researchers must ensure that three ethical principles are adhered to: (a) respect for persons, (b) beneficence/non-maleficence, and (c) justice, where respect for persons includes the respect for autonomy and the protection of dependent and vulnerable persons, beneficence/non-maleficence is defined as the ethical obligation to maximize benefits and to avoid or minimize harms, and justice is a fair distribution of the burdens and benefits of research.

The EMA explicitly states the requirement that an independent ethics committee, or its equivalent such as an independent review board required by the FDA, meets all local requirements relative to study oversight.

Other sections explicitly covered in EMA/712397/2009 are:

- information/consent procedure
- confidentiality
- fair compensation
- vulnerable populations
- placebo and active comparator
- access to treatment post-trial
- applicability to EU populations
- study feasibility
- inspections.

Even though there have been strides made toward harmonization, there are noteworthy differences between the requirements of the FDA and the EMA. In the EU, GCPs and GMPs are more interrelated than in the USA. The interrelation begins with the reference to the EU GMPs and Annex 13 in the clinical trial directive. While the FDA has not finalized guidance documents relative to GMPs for clinical trial materials, both the FDA and the EMA require product information before the submission of new drug applications. For the FDA, the chemistry, manufacturing, and control (CMC) section of the submission addresses most if not all the requirements of the EMA's product characterization section of its MAA.

Another area of difference is quality assurance (QA) oversight. In addition to QA groups in the EU reviewing batch recodes against specification as in the USA, there is the requirement that a qualified person (QP) must certify that the clinical trial material meets the appropriate level of GMP requirements. This means that the QP must have audited all the third party clinical trial material sources.

In the USA, changes to original IND submissions that have safety implications are handled by amendments followed by verbal or written confirmation that the changes are acceptable. Other updates to the IND may be made without prior approval in an annual report or an information amendment such as the postapproval changes being effected (CBE) process.

In the EU, updates are made through amendments, which can be classified as major or minor. Since the 2004 changes, all major amendments require prior approval from the competent authorities; however, there is no consistency among the 27 member states about what constitutes a major amendment. For example, in the UK, the Medicines and Healthcare Products Regulatory Agency (MHRA) expects to review formally and approve the original and any extensions to the provisional or tentative expiry (or shelf-life or retest) date, and has requested the actual stability data that were used by the sponsor to project the expiry date. In contrast, this is a notification in other EU member states.

Another area of significant differences is in reporting serious adverse events during clinical trials. In the USA, all serious adverse events that occur during a study are reported, while in the EU if the adverse event is caused by a known or expected side-effect, the reporting need not be done in the same period as those events that are not.

4.7. MEDICAL DEVICE CLINICAL TRIAL REQUIREMENTS

With respect to medical devices, in addition to other requirements, the EU member states have agreed to comply with a set of standards entitled "EN 540: Clinical Investigation of Medical Devices for Human Subjects" that was produced by the European Committee for Standardization (CEN). Member nations of CEN include all the EU countries as well as Iceland, Norway, and Switzerland. EN 540 provides a framework for voluntary compliance with the terms of the "clinical evaluation" and "clinical investigation" sections of the EU's medical device directive 90/385 for active implantable devices, and 90/683 for all other devices.

EN 540 is similar in purpose to the FDA's regulations for informed consent and the FDA's investigational device exemption (IDE) regulations (21 CFR Part 50, 21 CFR Part 812). Both EN 540 and the IDE rules are designed to protect study subjects being treated with investigational devices while scientific data are collected to support a medical device application for approval. While FDA's IDE regulations cover all studies on humans for the purpose of investigating the safety and effectiveness of devices, the purpose of EN 540 is to "protect subjects and ensure the scientific conduct of the clinical investigation". However, similarities include provisions for investigational plans, study monitoring, record-keeping, and informed consent.

Other general requirements set forth in EN 540 address patient confidentiality; general qualifications for investigators, monitors, and other persons involved in a clinical study; and the termination of studies if unforeseen or increased risks occur. An entire section of the standard is devoted to methodology, including the respective roles of sponsors, monitors, and investigators, and documentation of the following:

- justification for the study
- plan for the clinical investigation

- agreements between sponsor, monitor, and investigator
- provisions made to compensate subjects in the event of injury arising from participation in a clinical investigation.

EN 540 also requires that a final report of the clinical investigation be drafted and all investigators sign it.

Under EN 540, an ethics committee operating under the rules of the particular country must approve or express an opinion about a clinical study. In the EU, individual countries also may have different levels of study review depending on the device and the amount of risk it may present.

Requirements in EN 540 and the IDE regulations regarding confidentiality, qualification of those participating (for example, investigators), and study termination are analogous. The methodology section of EN 540 states that the role of the sponsor includes the selection of qualified investigators and monitors (or assuming the monitor's responsibilities) and record-keeping. The provisions in 21 CFR Part 812 regarding the roles of sponsors are similar.

Monitors have responsibilities similar to those set forth in the FDA's regulations, including the reporting of adverse events and adverse device effects to the sponsor. Investigators' responsibilities are similar as well. Both sets of clinical standards provide for relief from informed consent in emergency situations as well as for termination of studies in the case of unanticipated adverse events. However, the FDA's regulations provide more detail regarding the situations in which informed consent may not be necessary, hence limiting the number of situations in which relief from informed consent can be obtained (21 CFR Part 50).

While many similarities between the two standards exist, there are significant differences. The most significant difference is that, while the IDE regulation exempts manufacturers from all requirements for approved devices under the Federal Food, Drug, and Cosmetic Act, the CEN standard does not exempt manufacturers from these other obligations.

Also, the IDE exemption permits unapproved devices to be distributed in interstate commerce to qualified investigators to evaluate their safety and effectiveness. It appears that European investigational devices are subject to all requirements applicable to approved or cleared devices, unless they are individually exempted.

Another major difference between the EU's and FDA's requirements is that EN 540 specifically exempts in vitro diagnostics (IVDs) from its purview and states that the standard applies only to a medical device when its clinical performance needs assessment prior to being placed on the market.

It is unclear from the EN 540 standard exactly what criteria are used to determine whether or not a device's clinical performance needs assessment, but it can be assumed that decisions are made according to the rules of ethics committees, which range from country to country and are similar to institutional review boards in the USA. The involvement of such ethics committees, which are subject to individual countries' sets of rules, introduces a random element to the European clinical trials standard for medical devices.

Under EN 540, the sponsor is responsible for ensuring that provisions for compensation in the event of injury arising from participation in the investigation have been made. This provision has no analog in the FDA's regulations.

EN 540 lacks the IDE regulations' labeling requirements, although it does not exempt devices from normally applicable labeling and other requirements. Also, it does not prohibit the promotion, commercialization, prolonged study of, or false representation of an investigational device. Such activities are prohibited under the IDE regulation (21 CFR Section 812.7, see Appendix II).

Although EN 540 and the FDA's regulations attempt to minimize or eliminate the inappropriate entry of subjects into studies, EN 540 shows a particular concern for improper patient inducements, including payments for entry into a study.

Unlike the FDA regulations, EN 540 contains no provisions that state whether or not patients may be charged for investigational devices.

In the EU, in addition to meeting all other clinical trial requirements, it is illegal to market a product that does not have the CE Mark – the Communauté Européenne's official seal of approval – that certifies a product meets a set of "essential requirements", which are listed in annexes (appendices) to EU directives (regulations) and in guidelines issued by groups working to develop the EU's single internal market such as the CPMP. These directives and their annexes are published in the EU Official Journal. Copies of the directives and annexes can be obtained from the following locations:

US Department of Commerce
International Trade Administration
Office of EU Affairs
Single Internal Market Information Service (SIMIS)
Room 3036
14th & Constitution Avenue, NW

Washington, DC 20230
Tel: +1 202 482 5276

or

Principal Administrator
Commission of the European Union
Rond-Point Schuman #3
Second Floor #8
B-1049 Brussels, Belgium
Tel: 32 2 235 5956
Fax: 32 2 235 7191

REFERENCES

[1] http://europa.eu/abc/history/
[2] http://www.eme.europa.eu/ema
[3] Sherwood T. Generic Drugs: Overview of ANDA Review Process. CDER Forum for International Drug Regulatory Authorities. Food and Drug Administration, Office of Pharmaceutical Science. http://www.fda.gov/downloads/Drugs/NewsEvents/UCM167310.pdf; 2008 (accessed November 30, 2010).
[4] http://www.hma.eu/
[5] http://www.ich.org/
[6] http://www.ema.europa.eu/docs/en_GB/document_library/Other/2009/12/WC500016817.pdf
[7] OIG Report ID, OEI-01-08-00510, June 2010.
[8] http://www.ema.europa.eu/docs/en_GB/document_library/Other/2009/12/WC500016819.pdf
[9] http://www.ema.europa.eu/docs/en_GB/document_library/Regulatory_and_procedural_guideline/2010/06/WC500091530.pdf

Japanese Regulations

Mary Bareilles*, Suzanne Gagnon, Masanobu Kimura*, Hiroshi Naito* and Hiromi Wakita***

*ICON Clinical Research, Japan
**Medical, Safety and Scientific Services, ICON Clinical Research, USA

Contents

Global Clinical Trials
ISBN 978-0-12-381537-8, Doi:10.1016/B978-0-12-381537-8.10005-6

5.1. INTRODUCTION

Japan's current Pharmaceuticals and Medical Devices Agency (PMDA) (Figure 5.1) was established in April 2004 and as a result of the revised Pharmaceutical Affairs Law of 2005, the Japanese regulatory review process has moved into a new era. Various guidelines have been harmonized and there is now a common international understanding regarding clinical development. This chapter summarizes the current regulatory requirements for clinical development in Japan and offers some help for companies that plan to conduct clinical trials in Japan.

5.2. DOCUMENTS REQUIRED FOR APPLICATION

5.2.1. New Drug Classification

Figure 5.2 shows the classification of new drugs based on the review category specified for ethical drugs. Those drugs listed above the dotted line are regarded as new drugs.

- **Drugs containing new active ingredients**: Drugs with active ingredients which are neither contained in any drug currently approved for marketing in Japan nor any drug listed in the Japanese Pharmacopoeia (JP, hereafter referred to as "already-approved drugs"). "New active ingredients" may include active ingredients already used for other purposes, but not as pharmaceuticals. For example, ingredients that are commonly used as additives but not as active pharmaceutical ingredients are regarded as "new active ingredients" if they are used as such for the first time.
- **New ethical combination drugs**: Drugs whose active ingredients and/or their combination ratio differ from those for combination drugs

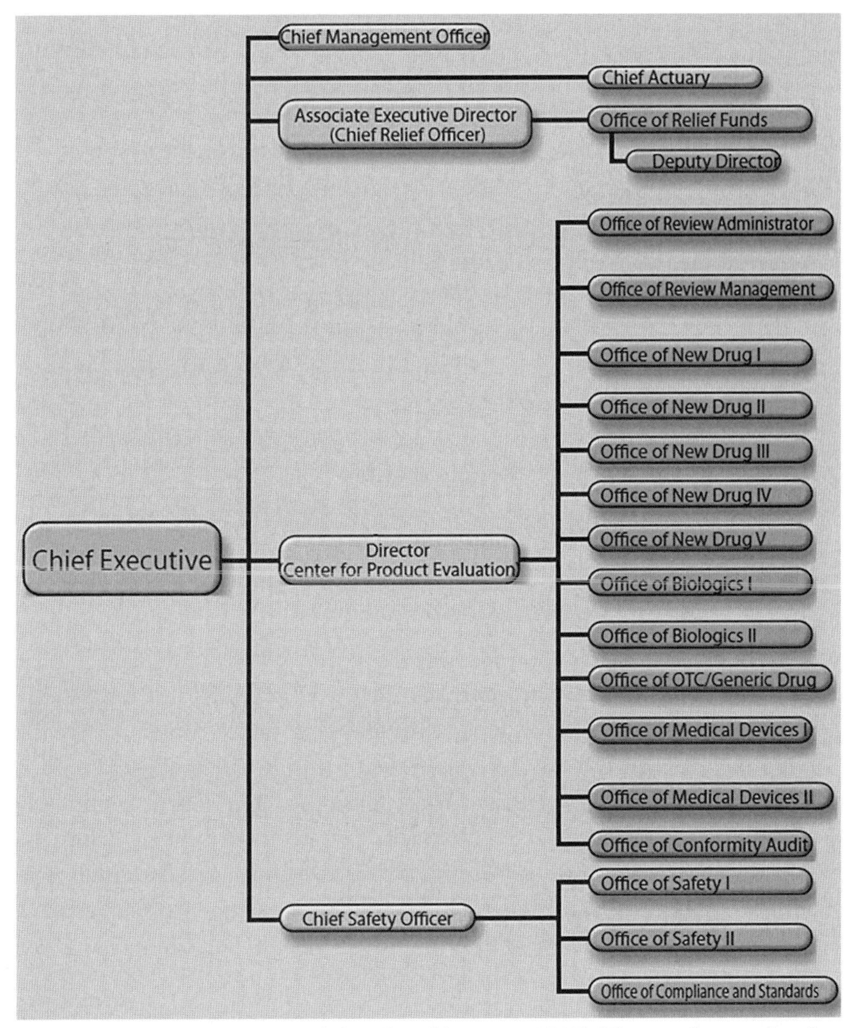

Figure 5.1 PMDA organizational chart (as of August 1, 2009). (Please refer to color plate section)

included in the JP or combination drugs approved for marketing as ethical drugs.

- **Drugs with a new administration route**: Drugs containing the same active ingredients as already-approved drugs, but which are administered through a different route (oral, subcutaneous, intramuscular, intravenous, percutaneous, intrarectal, transvaginal, eye drops, ear drops, nose drops, inhalation drugs, etc.).

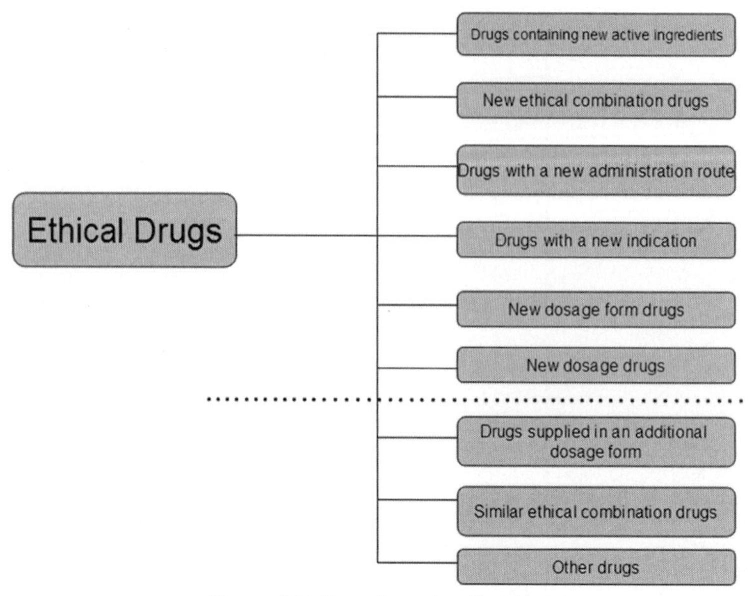

Figure 5.2 New drug classification

- **Drugs with a new indication**: Drugs containing the same active ingredients and administered through the same route as already-approved drugs, but currently approved for different indications.
- **New dosage form drugs**: Drugs with the same active ingredients, administration route, and indications as already-approved drugs, but used in different treatment regimens because of pharmaceutical changes.
- **New dosage drugs**: Drugs with the same active ingredients and administration route as already-approved drugs, but used in different doses.

5.2.2. Data to be Attached to Drug Approval Applications

The following information is required to be submitted with a drug approval application: origin or background of discovery, characteristics and efficacy, PMDA consultation records, a list of drugs of the same type and same indications, conditions of use in foreign countries, and package inserts.

- **Manufacturing methods, standards, and test methods**: Manufacturing method – must follow good manufacturing practice (GMP); bulk drugs and formulations – to be prepared on the basis of the guidelines (this is the most important item in the case of biological products).

- **Stability**: Bulk drugs and formulations – long-term stability studies, stress-stability studies (photostability, thermostablity, and humidity stability), and accelerated-stability studies (refer to the guidelines).
- **Pharmacological action**: Documentation to support efficacy, trial results of secondary/safety pharmacology and other pharmacology [good laboratory practice (GLP) applies to safety pharmacology].
- **Absorption, distribution, metabolism, and excretion**: Absorption, distribution, metabolism, excretion, and other pharmacokinetics.
- **Acute, subacute, and chronic toxicity, teratogenicity, and other types of toxicity**: Trial results of single-dose toxicity, repeated dose toxicity, genotoxicity, carcinogenicity, reproductive toxicity, local irritation, and other toxicities (refer to the guidelines and GLP).
- **Clinical studies**: Phase I, II, and III studies [must follow good clinical practice (GCP)].

5.2.3. NDA Application Procedures

The contents of this application form and a summary of the considerations for completion are provided in Figure 5.3.

5.2.3.1. Items for Drug Approval Applications

- **Non-proprietary name**: This item is needed in the case of drugs listed in "Minimum Requirements for Biological Products". A JAN (Japan Accepted Name) should be obtained approximately one year before the approval application is to be filed.
- **Brand name**: In principle, the brand name (registered trademark), formulation, and content of active ingredients are required.
- **Ingredients, composition, and essence**: Active ingredients per unit (e.g. 1 vial), the name of excipients and their specifications should be entered here. Regarding excipients, the purpose of the combination should be provided. A safety document is required for excipients that have not been used previously.
- **Manufacturing methods**: Manufacturing methods should be provided based on the plant where the active ingredients are manufactured (bulk drugs) from the starting materials to the final product. The means of process control and important parameters in the process should be included. (Information on and the process flowchart for both manufacturing plants and outsourcing laboratories are required.) Information on bulk drugs (names, storage conditions and expiry dates, standards, and test methods) is required. Packages and containers are also to be entered.

Form 22-1 (Article 38)

Revenue stamp

Application Form for Approval to Market Drugs

<table>
<tr><td rowspan="2">Name</td><td colspan="2">Non-proprietary name</td><td colspan="3"></td></tr>
<tr><td colspan="2">Brand name</td><td colspan="3"></td></tr>
<tr><td colspan="3">Ingredients, composition and essence</td><td colspan="3"></td></tr>
<tr><td colspan="3">Manufacturing methods</td><td colspan="3"></td></tr>
<tr><td colspan="3">Dosage and administration</td><td colspan="3"></td></tr>
<tr><td colspan="3">Indications</td><td colspan="3"></td></tr>
<tr><td colspan="3">Storage conditions and expiration date</td><td colspan="3"></td></tr>
<tr><td colspan="3">Standards and test methods</td><td colspan="3"></td></tr>
<tr><td colspan="2" rowspan="2">Manufacturing plant for marketing /distributing drug</td><td>Name</td><td>Address</td><td>Division of permit or accreditation</td><td>Number of permit or accreditation</td></tr>
<tr><td></td><td></td><td></td><td></td></tr>
<tr><td colspan="2" rowspan="2">Manufacturing plant for bulk drugs</td><td>Name</td><td>Address</td><td>Division of permit or accreditation</td><td>Number of permit or accreditation</td></tr>
<tr><td></td><td></td><td></td><td></td></tr>
<tr><td colspan="3">Remarks</td><td colspan="3"></td></tr>
</table>

As indicated above, we hereby apply for approval to market drugs.

Date

Name (Name and name of representative in case of a corporation)

Address (Head of office in the case of a corporation) Seal

Minister of Health, Labour and Welfare
Prefectural Governor

Figure 5.3 Form 22-1: application form for approval to market drugs

- **Dosage and administration**: This information is to be entered on the basis of the results of clinical studies.
- **Indications**: Diseases or conditions should be entered using medical terminology.
- **Storage conditions and expiry date**: Storage conditions that assure quality until the date of expiry is to be established, and a stability study is to be performed under such conditions so as to establish the expiry date. It is not necessary to enter products that are stable for more than three years.
- **Standards and test methods**: Standards are to be set in keeping with the guidelines; the values of the standards is to be established from the observed values.
- **Information on the manufacturing plant**: Information includes name, address, division, and authorization number (for overseas plants,

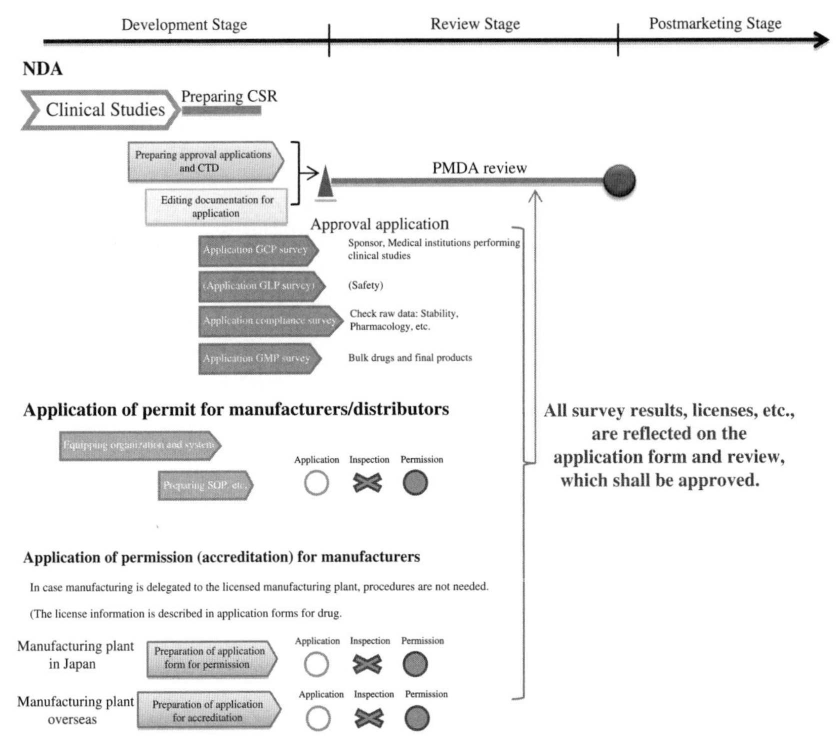

Figure 5.4 Pharmaceutical procedures in initial approval applications. (Please refer to color plate section)

this is their accreditation) for all plants manufacturing formulations and bulk drugs.

- **Remarks**: Package unit and request for entry in the Drug Price List. Details are entered for cases where stability testing is ongoing.

5.2.3.2. Pharmaceutical Procedures for Drug Approval Applications

Figures 5.4 and 5.5 are schematics of the procedures involved in submission of an initial new drug application (NDA). After completion of non-clinical and clinical trials, an approval application form and common technical document (CTD) are prepared and edited, and an application for approval of the new drug is made. GCP, GLP, compliance, and GMP surveys are required and should be applied for immediately after NDA submission. In addition, no license for application of approval will be granted unless the manufacture and/or distribution of a drug is authorized. If a permit for the manufacture and/or distribution of the drug is not already authorized, approval for permission should be applied for simultaneously

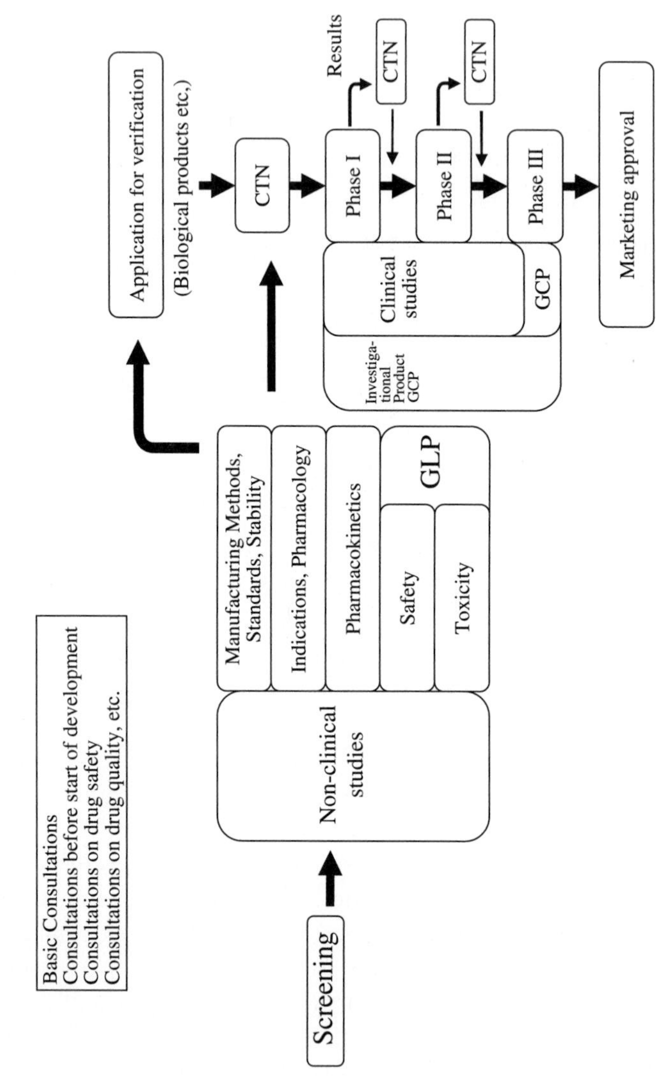

Figure 5.5 Process for new drug development in Japan. CTN: clinical trial notification

with NDA submission. If the manufacturer of a drug is not already authorized, such authorization must also be applied for because this information is requested in the application. If the manufacturing plant is located overseas, accreditation of the manufacturer must still be obtained.

5.3. REVIEW PROCESS FOR NEW DRUG APPLICATIONS

Figure 5.6 provides an overview of the review process by PMDA for NDAs in Japan.

5.3.1. Team Review

Frequently, the PMDA will perform a team review. The reviewing team is composed of experts in quality, non-clinical and clinical trials, biostatistics, and other fields (Figure 5.7).

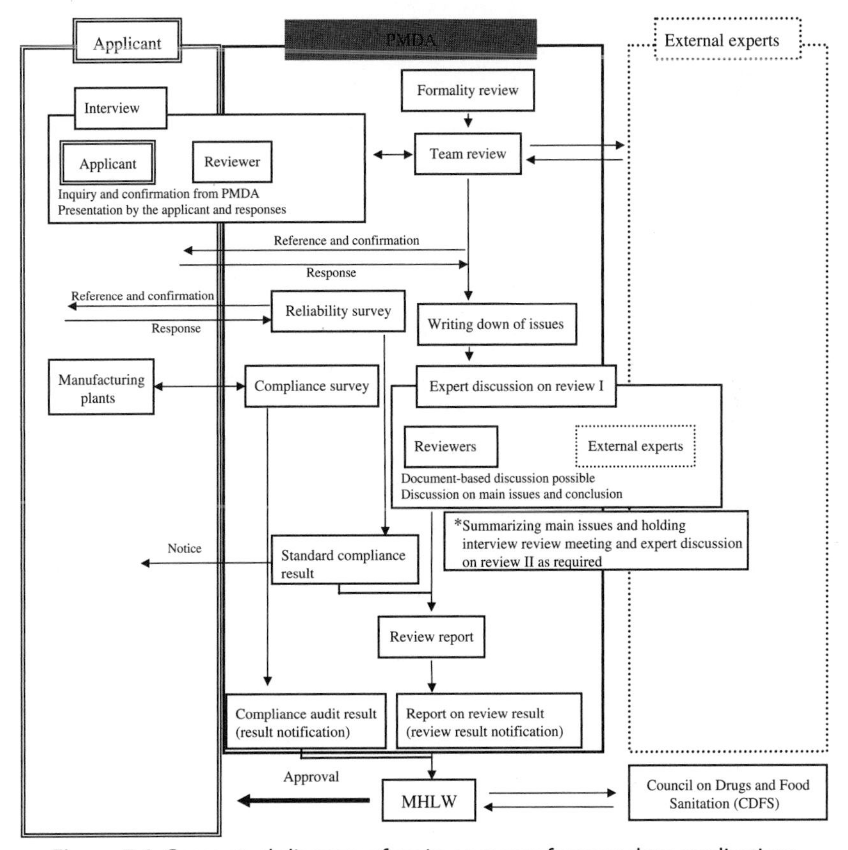

Figure 5.6 Conceptual diagram of review process for new drug applications

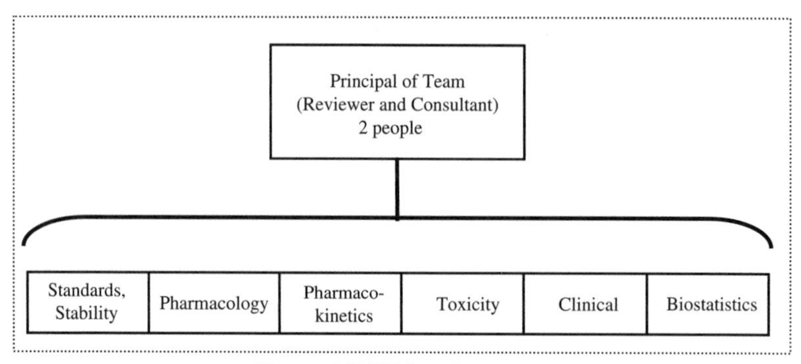

Figure 5.7 NDA review team

5.4. REQUIRED SURVEYS

5.4.1. GCP (Compliance) Survey

In compliance with GCP (Ordinance No. 28, March 27, 1997) concerning standards for preparing (sponsoring), managing, and conducting clinical trials, a survey is conducted to determine whether the clinical studies have been performed appropriately. This may include information on the facilities of both the sponsor and the investigators performing the clinical trial(s).

5.4.2. GLP (Compliance) Survey

Non-clinical studies which assess clinical safety must be performed in compliance with GLP (Article 14, Paragraph 3 of the Pharmaceutical Affairs Law). Therefore, the documentation used in the application must be obtained in accordance with GLP Inspection Guidelines. On-site inspections are usually required. However, in cases where satisfactory inspections have been performed within three years (for evaluation B, within two years) of the date on which confirmation of GLP compliance was issued, statements which assure the consistency of the study results and final reports can be attached.

5.4.3. Compliance Survey

In addition to the above surveys, other paper surveys or on-site inspections may be performed to confirm that the studies have been performed in accordance with standards for reliability. (Article 14, Paragraph 5 of the Pharmaceutical Affairs Law).

5.4.4. Good Manufacturing Practice (Quality Management System) Surveys

5.4.4.1. GMP (QMS) Compliance Survey

On-site and paper inspections are performed to check that the manufacturing process from bulk drugs to final product is in accordance with GMP standards (Article 14, Paragraph 6 of the Pharmaceutical Affairs Law). A GMP survey is also performed when a manufacturing plant or manufacturing process is changed.

5.4.4.2. GMP (QMS) Inspection of Manufacturing Plants and Application for Accreditation

When request for permission or accreditation is received to manufacture a new drug, the plans are checked to ensure compliance with standards. In the case of accreditation, a GMP survey is performed. This is revisited at least once every five years when permission is updated.

5.5. CONTENTS REVIEWED IN THE DRUG APPROVAL APPLICATION

- **Bulk drugs**: New active ingredients (bulk drugs) which have not previously been approved in Japan.
  - It is only possible to apply for approval of bulk drugs.
  - Before applying for the approval of a drug, it is necessary to follow the correct procedure to decide on a Japanese Accepted Name (JAN).
- **New excipients**: In cases involving the use of an excipient which has never been used as a drug or being used with different administration route (including adjuvant). The following are required for approval of applications requesting the use of a new excipient (Notification No. 0330009 of the Evaluation and Licensing Division, PFSB, dated March 31, 2005):
  - data on the origin or background of discovery and conditions of use in foreign countries
  - data on manufacturing processes, standards and test methods
  - data on stability (if there are unavoidable reasons, it is possible to use data on the stability of formulations)
  - data on toxicity (if there are reliable experiments, it is possible to use a published paper)

    single-dose toxicity

    repeated dose toxicity

reproductive toxicity
genotoxicity

  - New excipients cannot be applied for by themselves.
  - Note that there are some excipients that are not permitted to be dispensed as drugs.
- **Final products (formulations)**: Products that have actually been distributed as drugs – active ingredients, efficacy, and safety of formulations and quality.
- **Division of pharmaceutical regulations**: The postapproval division of the new drug (poisons, deleterious substances, prescription drugs, etc.) is reviewed and the division into bulk drugs and final products is decided on. Documents for the review of division designation are established. This is prepared and attached in the application for approval.
- **Package insert**: The draft package insert intended to be used in postmarketing is reviewed.
- **Basic protocol for postmarketing surveys**: Eligibility of the protocol for postmarketing surveys is reviewed.

5.6. DIFFERENCES BETWEEN J-GCP AND ICH-GCP

The Japanese regulatory authorities are particularly sensitive about how clinical trials are performed because of misconduct that occurred in the past. In order to maintain transparency of the processes used in performing clinical trials, Japanese (J)-GCP places greater responsibility on the institute and on the institutional head where a trial is performed, as follows:

- The contract is between the sponsor and the institute and not between the investigator and the sponsor.
- The head of the institute must consult the institutional review board (IRB) about the planned clinical trial.
- Each institute shall, in principle, have an IRB.

The following points specific to J-GCP also differ from International Conference on Harmonisation (ICH)-GCP:

- The items to be covered in the contract must be defined (Article 13).
- The sponsor must prepare a policy for compensation of injuries to a person's health incurred during the clinical trial, before submitting the necessary documents to the IRB (Article 14). This shall be included while obtaining informed consent (Article 51). In the case of injury, "the subject should be adequately compensated, *regardless of whether or not the injury is due to negligence* (authors' emphasis). In such cases, the subject

should not be burdened with proving a causal relationship, etc." (PFSB/ELD Notification No. 1001001, dated 1 October, 2008).

- In cases where the sponsor does not have an address in Japan, it must select an in-country caretaker (ICC). A subsidiary company in Japan can act as the ICC (Article 15).
- The standard for the clinical trial conducted by the sponsor–investigator is described in Section 2-2 (Article 15-2 to 15-9) and Section 3-2 (Article 26-2 to 26-12).
- Labels on the container or the wrapper of the investigational product shall be written in Japanese (Article 16).
- Safety information shall be reported periodically, i.e. every six months after submission of the initial clinical trial notification (CTN) for the study and within two months after completion of the study (Article 19).

The above is not all inclusive and it is important to consult a regulatory expert when planning any clinical trials in Japan.

5.7. NOTIFICATION OF THE INTENT TO PERFORM CLINICAL TRIALS

Persons or companies intending to sponsor clinical trials must notify the Minister in Charge of Clinical Study Protocols for Drugs in advance. This is considered to be especially important in understanding the conditions of clinical studies with regard to public health and hygiene.

5.7.1. Type of Clinical Trial Notification

5.7.1.1. Notification of a Clinical Trial Plan

It is necessary to notify the authorities of a clinical study protocol in the following situations:

(1) drugs with new active ingredients
(2) drugs with new routes of administration (excluding bioequivalence studies)
(3) new combination drugs, drugs with new indications and drugs with new dose regimens and dosages (prescription drugs only) (excluding bioequivalence studies)
(4) drugs which have the same ingredients as drugs with new active ingredients within the re-examination period (excluding bioequivalence studies)
(5) drugs anticipated to be biological products (excluding (1–4), and excluding bioequivalence studies)

(6) drugs manufactured using gene-recombination technology (excluding (1–5), and excluding bioequivalence studies).

5.7.1.2. Notification of a Change in a Clinical Trial Plan, Clinical Trial Completion, or Premature Termination of a Clinical Trial

When the person or company who submits a CTN changes the contents of the CTN, or completes or discontinues the study, notification thereof is to be submitted to the Minister. If the person or company who submits the CTN is an overseas manufacturer, a change of ICC must also be submitted.

5.7.1.3. Notification of Discontinuation of Development

When it is decided to cancel the development of drugs submitted in a CTN, the Director of the Evaluation and Licensing Division of the Pharmaceutical and Food Safety Bureau must be notified.

5.7.2. Attached Documentation

Documents to be attached to each type of notification are as follows.

5.7.2.1. Notification of Clinical Trial Plan

Initial notification:

- document stating the scientific justification for sponsoring the clinical trial (in cases where the person who conducts clinical trial submits the CTN, a document stating the scientific justification for performing the clinical trial)
- protocol
- written information and the form used for obtaining informed consent (in cases where these documents have the same contents and are used at two or more medical institutions, a single set of respective copies may be attached to the notification)
- sample case report form
- latest version of the investigator's brochure.

In cases where the person who conducts the clinical trial submits the CTN, the "Argument by IRB, approval by the head of a medical institution, etc." shall also be attached.

Second or subsequent notification(s):

- document stating the scientific justification for sponsoring the clinical trial, including a summary description of new trial results and information obtained after the previous submission

- protocol
- written information and the informed consent form used for obtaining informed consent (in cases where these documents have the same contents and are used at two or more medical institutions, a single set of respective copies may be attached to the notification)
- sample case report form
- latest version of investigator's brochure.

5.7.2.2. Notification of Changes in a Clinical Trial Plan

Documented changes, where required.

5.7.2.3. Notification of Premature Termination of a Clinical Trial

Documented reasons for premature termination, where required (which shall include information on trial cases treated up to said termination).

5.7.3. Reporting on Adverse Drug Reactions and Infections

5.7.3.1. Scope of Reporting

The following are required to be reported when performing a clinical trial:

- unexpected death or cases of adverse experiences potentially leading to death
- the following matters shall be reported (other than those specified in the preceding item):
  (a) other unexpected serious events
  (b) adverse experiences requiring hospitalization or prolongation of hospital stay for their treatment
  (c) disability
  (d) adverse experiences potentially leading to disability
  (e) adverse experiences essentially comparable with those specified in (a)–(c) above, as well as death, and cases of adverse experiences potentially leading to death
  (f) congenital disease or abnormality occurring in the subsequent generation
- measures taken in foreign countries:
  - measures taken to prevent the occurrence or spread of risk to public health and hygiene, including discontinuation of manufacturing, import or marketing of, withdrawal or disposal of, etc., an item

whose ingredients are equivalent to those of the test products (including revision of the precautions, accompanied by a letter to the distributing doctor)

- research reports
  - where there is the possibility of the drug causing cancer or other serious disease due to adverse reactions, etc., of the investigational product concerned
  - research reports showing a lack of the test product's anticipated efficacy or clinical benefit for the disorder
  - other
- when similar measures taken overseas are to be implemented in Japan.

5.7.3.2. Determination of "Expectedness" and Causal Relationship

- When reporting adverse drug reactions, determination of "expectedness" is to be based on the following:
  - Adverse events described in the investigator's brochure.
  - The time of "expectation" is the preparation or revision date in the investigator's brochure or the preparation date of contact papers. An adverse event noted by an investigator in a contact paper shall be "expected" even if the investigator's brochure has not been revised.
  - Even if described in the investigator's brochure, events which occur at a different frequency or severity, or under different conditions of occurrence than described in the investigator's brochure and do not correspond with its content shall be deemed "unexpected".
- Causal relationship:
  - A report shall be required when both the principal investigator and person/company sponsoring the clinical trial deny any causal relationship.
  - In the case of a foreign subject, if both the foreign subject or subject's family and health professionals, and person/company sponsoring clinical trials deny any causal relationship, the event shall not be reported.

5.7.3.3. Reporting Period

Reporting periods for adverse drug reactions and infections, measures in foreign countries, and research reports are shown in Table 5.1.

Table 5.1 Reporting period

(a) Adverse drug reactions and infections		**Japan**	**Overseas**	**The Pharmaceutical Affairs Law, Enforcement Ordinance Article 273**
Expectation	**Severity**			
Unexpected	Death or adverse experiences potentially leading to death	7 days	7 days	No. 1a, b
	Other serious event	15 days	15 days	No. 2a (1)–(5)
Expected	Death or adverse experiences potentially leading to death	15 days	15 days	No. 2b
	Other serious event	—	—	—
(b) Measures in foreign countries				
Ingredients equivalent to the test products			**Overseas**	**The Pharmaceutical Affairs Law, Enforcement Ordinance Article 273**
Measures to prevent the occurrence or spread of risk to public health and hygiene, including discontinuation of manufacturing, import or marketing of, withdrawal or disposal, etc.			15 days	No. 2c
(c) Research reports				
			Overseas	**The Pharmaceutical Affairs Law, Enforcement Ordinance Article 273**
Showing the potential risk for cancer or other serious disease to occur Showing that substantial changes have been observed in the tendency for occurrence, including the number of cases, incidence, and conditions of occurrence Showing a lack of the test product's anticipated efficacy or clinical benefit for the disorder			15 days	No. 2d

5.7.3.4. Other Information

The date of the receipt of information must be obtained. The "start date" is the date on which minimum information on the following four items was reported:

- information on individual subjects: age (child, infant, middle-aged, elderly), gender, etc.
- reporting source: doctor, pharmacist, other health professional, consumer, paper, agency, etc.
- event: the name of adverse reactions, infections, severity, etc.
- the name of suspected trial products.

The start date is considered day zero for reporting purposes, in order to be compliant with the reporting timeline. When the reporting date is not a normal business day, the next day shall begin the reporting period.

5.7.4. Consulting with PMDA During the Development Process

The PMDA is available to consult with during the following stages of development (Figure 5.8).

5.7.5. Japan-Specific Issues

5.7.5.1. In-Country Caretaker

J-GCP ordinance Article 15 (In-Country Caretaker) is required as follows:

> *In order to avoid any occurrence of harm to public health and hygiene with respect to the new investigational drug, and to enable the party to take any necessary actions to prevent such harm from spreading, all parties who do not have a presence in Japan, but wish to conduct clinical trials in this country, shall identify a party who has a presence in Japan, such as a representative of the Japan office of the foreign corporation, that has the capability of requesting clinical trials to be conducted on its behalf, and shall allow said party (hereafter referred to as the "In-Country Caretaker") to carry out the necessary procedures relating to the clinical trial.*

Simply speaking, if a sponsor does not have an office located in Japan, the sponsor must designate an ICC.

The ICC is responsible for the progress of the clinical trials and must be the primary contact for any regulatory interaction on behalf of the sponsor company. The ICC is responsible for the following tasks:

- all clinical development work on behalf of the sponsor
- submission of a CTN to the PMDA
- replies to questions raised by the PMDA when the CTN is submitted, etc.

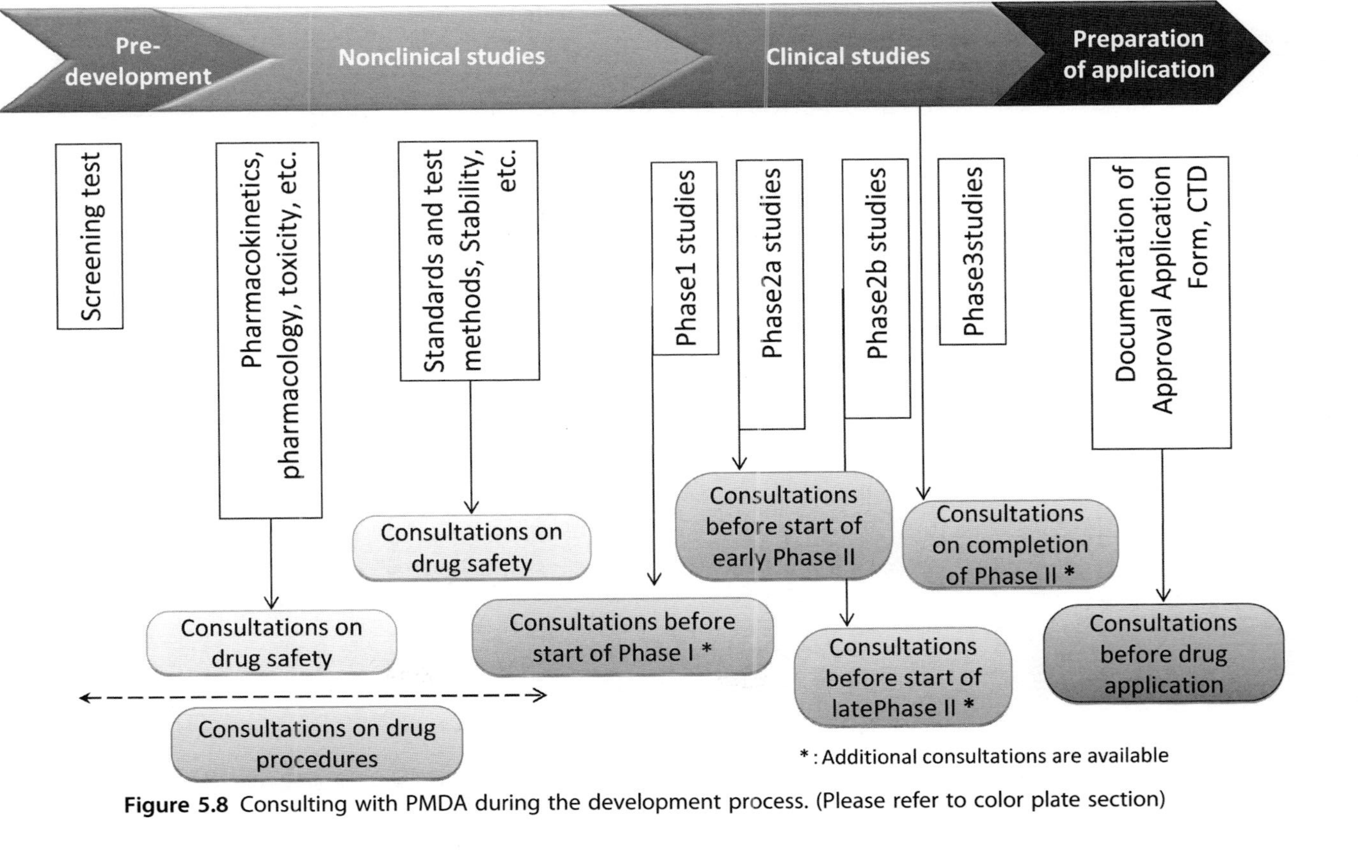

Figure 5.8 Consulting with PMDA during the development process. (Please refer to color plate section)

- supply of the investigational drug (customs clearance, requests made to the drug depot or the contract manufacturing organization, checking manufacturing records and quality control tests)
- monitoring of the clinical trial and preparation of the CSR
- collection of and reporting on adverse reactions, including information from overseas
- translation of documents and data (English to Japanese/Japanese to English).

5.7.5.2. Reimbursement of Concomitant Drugs

In Japan, a requirement regarding concomitant drugs in a clinical trial, including the screening phase, is as follows:

The concomitant drugs used during a clinical trial are defined as all the drugs expected to be used with the investigational products. The expenses for the drug products among the concomitant drugs, which have the same indications and effects as those of the investigational product, shall be borne by the sponsor, while those for all other concomitant drugs shall be covered by the Special Medical Care Scheme (Public Health Insurance).

Therefore, the sponsor or ICC must pay when requested to do so by the site.

5.7.5.3. Cross-Reporting to the PMDA

Safety reports from ongoing studies in other regions using the same compound must be submitted to the PMDA. In addition, when negative results are obtained from other studies, the sponsor or ICC is required to report them to the PMDA.

5.7.5.4. Aggregated Reporting to the PMDA

Every six months, periodic aggregated safety reports are required for products in Phases I–III. Unblinding is not necessary. The reports must be submitted in Japanese.

5.7.5.5. Contracts with Sites

Contracts for performing clinical research at sites are made between the sponsor (ICC) and the head of the medical institute where the trials are to be performed, and not with the principal investigators. This includes an agreement on clinical trial-related financing. Costs for the following items may be included (in the case of national hospitals): remuneration, travel expenses, research expenses for clinical studies (estimated by the point computation table), expenses for investigational product management, costs for fixtures, wages for staff (clinical research coordinator, etc.), expenses for the IRB, administrative costs, technical fees/machine depreciation fees/ building fees, and patient fees.

Expenses not covered under special medical care expenses shall be covered by the sponsor (ICC). A site may request the use of a site management organization (SMO) to support the conducting of the clinical study. In such cases, the sponsor (ICC) also covers the SMO's fee.

5.7.5.6. Investigator Grants

Investigator grants are not paid to the principal investigators directly but are paid to the medical institute in accordance with the contract. Types of payment arrangements may include the following:

- an advance payment in full, with no refund even if the contracted number of patients is not attained
- an advance payment in full, with partial refund if the contracted number of patients is not achieved
- a partial advance payment, with a performance-based payment for the remainder.

5.7.5.7. Importation

When new investigational drugs are imported (Figure 5.9), prior to customs clearance pharmaceutical inspectors stationed in the Regional

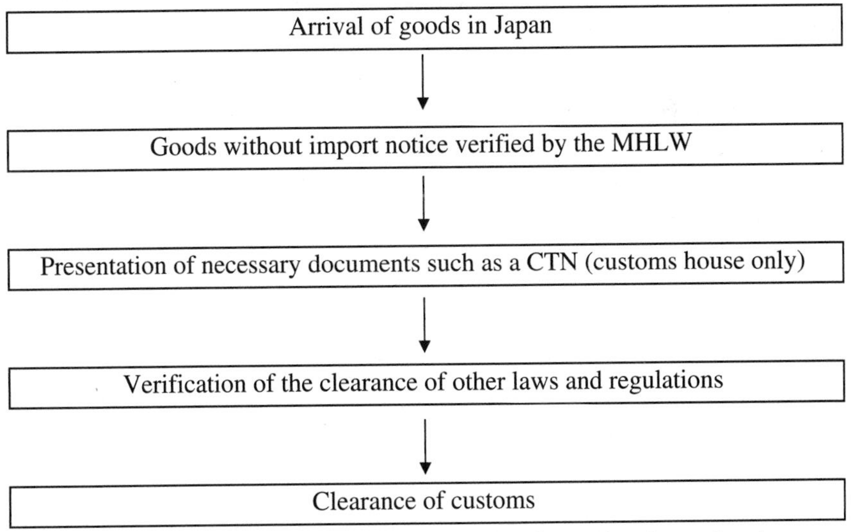

Figure 5.9 Flowchart of customs clearance procedures when importing drugs

Table 5.2 List of documents required for import procedures

Purpose of import		Document to be submitted				
		Import report form	Memorandum	CTN (copy)	Invoice (copy)	AWB or B/L (copy)
Clinical plan not submitted (import notice not available), (compliance certificate required)	Company-initiated	◎	○	○★ [1]	○	○
	Physician-initiated	◎	○	○★ [1]	○	○

When the documentation submitted does not include an invoice, a Contents Inspection Record with a stamp of the Customs House may be submitted instead.
When the documentation submitted does not include a BIL or AWB (because of the use of the postal mail service), an external postage notification postcard (copy) from the Customs House may be submitted instead.
Double circles (◎) indicate that the documents must be submitted in duplicate and single circles (○) that a single copy is sufficient. The ○★ mark indicates that the document must be submitted when required.
[1]When a drug is imported for use in two or more clinical trials, a clinical trial protocol estimate form showing the basis for calculation of the quantities imported should be attached instead of the clinical trial protocols for individual trials. In such cases, reports should be presented when clinical trials using the imported drugs are not started (cancelled), when a clinical trial notice is submitted, or when the clinical trial plan is changed, discontinued, or completed.

Bureaus of Health and Welfare confirm that such importation is appropriate, based on a review of the documentation submitted by the importer (Table 5.2). Once approved, the import report form is stamped "Verified by the Ministry of Health, Labor and Welfare" (MHLW) and transferred to the importer. This stamped report is called a Compliance Certificate and is required for customs clearance procedures. Drugs that have not yet arrived but for which an ocean Bill of Landing (BIL) or Airway Bill (AWB) has been issued are handled in the same manner.

REFERENCES

Websites

Pharmaceutical Medical Device Agency: http://www.pmda.go.jp
National Institute of Biomedical Innovation: http://www.nibio.go.jp/index.html

Regulations and Notifications

Article 54, Paragraph 3 of the Enforcement Regulations, Pharmaceutical Affairs Law.
Article 273 of the Enforcement Regulations, Pharmaceutical Affairs Law.

Notification No. 0331003 of the Evaluation and Licensing Division, PFSB, dated March 31, 2005.
Notification No. 0331009 of the Evaluation and Licensing Division, PFSB, dated March 31, 2005.
Notification No. 0815005 of the Evaluation and Licensing Division, PFSB, dated August 15, 2008.

Publication

Drug Approval and Licensing Procedures in Japan (2010). Jiho, Tokyo.

SECTION 2

Bioethical Considerations in Global Clinical Trials

Indian Regulatory Framework

S.D. Seth*, Surinder Singh and Mohua Maulik***
*Clinical Trials Registry – India, National Institute of Medical Statistics, Indian Council of Medical Research, New Delhi, India
**Central Drugs Standard Control Organization, Ministry of Health and Family Welfare, Government of India, New Delhi, India

Contents

Global Clinical Trials
ISBN 978-0-12-381537-8, Doi:10.1016/B978-0-12-381537-8.10006-8

6.1. INTRODUCTION

The Indian pharmaceutical sector has witnessed a remarkable growth, from around $1 billion in 1990 to $20 billion in 2009–2010, and has a current growth rate of 12–14 percent per annum, according to data available from the Central Drugs Standard Control Organization (CDSCO), Government of India. India has demonstrated its technological capabilities and manufacturing standards by having the highest number of Food and Drug Administration (FDA)-approved plants outside the USA [1].

Moreover, India has emerged as an attractive region for conducting clinical trials. The FICCI–Ernst & Young Survey Report 2008 notes that India is one of the fastest growing clinical research destinations, with a growth rate that is two and a half times the overall market growth [2]. India ranks second in Asia after Japan in its number of industry-sponsored Phase II–III clinical trial study sites and accounts for nearly 20 percent of all Asian study sites [2].

Data compiled by the CDSCO also document the growth in receipt as well as permissions granted for the conduct of global clinical trials in India (Figures 6.1 and 6.2). Phase III trials comprise the bulk of these trials (Figure 6.3). Currently, Indian companies are investing in new drug discovery research. Multinational companies are also showing interest in collaborating with Indian companies for such new drug discovery research.

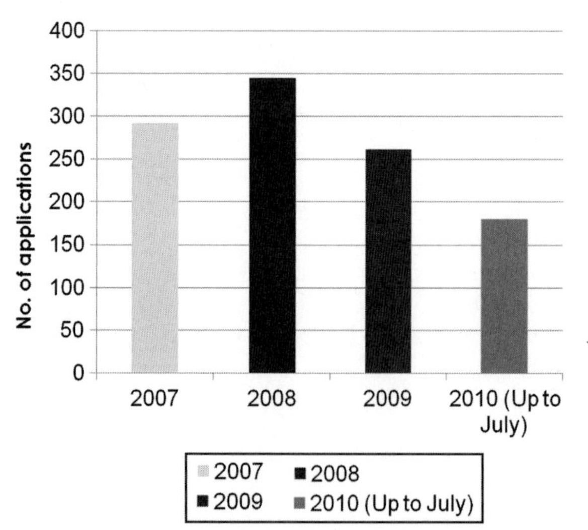

Figure 6.1 Number of global clinical trial applications received. (Please refer to color plate section)

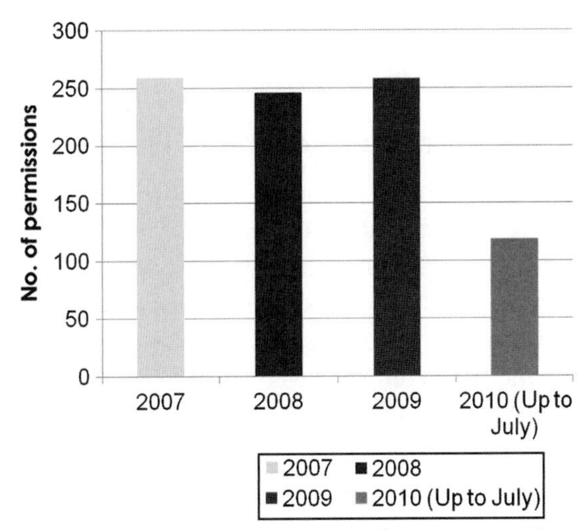

Figure 6.2 Number of global clinical trials granted permission. (Please refer to color plate section)

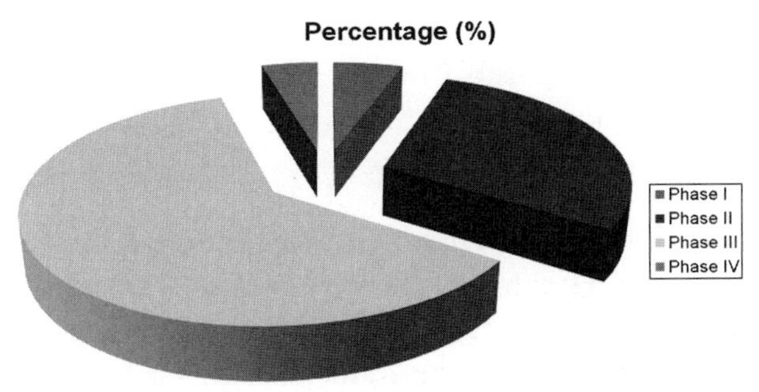

Figure 6.3 Phase-wise breakdown of clinical trials conducted in India. (Please refer to color plate section)

Some of the factors that make India a viable clinical trial destination are:

- a large genetically diverse and treatment-naïve patient population
- large patient pool with disease patterns reflective of both the developing world (e.g. malnutrition, diarrhea, and other infectious diseases) and the developed world (e.g. diabetes, heart disease)
- availability of specialty hospitals with state-of-the-art facilities and staffed by a significant number of Good Clinical Practice (GCP)-trained clinicians
- a large English-language proficient workforce
- high-quality Internet connectivity
- a highly developed information technology/information technology enabled services (IT/ITES) sector

- lower overall costs associated with the conduct of clinical trials
- a robust regulatory system.

6.1.1. Drugs and Cosmetics Act and Rules

Stringent but streamlined regulatory processes have been put in place that ensure not only the generation of quality data but also the safety of the Indian trial participant. Drugs are on the concurrent list of the Indian Constitution and the regulatory system in India is managed by both central and the state government under the Drugs and Cosmetics Act, 1940, and Rules, 1945, as amended from time to time. The Act regulates the import, manufacture, distribution, and sale of drugs and cosmetics [3]. Schedule Y of the Drugs and Cosmetics Act, 1940, deals with the rules and regulations applicable to the conduct of clinical trials in India [3].

In addition, the Indian GCP guidelines [4] and the Indian Council of Medical Research's Ethical Guidelines for Biomedical Research on Human Participants [5] lay down standards and norms for the conduct of clinical trials in India, and ensure that the Indian regulatory scenario is in tandem with the international regulatory norms.

6.2. REGULATORY HIERARCHY

The CDSCO, under the Directorate General of Health Services (Dte GHS), Ministry of Health and Family Welfare (MOHFW), Government of India, is the primary drug regulatory authority in India [6]. The Dte GHS is headed by the Director General of Health Services (DGHS), who is chair of the Drugs Technical Advisory Board (DTAB), the highest drug advisory body of the Government of India. The DTAB advises the central government on technical matters arising out of the administration of the Act and carries out the other functions assigned to it by the Act.

The Drugs Controller General of India (DCGI) presides over the functioning of the CDSCO and the Central Drugs Laboratory. The DCGI is also the Central License Approving Authority (CLAA) for specified categories of products. Clinical trials in India are regulated by the office of the DCGI (i.e. CDSCO) in accordance with the provisions of the Drugs and Cosmetics Act, 1940, and the Drugs and Cosmetics Rules, 1945, made thereunder. The structure and functions of the DCGI office are depicted in Figure 6.4. The geographical locations of CDSCO offices in India are shown in Figure 6.5.

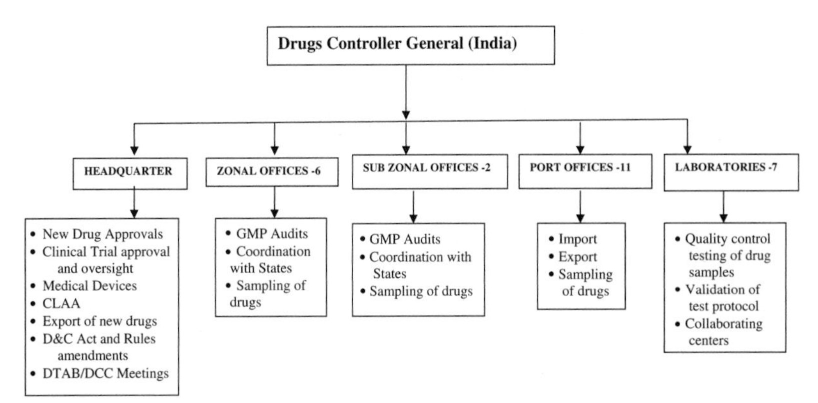

Figure 6.4 Structure and functions of the Drugs Controller General (India) office. CLAA: Central License Approving Authority; D&C: Drugs and Cosmetics; DTAB: Drugs Technical Advisory Board; DCC: Drugs Consultative Committee; GMP: good manufacturing practice

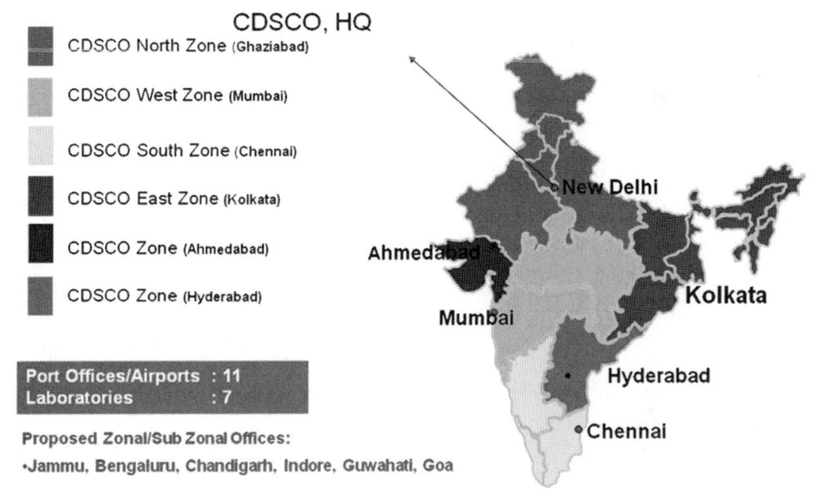

Figure 6.5 Geographical locations of Central Drugs Standard Control Organization (CDSCO) offices in India. (Please refer to color plate section)

The drug regulatory functions for the state are performed by the State Licensing Authorities (SLAs). The SLAs fall under the concerned ministries of the various state governments.

6.3. CENTRAL DRUGS STANDARD CONTROL ORGANIZATION

The CDSCO regulates and monitors not only the conduct of clinical trials but also entire spectra of activities related to drugs, devices, and substances to be used for therapeutic, preventive, and diagnostic purposes [6]. To strengthen the National Regulatory Authority (NRA), CDSCO has collaborated with the World Health Organization (WHO) and Health Canada. Furthermore, to increase knowledge and skill in clinical trial inspection, CDSCO has also collaborated with the US FDA.

6.3.1. Functions Undertaken by the CDSCO

- Statutory functions:
  - laying down standards of drugs, cosmetics, diagnostics, and devices
  - laying down regulatory measures, amendments to Acts and Rules
  - regulating market authorization of new drugs
  - regulating clinical trials in India
  - approving licenses to manufacture certain categories of drugs as CLAA, i.e. for medical devices, blood products, blood banks, large volume parenterals, and vaccines and sera
  - regulating the standards of imported drugs
  - work relating to the DTAB and Drugs Consultative Committee (DCC)
  - testing of drugs by Central Drugs Laboratories
  - banning of drugs and cosmetics
  - granting of Test Licenses, Personal Licenses, and No Objection Certificates (NOCs) for export
- Other functions:
  - coordinating the activities of the State Drugs Control Organizations to achieve uniform administration of the Act; and policy guidance
  - guidance on technical matters
  - participation in the WHO good manufacturing practice (GMP) certification scheme
  - monitoring adverse drug reactions (ADRs)
  - conducting training programs for regulatory officials and government analysts
  - screening of drug formulations available in the Indian market
  - overseas inspection.

6.3.2. Central Drugs Laboratory

The Central Drugs Laboratory, Kolkata, headed by the Director, is the National Statutory Laboratory of the Government of India for Quality Control of Drugs and Cosmetics under the direct administrative control of the DCGI, Directorate General of Health Services, Ministry of Health and Family Welfare. The organization plays a crucial role in ensuring quality control of imported drugs available in the Indian market as well as quality control of drug formulations manufactured within the country on behalf of the central and state drug control organizations [7].

6.4. NORMS AND REGULATIONS FOR THE CONDUCT OF GLOBAL TRIALS

Global clinical trials are those that are conducted as part of a multinational drug development program. Requirements and guidelines for conducting clinical trials in the country are prescribed under Rule 122DA, 122DAA, 122DB, and 122E, and Schedule Y to the Drugs and Cosmetics Rules [3]. The said Rules also provide specific application forms and relevant document requirements for seeking permission to conduct clinical trials.

As per Rule 122 DA of the Drugs and Cosmetics Rules, no clinical trial of a new drug can be conducted in the country without permission from the DCGI. As per Rule 122E of the Drugs and Cosmetics Rules, a new drug is defined as a drug, including bulk drug substance, that has not been used in the country to any significant extent under the conditions prescribed, recommended, or suggested in the labeling thereof, and has not been recognized as effective and safe by the DCGI, for the proposed claim.

In addition, the following categories of substances are considered as new drugs under Rule 122E:

- a drug already approved by the Licensing Authority mentioned in Rule 21 for certain claims, which is now proposed to be marketed with modified or new claims, namely, indications, dosage, dosage form (including sustained release dosage form), and route of administration
- a fixed dose combination of two or more drugs, individually approved earlier for certain claims, which are now proposed to be combined for the first time in a fixed ratio, or if the ratio of ingredients in an already marketed combination is proposed to be changed, with certain claims, namely, indications, dosage, dosage form (including sustained release dosage form), and route of administration
- all vaccines shall be new drugs unless certified otherwise by the Licensing Authority under Rule 21
- a new drug shall continue to be considered as new drug for a period of four years from the date of its first approval or its inclusion in the Indian Pharmacopoeia, whichever is earlier.

In India, permission from DCGI is required for conducting clinical trials including bioequivalence studies for:

- new chemical entities, including investigational new drugs (INDs) and drugs approved outside for marketing but not approved in India
- all vaccines and recombinant DNA (rDNA) products
- a fixed dose combination, considered as a new drug under Rule 122E
- global clinical trials
- new indication/dosage/dosage form including sustained release form and route of administration
- approved new drugs (within four years of approval)
- Phase IV clinical trials.

6.4.1. Schedule Y

Guidelines and requirements applicable to the conduct of a clinical trial in India are detailed in Schedule Y to the Drugs and Cosmetics Rules, 1945, as amended from time to time [3].

For new drug substances discovered in countries other than India, Phase I (first-in-human) studies are not permitted. However, repeat Phase I clinical trials can be permitted for which Phase I data generated outside should be submitted along with the application. After submission of Phase I data to the DCGI, permission may be granted to repeat Phase I trials and/or to conduct Phase II trials and subsequently Phase III trials concurrently with other global trials for that drug. For new drugs having potential for use in children, permission for clinical trials in the pediatric age group is usually granted after Phase III trials in adults are completed. However, if the drug is of value primarily in a disease of children, early trials in the pediatric age group may be allowed.

In Schedule Y, provisions have been made to ensure that the patients and volunteers participate in studies only after proper understanding of the study. An elaborate informed consent process has been prescribed in Schedule Y. The responsibilities of institutional ethics committees (IECs), clinical investigators, and sponsors have also been clearly defined [3]. Schedule Y has also made it mandatory that trials have to be conducted as per GCP guidelines published by the CDSCO [4].

In brief, the salient features of Schedule Y are:

- Phase I (first-in-human) study of new drug/substance discovered outside the country not permitted
- repeat Phase I studies of new drug/substance discovered outside the country may be permitted

- concurrent phase clinical trials permitted
- provides statutory support to Indian GCP guidelines
- stipulates responsibilities of ethics committees, investigators, and sponsor
- structure, contents, and formats for clinical trial protocols, reports, ethics committee approvals, informed consent forms (ICFs), serious adverse event (SAE) reporting, etc., are provided in the various appendices to schedule Y, e.g.:
  - Appendix I: Data to be submitted with Form 44
  - Appendix IA: Data required for grant of permission to market-approved new drug
  - Appendix II: Contents of clinical study reports
  - Appendix III: Animal toxicology
  - Appendix IV: Animal pharmacology
  - Appendix V: Informed consent checklist and form [patient information sheet (PIS)]
  - Appendix VI: Fixed dose combinations (FDCs)
  - Appendix VII: Investigator's undertaking
  - Appendix VIII: Ethics committee
  - Appendix IX: Stability testing of new drugs
  - Appendix X: Contents of clinical study protocol
  - Appendix XI: SAE reporting data elements.

6.4.2. GCP – India, 2001

Good Clinical Practice Guidelines of India (GCP – India) is a set of guidelines for biomedical studies which encompasses the design, conduct, termination, audit, analysis, reporting, and documentation of the studies involving human subjects [4]. The fundamental tenet of GCP is that in research on humans, the interests of science and society should never take precedence over considerations related to the well-being of the study subject. It aims to ensure that the studies are scientifically and ethically sound and that the clinical properties of the pharmaceutical substances under investigation are properly documented. The guidelines seek to establish two cardinal principles: protection of the rights of human subjects and authenticity of the biomedical data generated. These guidelines should be followed for carrying out all biomedical research in India at all stages of drug development, whether prior or subsequent to product registration in India [4].

Some of the important features of the GCP – India guidelines are highlighted below.

6.4.2.1. Foreign Sponsor

With regard to the conduct of a clinical trial in India by any foreign sponsor, as per GCP guidelines, the sponsor should appoint a local representative or Indian contract research organization (CRO) to fulfill the appropriate local responsibilities as governed by the national regulations. The sponsor may transfer any or all of their study-related duties and functions to a CRO, but the ultimate responsibility for the quality and the integrity of the study data shall always reside with the sponsor. Any study-related duty, function, or responsibility transferred to and assumed by a local representative or a CRO should be specified in writing [4].

6.4.2.2. Sponsor Responsibilities

The clinical trial sponsor is responsible for implementing and maintaining quality assurance systems to ensure that the clinical trial is conducted and data are generated, documented, and reported in compliance with the protocol and GCP – India guidelines issued by the CDSCO as well as with all applicable statutory provisions. The sponsor should establish detailed standard operating procedures (SOPs) and ensure compliance with GCP and applicable regulations. The sponsor and the investigator(s) should sign a copy of the protocol and the SOPs or an alternative document to confirm their agreement.

Sponsors are required to submit a status report on the clinical trial to the DCGI at the prescribed periodicity. In case of studies prematurely discontinued for any reason, including lack of commercial interest in pursuing the new drug application, a summary report should be submitted within three months. The summary report should provide a brief description of the study, the number of patients exposed to the drug, dose and duration of exposure, details of adverse drug reactions as per Appendix XI of Schedule Y [3], if any, and the reason for discontinuation of the study or non-pursuit of the new drug application.

Any unexpected SAE (as defined in GCP guidelines) occurring during a clinical trial should be communicated promptly (within 14 calendar days) by the sponsor to the DCGI and to the other investigator(s) participating in the study (see Appendix XI of Schedule Y).

No deviations from or changes to the protocol should be implemented without prior written approval of the ethics committee and the DCGI except when it is necessary to eliminate immediate hazards to the trial subjects or when the change involved is only logistic or administrative in nature. All such exceptions must be immediately notified to the ethics committee as well as to the DCGI. Administrative and/or logistic changes in the protocol should be notified to the DCGI within 30 days.

6.4.2.3. Investigator Responsibilities

All trial investigators should possess appropriate qualifications, training, and experience, and should have access to such investigational and treatment facilities as are relevant to the proposed trial protocol. A qualified physician (or dentist, when appropriate) who is an investigator or a subinvestigator for the trial should be responsible for all trial-related medical (or dental) decisions. Laboratories used for generating data for clinical trials should be compliant with Good Clinical Laboratory Practices. If services of a laboratory or facilities outside the country are to be availed, their name(s), address(es), and specific services to be used should be stated in the protocol to avail DCGI permission to send clinical trial-related samples to such a laboratory and/or facility. In all cases, information about laboratories/facilities to be used for the trial, if other than those at the investigation site, should be furnished to the DCGI prior to initiation of the trial at such site.

The investigator is responsible for the conduct of the trial according to the protocol and the GCP guidelines and also for compliance as per the undertaking given in Appendix VII of Schedule Y [3]. In India, as per Schedule Y, standard operating procedures are required to be documented by the investigators for the tasks performed by them.

During and following a subject's participation in a trial, the investigator should ensure that adequate medical care is provided to the participant for any adverse events. The investigator is expected to report all serious and unexpected adverse events to the sponsor within 24 hours and to the ethics committee that accorded approval to the study protocol within seven working days of their occurrence.

The investigator has the primary responsibility for the investigational product's accountability at the study site. A reconciliation statement of the investigational product should be maintained at the site at any given point of time. The investigator should maintain records of the product's delivery to the study site, the inventory at the site, the use by each subject, and the return to the sponsor or the alternative disposal of the unused product(s). These records should include dates, quantities, batch/serial numbers, expiry dates if applicable, and the unique code number assigned to the investigational product packs and study subjects. The investigator should maintain records that describe that the subjects were provided the dosage specified by the protocol and reconcile all investigational products received from the sponsor.

The investigator should ensure that the product(s) are stored under specified conditions and are used only in accordance with the approved protocol [4].

The investigator should ensure that the observations and findings are recorded correctly and completely in the case report forms (CRFs) and signed by the responsible person(s) designated in the protocol. The medical records of the trial participants should be clearly marked so as to indicate that the individual is participating in a clinical trial [4].

6.4.2.4. Informed Consent

The subject's consent must be obtained in writing using an ICF, prior to the subject's participation in the study, as per GCP – India (participation includes patient screening for eligibility). In all trials, freely given, informed, written consent is required to be obtained from each study subject. The investigator must provide information about the study verbally as well as using a patient information sheet, in a language that is non-technical and understandable by the study subject. In view of the multilingual nature of India's population, the PIS and ICF are to be drafted in English and local languages and then backtranslated to ensure the clarity of the information provided.

Both the PIS and the ICF should have been approved by the ethics committee and furnished to the DCGI. Any changes in the informed consent documents should be approved by the ethics committee and submitted to the DCGI before such changes are implemented [4].

Where a subject is not able to give informed consent (e.g. an unconscious person or a minor or those suffering from severe mental illness or disability), the same may be obtained from a legally acceptable representative (a legally acceptable representative is a person who is able to give consent for or authorize an intervention in the patient as provided by the laws of India). If the subject or his or her legally acceptable representative is unable to read and write, an impartial witness should be present during the entire informed consent process who must append his or her signature to the consent form.

A checklist of essential elements to be included in the study subject's informed consent document as well as a format for the ICF for study subjects is given in Appendix V of Schedule Y (3).

6.4.2.5. Ethical Considerations

All research involving human subjects should be conducted in accordance with the ethical principles contained in the current revision of the Declaration of Helsinki and should respect three basic principles, namely justice, respect for persons, beneficence (to maximize benefits and to minimize harms and wrongs), and non-maleficence (to do no harm), as defined by "Ethical Guidelines for Biomedical Research on Human Participants" issued by the Indian Council of Medical Research (ICMR), New Delhi [5], and any other laws and regulations of the country, which ensure greater protection for subjects [4].

In brief, the following principles are to be followed:

- principles of essentiality
- principles of voluntariness, informed consent, and community agreement
- principles of non-exploitation
- principles of privacy and confidentiality
- principles of precaution and risk minimization
- principles of professional competence
- principles of accountability and transparency
- principles of the maximization of the public interest and of distributive justice
- principles of institutional arrangements
- principles of public domain
- principles of totality of responsibility
- principles of compliance.

All clinical trials conducted in India must be ethically approved at the site level by an appropriately constituted and competent ethics committee. A trial may be initiated at each site only after obtaining such an approval for that site. The trial site(s) may accept the approval granted to the protocol by the ethics committee of another trial site or the approval granted by an independent ethics committee, provided that the approving ethics committee(s) is (are) willing to accept their responsibilities for the study at such trial site(s) and the trial site(s) is (are) willing to accept such an arrangement and that the protocol version is the same at all trial sites.

All clinical trials being conducted in the India must follow in letter and spirit the Ethical Guidelines for Biomedical Research on Human Participants (2006) developed by the ICMR, which is also known as the ICMR code [5]. These guidelines, drawing from the International Ethical Guidelines for Biomedical Research Involving Human Subjects, updated in 2002, elaborate on the ethical norms that should be followed while conducting clinical trials in India. The guidelines are primarily recommendatory in nature; however, all trials approved by the DCGI emphasize adoption of these guidelines.

The guidelines attempt to apply universal ethical principles to biomedical research in the multicultural Indian scenario with its vastly varying standards of healthcare systems, while keeping the national policies and the unique and diverse demands of the pan-Indian culture in the forefront. While, on one hand, research involving human participants must not violate any universally applicable ethical standards, on the other hand, a researcher needs to consider local cultural values when it comes to the application of the ethical principles to individual autonomy and informed consent. In India, one will have to consider autonomy versus harmony of the environment of the research participant. In research on sensitive issues, this will have to be properly addressed in the research protocol to safeguard the human rights of the dependent or vulnerable persons and populations [5].

Apart from the general principles of clinical research on human participants, the ICMR Code has separate dedicated sections for clinical research on vaccines, herbal products and other traditional systems of medicine, organ transplantation, stem cell research, assisted reproductive technologies, and biobanking. These guidelines reflect the current global thinking on these topics while underscoring their application in the Indian circumstances from ethical, legal, and social angles.

The composition and responsibilities of the IEC/independent ethics committees [IEC (Ind)] are laid down in the ICMR Code [5]. The responsibilities of an IEC can be defined as follows:

- to protect the dignity, rights, and well-being of the potential research participants
- to ensure that universal ethical values and international scientific standards are expressed in terms of local community values and customs
- to assist in the development and the education of a research community responsive to local healthcare requirements.

The ethics committees not only are entrusted with the initial review of the proposed research protocols before initiation of the projects, but also have the continuing responsibility for regular monitoring of the approved programs to ensure ethical compliance during the period of the project.

All documentation and communication of an IEC are to be dated, filed, and preserved according to written procedures. It is recommended that all records be safely maintained after the completion or termination of the study for at least a period of five years if it is not possible to maintain the same permanently [4].

In the near future, a Biomedical Research Authority will be set up under the proposed Bill on Biomedical Research on Human Participants (Promotion and Regulation) and would necessitate registration of all IECs with this Authority. It will also evaluate and monitor functioning of the IECs, and develop mechanisms for enforcing accountability and transparency by the institutions.

6.4.2.6. Insurance

According to paragraph 2.4.7 of the GCP guidelines, trial participants are entitled to compensation for accidental injury and it is the obligation of the sponsor to pay adequate compensation for any temporary or permanent impairment or disability, subject to confirmation from the IEC. In case of death, their dependants are entitled to material compensation [4].

Study subjects should be satisfactorily insured against any injury caused by the study. Each clinical trial is expected to include an inbuilt mechanism for compensation for the human participants either through insurance cover or by any other appropriate means to cover all foreseeable and unforeseeable risks by providing for remedial action and comprehensive aftercare, including treatment during and after the research or experiment, in respect of any effect that the conduct of research or experimentation may have on the human participant and to ensure that immediate recompense and rehabilitative measures are taken in respect of all affected, if and when necessary [5]. To ensure smooth and easy access to compensation, the insurance agency recruited for the purposes must have an Indian arm.

6.4.3. Data Safety Monitoring Board

The Indian GCP Guidelines (2001) and the ICMR Ethical Guidelines for Biomedical Research on Human Participants (2006) recommend the setting up of independent data monitoring committees to monitor both the data emerging from clinical trials and the safety of the participants in these trials [4,5].

The guidelines provide recommendations to sponsors of clinical research for the establishment and functioning of data safety monitoring boards (DSMBs) in India [8].

The guidelines describe the constitution, role, responsibilities, and operating framework for DSMBs. They are also intended to assist DSMB members, sponsors, investigators, members of ethics committees, regulatory authorities, and research participants and their organizations in understanding the role and functions of DSMBs. The guidelines are not intended to supersede national laws and regulations.

The sponsor should report the safety and efficacy data, as well as other relevant study information, to the DSMB for its review. The sponsor's report to the DSMB is often provided in two parts: an open part and a closed part. The full report should be made available to DSMB members at least one week before the meeting. The contents of the report are determined by the DSMB charter and discussed in advance during the organizational meeting. The charter should specify who will prepare and provide the open and closed parts of the report; two separate parties can provide these parts, if appropriate. The sponsor should establish a procedure for receiving and distributing the recommendation of a DSMB. The sponsor is responsible for distributing the recommendation, in a timely manner, to the steering committee, investigator(s), ethics committee, and regulatory authority of the study. Procedures for implementing the recommendations of the DSMB also need to be specified in advance [8].

6.5. OBTAINING PERMISSION FOR CONDUCTING A CLINICAL TRIAL

An application for grant of permission to undertake a clinical trial has to be made as per the format of Form 44 of the Drugs and Cosmetic Rules, accompanied by data in accordance with Appendix I to Schedule Y to the DCGI [3].

Fees, as per the provisions of the Drugs and Cosmetics Act and Rules, in the form of a TR6 Challan issued by Bank of Baroda, are to be deposited at notified branches of Bank of Baroda under the head of account "0210 – Medical and Public Health, 04 – Public Health, 104 – Fees and Fines" adjustable to Pay and Account Officer, DGHS, New Delhi, in the form of a Treasury Challan. A fee of INR (Indian rupees) 50,000/– (equivalent to approximately US $1000) for Phase I, and INR 25,000/– (equivalent to approximately US $500) for Phases II and III is required to be submitted by the industry for processing of the application. However, no fee is required to be paid along with the clinical trial application by central government or state government institutes of India.

In case the investigational product is to be imported, an application for import of the investigational product as per Form 12 of the Drugs and Cosmetic Rules, including justification, along with requisite fees is to be submitted. Simultaneously, an application for an NOC for export of biological samples for testing purposes, if any, may be submitted along with justification.

In India, an IND refers to a new chemical entity or a product having a therapeutic indication but which has never been previously tested on humans. In India, an IND application refers to a substance that is developed in the country. Such an application would be evaluated by the IND Committee, specially constituted by the MOHFW, Government of India, to advise DCGI. Such a substance has to undergo Phase I, Phase II, and Phase III trials before a new drug application (NDA) can be made to the DCGI for marketing authorization.

Applications for global clinical trials are evaluated at CDSCO headquarters, New Delhi, by the New Drug Division, Biological Division or Medical Device Division, as the case may be. If needed, subject experts from reputed Indian institutes are consulted. The targeted first response for such applications is around 45 working days.

6.6. HEALTH MINISTRY'S SCREENING COMMITTEE CLEARANCE

For clinical trials to be conducted in India with foreign collaboration, approval of the Government of India through the Health Ministry's Screening Committee (HMSC) is mandatory. The application is to be submitted to the International Health Division (IHD) of ICMR, which is the secretariat of HMSC.

The following information is to be provided while submitting proposals for foreign collaboration/assistance after identification of the foreign collaborator and its role [9]:

- role/status/expertise of the Indian principal investigator (PI)
- availability of infrastructure and staff in the institution
- justification for foreign collaboration/funding
- relevance to India's national health priorities
- role, consent, and biodata of the foreign collaborator
- budget with justification and year-wise breakdown in a single currency, i.e. Indian currency, including training as well as foreign exchange component, if any.

Apart from the technical details such as rationale for the studies, objectives, review of the literature, materials and methods, techniques to be used, and so on, the following additional information is required for any Indo-foreign collaborative project. This information is to be submitted in the specified format available at http://icmr.nic.in/guide.htm:

- nature of work to be done in the Indian laboratory/institution and the foreign collaborator's laboratory/institution
- number of international collaborative projects (approved by HMSC) being undertaken by the Indian PI and the outcome of such approved projects (publications, patents, etc.)
- whether there would be transfer of technology as an outcome of the project
- whether there would be transfer of human biological material from India to the foreign laboratory, or vice versa, and if so the requisite details for the same, such as the nature and quantity of material to be sent abroad; purpose/need of transfer; nature of investigation to be done utilizing the material; institution(s)/scientist(s) to whom material to be sent; along with their addresses; copy of material transfer agreement (MTA)
- information pertaining to likely visits (year-wise) by Indian and foreign scientist(s), including duration and purpose of each visit
- institutional ethical clearance to be submitted at the time of submission of the proposal to ICMR
- appropriate clearances for research involving human subjects, radio-tagged material (for clinical and/or experimental purposes), or rDNA/genetic engineering work
- the proposals involving ICMR institutes/centers should be submitted with the recommendations of the scientific advisory committee (SAC) of the concerned institute/center
- mutual agreement on intellectual property rights (IPR) claims.

Hard copies of the proposals and associated documents as well as electronic submissions are to be submitted to the IHD of the ICMR by the Indian PI. Details of the number of copies and the documents required are available at http://icmr.nic.in/guide.htm

While proposals may be submitted throughout the year, HMSC meetings are generally held at intervals of three months. The proposals should be submitted to the IHD at least one month ahead of the scheduled HMSC meeting. Submissions should include all relevant approvals such as IEC approval, DCGI approval, if applicable; appropriately filled in and duly signed MTA, in case of transfer of biological material or any other relevant clearance/document. In general, six to nine months are taken for peer review and processing by the ICMR. However, if there is a specified and written deadline of an international funding agency and this is indicated by the PI at the time of submission of the proposal, there is provision for expedited review of the proposal [10].

6.7. CLINICAL TRIALS REGISTRATION

To enhance the transparency, accountability, and accessibility of clinical trials, registration of a clinical trial in the ICMR's Registry, the Clinical Trials Registry – India (CTRI), was made mandatory by the DCGI's office with effect from June 15, 2009 [11]. Registration of clinical trials in the CTRI is a free and online process [12]. The CTRI data set includes all the 20 data set points of the WHO [13], as well as certain items such as contact details of Indian PIs, ethics, and DCGI approval (including submission of

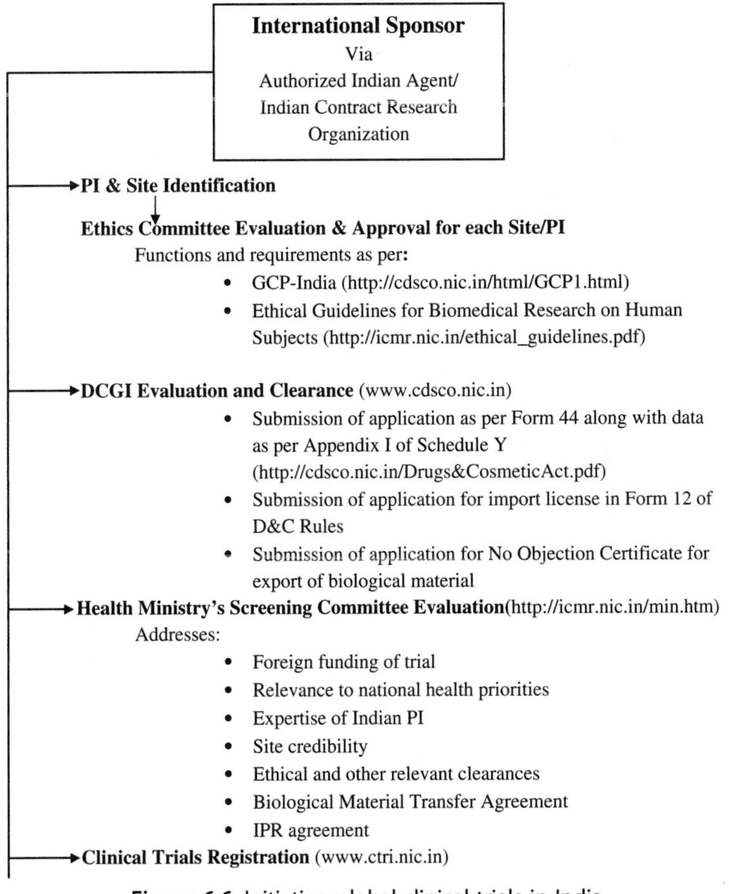

Figure 6.6 Initiating global clinical trials in India

approval documents). For global clinical trials, specific information regarding the number of patients being recruited and date of first enrollment in India is also recorded.

The key elements of the process of obtaining approval for the conduct of clinical trial in India are summarized in Figure 6.6.

6.8. SPECIAL CONSIDERATIONS

6.8.1. Biologicals

As per Rule 122E of the Drugs and Cosmetics Rules, all vaccines and rDNA products are considered as new drugs.

The guidelines for conducting a clinical trial on investigational vaccines are similar to those governing a clinical trial. The phases of these trials differ from drug trials as given below [4]:

- **Phase I:** This refers to the first introduction of a vaccine into a human population for determination of its safety and biological effects, including immunogenicity. This phase includes study of dose and route of administration and should involve low-risk subjects. For example, immunogenicity to hepatitis vaccine should not be determined in high-risk subjects.
- **Phase II**: This refers to the initial trials examining effectiveness (immunogenicity) in a limited number of volunteers. Vaccines can be prophylactic and therapeutic in nature. While prophylactic vaccines are given to normal subjects, therapeutic or curative vaccines may be given to patients suffering from a particular disease.
- **Phase III**: This focuses on assessments of safety and effectiveness in the prevention of disease, involving a controlled study on a larger number of volunteers (in thousands) in multicenters.

The risks associated with vaccines produced by rDNA techniques are not completely known. However, for all recombinant vaccines and products the guidelines issued by the Department of Biotechnology should be strictly followed [14].

Vaccine trials should be conducted by an investigator with the requisite experience and having the necessary infrastructure for the laboratory evaluation of seroconversion. Protocols for such trials should include appropriate criteria for the selection of subjects, and a plan of frequency of administration of the test vaccine in comparison with the reference vaccine. It should accompany detailed validation of the testing method to detect the antibody titer levels as well as specify the methodology to be adopted for prevention of centrifuged serum for the purpose of testing.

Protocols for testing a new vaccine should contain a section giving details of steps of manufacture, in-process quality-control measures, storage conditions, stability data, and a flowchart of various steps taken into consideration for manufacture of the vaccine. It should also contain detailed methods of quality control procedures with relevant references.

The investigator should be provided with quality control data of the experimental batch of the vaccine made for the purpose of clinical trials. The sponsor should provide the independent ethics committee approval of the nodal body (or bodies) to carry out clinical trials with the vaccine.

The generic version of new vaccines already introduced in the other markets after step-up clinical trials, including extensive Phase III trials, should be compared with the reference vaccine with regard to seroconversion in a comparative manner.

Postmarketing surveillance (PMS)/bridging studies are required following seroconversion studies. PMS data should be generated to detect side-effects and address other safety issues [4]. The purpose of PMS and bridging studies is to monitor the performance of a vaccine in the target population under conditions of routine use. The objectives are to detect adverse reactions and to monitor efficacy.

PMS programs should be appropriate to the disease epidemiology, infrastructure, and resources in the target area. Essential standards of safety, efficacy and quality should be defined before initiating a PMS program, without exception.

Bridging studies are intended to support the extrapolation of efficacy, safety, and immunogenicity data from one formulation, population, and dose regimen to another. The need for performing bridging studies should be considered carefully and justified in the protocol. The endpoints for clinical bridging studies are usually the relevant immune responses and clinical safety parameters.

Situations where bridging studies could be required are:

- for manufacturing change
- for a new dosing schedule
- for a new population
- for safety.

In the case of a clinical trial with indigenously developed recombinant products, permission from the Institutional Biosafety Committee (ISBC) and Review Committee of Genetic Manipulation (RCGM) under the Department of Biotechnology is essential before initiation of the trial [14].

For rDNA-based products developed and manufactured in another country, these approvals are not essential for the conduct of clinical trials in the country and application in Form 44 may be submitted directly to the DCGI. However, if the product contains a live modified organism (LMO), clearance from the Genetic Engineering Approval Committee (GEAC), under the Ministry of Environment and Forests, is essential before conducting the trial in the country.

While the requirements for the conduct of a clinical trial for biologicals remain the same as per Schedule Y of Drugs and Cosmetic Rules, 1945, the requirements in respect of chemistry and pharmaceutical information are distinct and details are available on the CDSCO website [15].

The document also contains detailed information regarding:

- requirements for permission of new drugs approval
- post-approval changes in biological products: quality, safety and efficacy documents
- preparation of the quality information for drug submission for new drug approval: biotechnological/biological products.

6.8.2. Medical Devices

In India, the import, manufacture, sale, and distribution of medical devices are regulated under the Drugs and Cosmetics Act, 1940, and Rules, 1945. Those medical devices that are notified by central government in the official gazette are regulated under the said Act [6].

Disposable hypodermic syringes, disposable hypodermic needles, and disposable perfusion sets were notified by the Government of India in 1989. Subsequently, in vitro diagnostic devices for human immunodeficiency virus, hepatitis B surface antigen, and hepatitis C virus were notified in 2002. In 2005, the following categories of medical devices were notified:

- cardiac stents
- drug-eluting stents
- catheters
- intraocular lenses
- intravenous cannulae
- bone cements
- heart valves
- scalp vein sets
- orthopedic implants
- internal prosthetic replacements.

The objectives of a clinical trial for a medical device are:

- to evaluate the safety and performance of the device in question and establish its superiority or inferiority over existing treatment options in a sufficiently large number of subjects that the results can be extrapolated in the real-world population
- to define method of use, indications (on-label and off-label use), contradictions, hazards, and any other device-related issue that must be brought to the notice of physician and patients
- to test the utility of the medical device to minimize invasive procedures in defects and conditions requiring compulsory surgical treatment
- to enhance the accuracy and quality of diagnostic tests, clinical outcomes, and procedural success with newer available devices
- to establish treatment protocols (especially if a certain disease or condition has multiple options or therapeutic combinations as treatment).

Regulatory control may include the operation of a quality system (recommended for all devices), technical data, product testing using in-house or independent resources, documentation of clinical evidences to support a manufacturer's claim, the need for independent external audits of the manufacturer's quality system, and independent external review of the manufacturer's technical data. Regulatory requirements for medical devices depend on the risk associated with the device and are under active consideration (Table 6.1).

6.8.2.1. Feasibility Studies and Pivotal Trials

- **Feasibility/safety study**: Feasibility studies are usually single-center studies of a limited number of patients designed to accomplish any number of objectives with a clinical testing program. The first study of the novel investigational device in humans allows the sponsor to collect data on a series of patient outcomes that may be related to device performance, thereby contributing to the identification and selection of clinically significant measures for use as effectiveness endpoints in subsequent pivotal clinical trials.
- **Pivotal trials**: For high-risk medical devices (see Table 6.1), data from well-randomized and statistically qualified trials should be submitted.

Requirements for pivotal trials:

- The trial should be prospective randomized, multicentric, in a significantly large patient population to achieve the desired endpoints with statistically significant results.

Table 6.1 Draft requirements for clinical trials with a medical device

Device class	Risk analysis/examples	Design control	Biocompatibility study	Animal study*	Feasibility/safety study	Pivotal trials
Class A	Minimum-risk devices, e.g. thermometer, tongue depressor	√				
Class B	Minimum- to moderate-risk devices, e.g. hypodermic needles, suction equipment	√	√			
Class C	Moderate- to high-risk devices, e.g. ventilator, bone fixation plate	√	√	√	√	**
Class D	High-risk devices, e.g. heart valve/implantable defibrillator	√	√	√	√	√

* If the active component of the device is defined as a drug (e.g. drug-eluting stent), the data for its animal studies should be submitted as per Schedule Y.
** To be decided on a case-by-case basis.

- The study must establish superiority or at least non-inferiority to a control group with adequate statistical significance.
- The trial must follow international safety standards.
- Endpoints, both primary and secondary, should be based on objective and quantitative performance criteria which define the success or failure of the device in achieving the desired effect on the disease.
- The mean follow-up of the device shall be at least six months, or more, depending on the statistical significance of the endpoints as mentioned above.

The fee for clinical trials on medical devices is to be paid is as follows:

- for a feasibility study: INR 50,000/– (equivalent to approximately US $1000)
- for a pivotal study: INR 25,000/– (equivalent to approximately US $500).

Medical devices with relatively low risks within a defined class may be accorded some relaxation or exemptions depending on their degree of resemblance to already approved devices or the quality of available data, or both. However, such a claim for relaxation needs to be thoroughly documented and justified.

A guidance document, "Requirements for Conducting Clinical Trial(s) of Medical Devices in India", is available on the CDSCO website [16]. This document is intended to provide non-binding guidance for conducting clinical trials of medical devices in India. In brief, in conducting clinical trials on medical devices in India, the following documents/information should be provided:

- covering letter
- Form 44, containing relevant preclinical and clinical data
- TR6 Challan receipt (original)
- delegation of responsibility
- protocol
- ethics committee approval
- informed consent form
- case report form (CRF)
- patient record form
- investigator undertaking
- relevant published literature
- global regulatory status of the device
- clinical study report, if any

- suspected unexpected serious adverse reactions (SUSARs)
- affidavit from the sponsor.

Global regulatory status of the device: As regards the global regulatory status of the device, the sponsor should submit information regarding the regulatory status of the device in other countries, particularly in five Global Harmonization Task Force (GHTF) countries, i.e. the USA, Australia, Japan, Canada, and the European Union:

- approved
- marketed (if marketed a copy of package insert)
- withdrawn, if any, with reasons
- free sale certificate or certificate of analysis, as appropriate
- International Organization for Standardization (ISO) certificates and/or Conformité Européenne (CE) certificate (if available).

In addition, the status and details of the clinical trial protocol (regulatory approval, IEC/institutional review board approval, number of patients, etc.) from each participating country should be submitted.

SUSARs from other participating countries, if any have been reported, and a summary of any reported problems should be submitted. Data elements for reporting SAEs occurring in a clinical trial are provided in Annexure VI of the guidance document [16].

An affidavit from the sponsor is required, declaring that:

- the study has not been discontinued in any country
- in case of discontinuation the reasons for such a discontinuation
- in case of future discontinuation, the applicant would further communicate the same to DCGI
- the information in the investigator's brochure is based on facts.

6.8.3. Stem Cell Research

Guidelines for conducting research on stem cells have been developed jointly by the ICMR and the Department of Biotechnology (DBT) [17]. There is a separate mechanism for review and monitoring for research and therapy in the field of human stem cells, one at the national level, called the National Apex Committee for Stem Cell Research and Therapy (NAC-SCRT), and the other at the institutional level, called the Institutional Committee for Stem Cell Research and Therapy (IC-SCRT).

All institutions and investigators, both public and private, carrying out research on human stem cells should be registered with the NAC-SCRT through the IC-SCRT. All research studies using human stem cells require prior approval of IC-SCRT for permissive research and of the NAC-SCRT for restricted research. The details of these are available at Guidelines for Stem Cell Research and Therapy [17].

All established human stem cell lines from any source, imported or created in India, should be registered with IC-SCRT and NAC-SCRT. Permission for import/procurement from other Indian laboratories is to be obtained from IC-SCRT. It is the responsibility of the investigator to ensure that the cell line has been established in accordance with the existing guidelines of the country. An appropriate MTA shall be adopted for the purpose.

In the area of stem cell research and therapy, the following activities are prohibited in India:

- any research related to human germline genetic engineering or reproductive cloning
- any in vitro culture of the intact human embryo, regardless of the method of its derivation, beyond 14 days or formation of primitive streak, whichever is earlier
- transfer of human blastocysts generated by somatic cell nuclear transfer (SCNT) or parthenogenetic or androgenetic techniques into a human or non-human uterus
- any research involving implantation of human embryo into the uterus after in vitro manipulation, at any stage of development in humans or primates
- animals in which any human stem cells have been introduced at any stage of development should not be allowed to breed
- research involving directed non-autologous donation of any stem cells to a particular individual is also prohibited.

6.8.3.1. Use of Stem Cells for Therapeutic Purposes

Currently, there is no approved indication for stem cell therapy as a part of routine medical practice, other than bone marrow transplantation (BMT). Hence, stem cell therapy other than BMT (for accepted indications) is treated as experimental. It should be conducted only as a clinical trial after approval of the IC-SCRT/IEC and DCGI (for marketable products). All experimental trials are required to be registered with the NAC-SCRT.

Cells used in such trials must be processed under good tissue practices/ good manufacturing practices standards and the injectable product should meet pharmacopeia specifications for parenteral preparations.

The centers carrying out stem cell clinical trials and the agency/source providing such cells for the trial should be registered with the NAC-SCRT through IC-SCRT/IEC. In cases of international collaboration, the public funding agency evaluating the study/NAC-SCRT shall ensure that the certification provided by the collaborating country fulfills the requirements laid down in these guidelines.

For international collaborations, national guidelines of the respective countries should be followed. Exchange of biological material will be permitted as per the existing procedures of the funding agencies (Department of Science and Technology, DBT, ICMR, etc.) or the HMSC (as per Government of India Guidelines), even if no funding is involved after the joint proposal with appropriate memorandum of understanding is approved by NAC-SCRT.

If there is a conflict between the scientific and ethical perspectives of the international collaborator and the domestic side, then the Indian ethical guidelines or law shall prevail.

Research on stem cells/lines and their applications may have considerable commercial value. Appropriate IPR protection may be considered on the merits of each case. If the IPR is commercially exploited, a proportion of benefits shall be plowed into the community that has directly or indirectly contributed to the IPR. The community includes all potential beneficiaries such as patient groups, research groups, etc [17].

In brief, all clinical trials with any stem cells require prior approval of IC-SCRT, IEC, and DCGI for marketable products, and should be registered with the NAC-SCRT. For approval of clinical trials and market authorization of therapeutic products derived from stem cells, human gene manipulations, and xenotransplant technology, the DCGI is advised by an expert committee, namely, the Cellular Biology Based Therapeutic Drug Evaluation Committee (CBTDEC). International collaborations should also have the prior approval of NAC-SCRT and the respective funding agency as per its procedure, or HMSC.

6.9. NEW INITIATIVES

Currently, several new initiatives are in the pipeline, which will further strengthen the regulatory process in India. Notable among these is the proposal for converting CDSCO to the Central Drug Authority (CDA), an autonomous body under MOHFW, which is under active consideration. Other initiatives include the mandatory registration of stakeholders, i.e. CROs (Schedule Y1), ethics committees, investigator sites and investigators, monitoring and oversight of clinical trials (inspection), overseas inspection of manufacturing facilities as well as clinical trial sites, imposition of penal provisions for misconduct and fraud in trials, and e-governance of the entire drug approval process. CDSCO is in the process of creating a national database of all drug manufacturers and drug products manufactured and marketed in India. CDSCO is also considering putting a biometric system in place. This would help to identify prior enrollment of trial subjects in any clinical trial.

6.10. CONCLUSION

The above information is a brief summary of the technical requirements for the conduct of clinical trials in India as per details provided in the Drugs and Cosmetics Act and Rules, as amended from time to time [3]. The Act and Rules are liable to undergo amendments. Furthermore, in view of the number of new initiatives and policies under active consideration, interested sponsors are advised to go through the detailed version of the Drugs and Cosmetics Act and Rules as well as the amendments that are notified from time to time and displayed on the CDSCO website [6].

ACKNOWLEDGMENTS

The authors wish to thank A.K. Pradhan, A. Kukrety, and S.E. Reddy for their valuable suggestions.

REFERENCES

[1] India: An emerging knowledge superpower. In: Enhancing healthcare though quality products. Focus Reports, September 2006, http://www.pharmexcil.com/v1/docs/Brief_report_on_Indian_Pha_maceutical_Industry.pdf
[2] The glorious metamorphosis: compelling reasons for doing clinical research in India, http://www.ficci.com/Clinical-Research-Report.pdf
[3] Government of India; Ministry of Health and Family Welfare (Department of Health). The Drugs and Cosmetics Act and Rules. The Drugs and Cosmetics Act, 1940 (23 of 1940 as amended up to June 30, 2005) and The Drugs and Cosmetics Rules, 1945 (as amended up to June 30, 2005), http://cdsco.nic.in/Drugs&CosmeticAct.pdf
[4] Central Drugs Standard Control Organization, Directorate General of Health Services, Ministry of Health and Family Welfare, Government of India. Good Clinical Practices for Clinical Research in India, http://cdsco.nic.in/html/GCP1.html
[5] Ethical Guidelines for Biomedical Research on Human Participants. Indian Council of Medical Research, New Delhi, 2006, http://icmr.nic.in/ethical_guidelines.pdf
[6] Central Drugs Standard Control Organization, Directorate General of Health Services, Ministry of Health and Family Welfare, Government of India, www.cdsco.nic.in

[7] Central Drugs Laboratory, Kolkata, http://mohfw.nic.in/CDL%20Kolkata.pdf
[8] National Guidelines for Data Safety Monitoring Boards. CDSCO, Ministry of Health and Family Welfare, Government of India. World Health Organization, Country Office for India, July 25, 2007, http://www.whoindia.org/LinkFiles/Clinical_Trials_National_Guidelines_for_Data_Safety_Monitoring_Boards_.pdf
[9] Guidance for International collaboration for research in biomedical sciences, http://icmr.nic.in/guide.htm
[10] Indian Council of Medical Research, New Delhi. Office Memorandum, April 29, 2009, http://icmr.nic.in/icmrnews/OM_IHD.pdf
[11] Registration of Clinical Trials. Central Drugs Standard Control Organization, Directorate General of Health Services, Ministry of Health and Family Welfare, Government of India, http://cdsco.nic.in/index.html
[12] Clinical Trials Registry – India, National Institute of Medical Statistics (ICMR), www.ctri.nic.in
[13] WHO trial registration data set (Version 1.2.1), International Clinical Trials Registration Platform, World Health Organization, http://www.who.int/ictrp/network/trds/en/index.html
[14] Biosafety: Regulatory Mechanisms for GMO and Products Thereof. Department of Biotechnology, Ministry of Science and Technology, http://dbtindia.nic.in/unique page.asp?id_pk=110
[15] Guidance for Industry. Central Drugs Standard Control Organization, Ministry of Health and Family Welfare, Government of India, http://cdsco.nic.in/CDSCO-GuidanceForIndustry.pdf
[16] Guidance Document: Requirements for conducting clinical trial(s) of medical devices in India, dated 04-08-2010; Central Drugs Standard Control Organization (Medical Devices Division), Directorate General of Health Services, Ministry of Health and Family Welfare, Government of India, http://cdsco.nic.in/Requirements%20for%20Conducting%20Clinical%20Trial%28s%29%20of%20Medical%20Devices%20in%20India.PDF
[17] Guidelines for Stem Cell Research and Therapy. Department of Biotechnology and Indian Council of Medical Research, 2007, http://www.icmr.nic.in/stem_cell/stem_cell_guidelines.pdf

CHAPTER 7

Clinical Trials in India

Nermeen Varawalla* and Rajesh Jain**
*ECCRO, London, UK
**ECCRO, Mumbai, India

Contents

Global Clinical Trials
ISBN 978-0-12-381537-8, Doi:10.1016/B978-0-12-381537-8.10007-X

7.1. INTRODUCTION

Ever increasing requirements for clinical trial participants have forced pharmaceutical sponsors to access clinical trial participants from a number of emerging countries. India represents one of the more recent additions to this group, with the potential to enroll large numbers of patients for global clinical trials in a cost-effective manner. Although a new entrant to the international clinical trials marketplace, India has swiftly achieved fair progress in this competitive industry. Backed by compelling fundamentals and supported by activities initiated by various stakeholders, India is attracting increasing numbers of international clinical trials. This is further helped by India's limited but growing track record of quality clinical data contributions for pivotal product registration studies submitted to international regulatory authorities. India is ranked third across all countries after the USA and China in terms of its overall attractiveness as a clinical trial destination according to a recent global survey [1]. India represents one of the more recent additions to this group with the potential to enroll large numbers of patients at relatively low cost for global clinical trials. India has large numbers of treatment-naïve heterogeneous Caucasian patients with diseases of both the tropical and industrialized worlds.

7.2. INDIA: BACKGROUND

India occupies a total area of over 3 million square kilometers, making it the second largest country in Asia. India is divided into 25 states and seven union territories; the latter include New Delhi, the capital city. India is a constitutional democracy with a bicameral parliament. It is a secular country with a rich diverse culture represented by numerous religions, languages, and cultural traditions. The country's infrastructure is stretched at present, and although progress is being made in developing this, the new build cannot keep up with increasing demand.

The Indian economy is estimated to be roughly $1 trillion. The annual gross domestic product (GDP) growth rate has been between 8 and 9 percent for the past few years. The Indian economy is expected to continue growing by at least 5 percent annually for the next 45 years, making it perhaps the only economy in the world to maintain such a robust and sustained pace of growth. India has an ambitious, optimistic middle class of 300 million people that is increasing by approximately 5 percent per year. This section of the population contributes to the demand for better goods and services, driving consumer spending and, in turn, GDP growth. It remains a country with strong contrasts between rural and urban areas and between socioeconomic classes, with 22 percent of Indians living under the poverty line [2].

India traditionally has been a rural, agrarian economy. At the last Census of India in 2001, nearly three-quarters of the population lived in rural areas [3]. However, the growing economy has resulted in increasing the pace of urbanization, such that it is estimated that today about 35 percent of the population resides in urban areas. The urban literacy rate is 81 percent for males and 64 percent for females. In rural areas, literacy is about 76 percent for males and 55 percent for females. The 2001 Census, however, indicated a 1991–2001 decadal literacy growth of 12.63 percent, which is one of the fastest on record [3].

7.3. DEMOGRAPHICS

India's population is currently 1.1 billion and increasing at an annual rate of 1.4 percent. By 2030, India is expected to surpass China as the world's most populous nation. By 2050, the population is projected to reach 1.6 billion. The birth rate is 22 births per 1000 population and the death rate is 8 deaths per 1000 population, per year. Average life expectancy is rapidly approaching the levels of the Western world; at present it is 63 years, having arisen significantly from 32 years in 1947, the year of India's independence, and is expected to increase further to 72 years by 2020 [4]. By 2025, an estimated 189 million Indians will be at least 60 years of age, triple the number in 2004, a result of greater affluence and better hygiene. India's population increase is due in part to a decline in infant mortality, the result of better healthcare facilities and the government's emphasis on eradicating diseases such as hepatitis and polio among infants. The infant mortality rate is 55 deaths per 1000 live births.

India is one of the world's youngest countries, with more than 50 percent of its population below the age of 25 and more than 65 percent below the age of 35. Today, the mean age of the population is 24.9 years. It is expected that, in 2020, the average age of an Indian will be 29 years, compared to 37 for China and 48 for Japan.

All five major racial types, Australoid, Mongoloid, Europoid, Caucasian, and Negroid, are represented among the people of India, with Caucasian being the most prevalent.

7.4. DISEASE PROFILE

India has 16 percent of the global population but 20 percent of the global disease burden. In addition to widely prevalent infectious and tropical diseases, rapid and extensive urbanization has resulted in a disease prevalence similar to that found in developed countries. The share of non-communicable diseases, such as cardiovascular disease, cancer, and diabetes, in the total disease burden is expected to double in the 30 year period from 1990 to 2020. There are 30 million hospitalizations each year, the top three contributors being gastrointestinal diseases, febrile illness, and injuries.

7.4.1. Cancer

There are around 3 million patients with cancer in India with about 1 million new cases detected every year. The lower medical surveillance in India makes it probable that the true extent of the cancer caseload is much higher. Although cancers of the oral cavity, lung, and cervix form over 50 percent of cases, there is also an increasing incidence of breast, head and neck, and pancreatic cancers. The estimated age-standardized rate for 100,000 cases of cancer in India is 96.4 for males and 88.2 for females, which is among the highest in the world [5].

Oral cancer is the most common cancer in India owing to widespread habits of tobacco chewing and smoking. It is estimated that about 40 percent of men and 15 percent of women regularly abuse tobacco by smoking and/or chewing. Tobacco is deemed responsible for 52 percent of all cancers in Indian men and 25 percent in Indian women. Esophageal cancer is more prevalent among older men of low socioeconomic status. This is related to specific dietary deficiencies, tobacco intake, and consumption of areca nut, betel leaf, alcohol, and spicy food. The risk factors for stomach cancer in India are believed to be the consumption of spicy food, alcohol, and the intake of food at high temperatures. Over the past decade, with a general improvement in socioeconomic conditions, there has been a decline in the incidence of esophageal and stomach cancers with a concurrent increase in the incidence of cancer of the colon, rectum, pancreas, liver, and gall bladder [6]. Head and neck cancer accounts for about 25 percent of cancer in the Indian male, and the proportion of advanced disease states with a poor prognosis is high [7].

As India has about 20 percent of the world's smokers, it is not surprising that with an annual incidence of 33,000 it has one of the world's leading incidences of lung cancer. The disease pattern varies from that in Western Europe and the USA in that squamous cell carcinoma is the most common type, instead of adenocarcinoma, and occurs in a younger age group.

The incidence of breast cancer is rising and is the most common cancer among urban Indian women. Some 75,000 new cases are diagnosed each year. With 100,000 new cases of cervical cancer each year, India accounts for 20 percent of the global incidence of this disease.

7.4.2. Diabetes Mellitus

India has 60 million diagnosed cases of type 2 diabetes mellitus, with a five-fold greater prevalence in urban areas. The country has the largest number of individuals in the world with a combination of insulin resistance, hyperlipidemia, and obesity. This is expected to increase further owing to a combination of genetic and lifestyle factors.

7.4.3. Cardiovascular and Respiratory Disease

Eighty million individuals suffer from cardiovascular disease, 15 percent of the population is hypertensive, and 2 million deaths occur each year from cardiac disease [8]. There are 50 million asthmatics, many of whom are steroid naïve with uncontrolled symptoms.

7.4.4. Infection and Inflammation

Malaria, tuberculosis, tetanus, kala-azar, leprosy, influenza, diphtheria, genitourinary infections, and infective diarrheal syndromes form a major component of the infectious disease caseload. About 10 million Indians are known to be HIV positive, with a substantial antiretroviral-naïve patient segment.

7.4.5. Psychiatric and Neurological Diseases

One percent of the population is estimated to suffer from schizophrenia, mania, and/or bipolar disorders. This is probably an underestimation in view of the low psychiatric surveillance and still prevalent social taboos. India has 8 million epileptics, 1.5 million individuals with Alzheimer's disease, and 1 million patients with Parkinson's disease.

7.5. HEALTHCARE SYSTEM

Healthcare is one of India's largest sectors, in terms of revenue and employment, and the sector is expanding rapidly. During the 1990s, Indian healthcare grew at a compound annual rate of 16 percent. Today, the total value of the sector is more than $34 billion. This translates to $34 per capita, or roughly 6 percent of GDP. By 2012, India's healthcare sector is projected to grow to nearly $40 billion. The private sector accounts for more than 80 percent of total healthcare spending in India [4], an extremely high proportion by international standards.

A mixture of private and state-subsidized healthcare exists in India. Of the 15,000 hospitals in India in 2002, roughly two-thirds were public. After years of underfunding, most public health facilities provide only basic care. With a few exceptions, public health facilities are inefficient, inadequately managed and staffed, and have poorly maintained medical equipment. The number of public health facilities is also inadequate. For instance, India needs 74,150 community health centers per million population, but has less than half that number. The principal responsibility for public health funding lies with the state governments, which provide about 80 percent of public funding. The federal government contributes another 15 percent, mostly through national health programs.

Private firms are now thought to provide about 60 percent of all outpatient care in India and as much as 40 percent of all inpatient care. It is estimated that nearly 70 percent of all hospitals and 40 percent of hospital beds in the country are in the private sector. Private and corporate hospitals often provide subsidized and charitable treatment to patients who are poor. However, the majority of patients are private paying patients, an increasing proportion of whom are covered by health insurance.

When it comes to healthcare, there are two Indias: the country that provides high-quality medical care to middle-class Indians, the urban population, and medical tourists; and the India in which the majority of the population lives with limited or no access to quality care. Today, only 25 percent of the Indian population has access to Western medicine, which is practiced mainly in urban areas, where two-thirds of India's hospitals and health centers are located. Many of the rural poor must rely on alternative forms of treatment, such as ayurvedic medicine, unani, and acupuncture. Only 11 percent of the population has any form of health insurance coverage, further compounding the healthcare challenges that India faces.

The federal government has begun taking steps to improve rural healthcare. Among other things, the government launched the National Rural Health Mission 2005–2012 in April 2005. The aim of the Mission is to provide effective healthcare to India's rural population, with a focus on 18 states that have low public health indicators and/or inadequate infrastructure.

India has 16,000 hospitals, 942,000 hospital beds, 14,000 diagnostic laboratories and 210 medical colleges. However, only 41 hospitals are accredited by the National Accreditation Board for Hospital and Healthcare Providers, while 84 hospitals are currently in the accreditation process. Over the past decade there has been a substantial growth in the number of private corporate hospital groups which have a chain of hospitals located in India's large metropolitan cities and second and third tier cities and towns.

India has over 500,000 qualified physicians, with 17,000 physicians graduating each year from the country's 210 medical colleges [9]. Medical training is delivered in English based on a curriculum incorporating all the elements of a Western medical system. Many of India's leading specialists have received part of their postgraduate medical training in the USA or the UK.

7.6. INDIAN PHARMACEUTICAL MARKET

The Indian pharmaceutical market is worth US $25 billion, making it the world's fourth largest market in unit sales, but it ranks 13 in value terms [10]. It is a highly fragmented market, with the largest company having only a 7 percent market share. Indian companies, predominantly generic players, have a 70 percent market share. The discrepancy between the volume and value of this market is due to the fact that there is strong state-imposed price control so as to enable wider access to medicine. The market has enjoyed a 15 percent annual growth rate over the past decade, which appears to be set to continue, driven by India's rapidly growing middle class with the financial resources to purchase expensive drugs. The present per capita pharmaceutical expenditure is only US $4.5 per year, providing a low base for much future growth. Revenues from new drugs account for 75 percent of annual revenue growth, a reflection of the enthusiasm for new treatments. Six Indian pharmaceutical companies have been acquired by foreign multinationals in the past four years. Most recently, Abbott Laboratories acquired the pharmaceutical business of Piramal Healthcare for US $3.7 billion, which will give it a 7 percent market share in India. These acquisitions threaten the Indian government's efforts to control pharmaceutical prices.

Counterfeit drugs are a serious problem in India; recently, the Central Drugs Standards Control Organization (CDSCO) estimated that 11 percent of the drugs in India's pharmaceutical market were fake. To clamp down on the burgeoning illegal trade, the health ministry launched a reward program this year offering $55,000 to citizens who provide information about fake-drug syndicates. In addition, the ministry also strengthened its drug law to speed up court trials. Suspects found guilty of manufacturing and selling fake drugs can be sentenced to life imprisonment.

7.7. INDIA'S CHANGING INTELLECTUAL PROPERTY RIGHTS ENVIRONMENT

India's stance in relation to intellectual property (IP) protection has played a vital role in the evolution of the domestic pharmaceutical industry. The Indian Patents Act (1970) that replaced British colonial laws related to intellectual property rights did not recognize pharmaceutical product patents. Instead, the manufacturing process patents were recognized for a seven-year period. The Act's objectives were to encourage the development of an indigenous pharmaceutical industry and to provide Indian consumers with low-cost medicines. During the 1970s and 1980s, the Indian pharmaceutical industry enjoyed tremendous growth to become the world's largest producer of formulations and bulk drugs.

However, India's IP environment is swiftly transforming, forcing India's generic pharmaceutical industry to evolve. India is in complete concordance with the World Trade Organization (WTO) Trade-Related Aspects of Intellectual Property Rights (TRIPS) agreement and has legislation in place to enforce all patents registered after January 1, 1995. Clinical data exclusivity for five years is being enforced. The Indian government is determined to foster a research-driven life science industry, including a thriving pharmaceutical research services and contract research organization (CRO) sector, and recognizes that robust IP protection laws are critical to do so. Furthermore, India's IT and media sectors have driven the progress in the legal enforcement of IP protection so as to protect the IP that they continually generate [11].

The ruling of the Chennai high court in August 2007 that dismissed the challenge from Novartis regarding its product patent for imatinib mesylate (Gleevec®) has attracted substantial international attention. Via this ruling India has clearly declared its intention to adopt a patent regime that enables the country to meet its aspirations to ensure inexpensive access to medicines by prohibiting "ever greening" of patents, in keeping with the spirit of the WTO TRIPS agreement. India has demonstrated that it does have an effective legal process for patent grant, review, and challenge; however, it has set a different, perhaps higher standard of patent law for itself. International sponsors choosing to conduct clinical trials in India will be those who conclude that the advantages of doing so outweigh the possible disadvantages of operating in a research and commercial environment with a higher bar to secure for a product patent.

7.8. REGULATORY ENVIRONMENT

The Ministry of Health and Family Welfare oversees all medical and public health matters, including the control of drugs. The secretary to the Government of India heads this department and is supported by two deputy secretaries; they, in turn, are supported by the Drugs Controller General of India (DCGI).

The DCGI's office is responsible for the approval of manufacture, registration, marketing, and clinical trial applications for pharmaceutical products. Its functions include the granting of blood bank, import and export licenses, regulation of drug packaging, narcotics control, granting of permission for clinical research, supporting the compliance of good manufacturing practices (GMP) and good clinical practice (GCP), regional monitoring and implementation of the Drugs and Cosmetics Act, supporting domestic pharmaceutical exports, GMP inspection at manufacturing plants, licensing of biologicals, and coordination with the Department of Biotechnology (DBT), National Institute of Biologicals (NIB), and Indian Council of Medical Research (ICMR).

The DCGI has zonal offices located at Mumbai, Kolkata, Ghaziabad, and Chennai, and subzonal offices, the main ones being at Ahmedabad, Hyderabad, and Patna. They are responsible for coordination between the state and the DCGI and conduct joint inspections of manufacturing plants, blood banks, and pharmacies. Port offices are responsible for clearance of products that are imported into the country either for marketing purposes or for clinical research.

The Central Drugs Laboratory (CDL) is the national statutory laboratory of the Indian government. It is responsible for the quality control of drugs and cosmetics. India has six national laboratories, situated in Kolkata, two in Mumbai, Ghaziabad, Chennai, and Ranchi. Their functions include analysis and quality control of drugs and cosmetics, including those imported into the country. The CDL collaborates with the World Health Organization (WHO) in the preparation of international standards and specifications for international pharmacopeias.

7.8.1. Regulations

India's regulatory environment for clinical trials is stable, progressive, and evolving to meet the needs of India's international clinical research sector. The regulations balance the need to be industry friendly by making India an attractive clinical trial destination with the requirement to regulate the sector, ensuring subject protection.

Clinical trials in India are conducted under the legislation contained in Rule 122-DA, Drugs and Cosmetics Act, 1940 and Drugs and Cosmetics Rules, 1945. Schedule Y details the requirement and guidelines on Clinical Trials for Import and Manufacture of New Drugs. This came into force in 1988 essentially to serve a predominantly generic pharmaceutical industry. The 1945 Rules have been amended up to the Seventh Amendment, 2008. The most recent amendment, a guidance document on Registration of Medical Devices, is currently in circulation in draft form.

Indian GCP guidelines for research with human subjects published by the CDSCO are based on the Declaration of Helsinki, WHO, and the International Conference on Harmonisation (ICH) guidelines for GCP. Ethical Guidelines for Biomedical Research on Human Subjects issued by the ICMR, 2006, refers to the Council for International Organizations of Medical Sciences (CIOMS) and ICH-GCP, and is currently awaiting clearance by parliament to become law.

In order to carry out clinical trials in India for new drug substances discovered in countries other than India, submission of Phase I data generated outside India is required. First-in-human, Phase I clinical trials for these new drug substances discovered and developed in a foreign country are not allowed in India. Permission to carry out clinical trials in India is generally given in stages, considering the data emerging from earlier phase(s). Trials with drugs that have completed four years in India after first marketing or clinical trial approval are not considered as new drugs and do not require approval from the DCGI. However, ethics committee approval is required for trials involving such drugs. For systemically absorbed drugs approved elsewhere in the world, bioequivalence studies with the reference formulation are required to be carried out. For new drug substances discovered in India, clinical trials are required to be carried out in India from Phase I onwards.

Application to the DCGI for permission to conduct a clinical trial in India must include the following information:

- name of the drug
- objective of the study
- phase of study
- names of the participating countries and investigative sites
- total number of patients to be enrolled globally
- number of investigator sites to be utilized in India
- number of patients to be enrolled in India
- regulatory and ethics committee approvals from participating countries with their English translation
- status of the study in other countries, including the number of patients enrolled, the number of patients who have completed the study, and the number of discontinued patients
- suspected unexpected serious adverse reactions (SUSARs) from other participating countries
- an affidavit from the sponsor stating that the study has not been discontinued in any country; in case of discontinuation, the reasons for this and an undertaking that the sponsor would communicate future discontinuations to the DCGI
- a detailed product dossier with the following information:
  - a brief description of the drug and the therapeutic class to which it belongs
  - chemical and pharmaceutical information
  - information on active ingredients, including their generic and chemical names
  - physicochemical data: chemical name, structure and physical properties
  - analytical data, including spectrometry data
  - complete monograph specification
  - validations including assay methods

- stability studies
- formulation data, including dosage form, composition, stability evaluation, and packing specifications
- composition
- animal pharmacology data, including pharmacokinetics
- animal toxicology data, including male fertility, reproductive and developmental studies, allergenicity, genotoxicity, and carcinogenicity
- human/clinical pharmacology (Phase I) data, including pharmacokinetics, pharmacodynamics, and early measurement of drug activity
- therapeutic exploratory trials (Phase II) that evaluate the effectiveness, side-effects, and risks of a drug for a particular indication(s) in patients with the condition under study
- therapeutic confirmatory trials (Phase III) that demonstrate or confirm therapeutic benefits(s), drug safety, and efficacy in the intended indication (if available)
- Phase IV postmarketing trials performed after drug approval and related to the approved indication(s); if available, additional information on drug–drug interaction(s), dose–response, or safety studies
- rationale for selecting the proposed dose(s) and indication(s)
- reports of completed clinical studies, each of which to be certified by the principal investigator or, in the absence of a designated principal investigator, then by each of the investigators participating in the study
- special studies, if available, to include bioavailability/bioequivalence, studies in special populations such as geriatrics, pediatrics, pregnant or nursing women
- regulatory status in other countries: list of countries where the drug is marketed, approved, investigational new drug (IND) application filed, approved as IND and/or withdrawn, with reasons for the same
- restrictions on use, if any, in countries where marketed/approved
- free sale certificate or certificate of analysis, as appropriate
- prescribing information: proposed full prescribing information and drafts of labels and cartons
- samples of pure drug substance and finished product.

7.8.2. Documents to be Submitted to the DCGI for a Clinical Trial Application

- Completed Form 44 and Treasury chalan
- Form 12 application for a Test License (TL) to import a drug for clinical trials and Treasury chalan

- details of biological specimens to be exported
- protocol
- informed consent documents
- case report form
- investigator's brochure duly supported by an affidavit that the summarized information submitted is based on facts
- undertakings by the investigators
- ethics committee approvals (if already available)
- fee of Indian rupee (INR) 50,000 (US $1020) for Phase I trials, INR 25,000 (US $510) for Phase II, and INR 25,000 (US $510) for Phase III trials
- authorization letter from the sponsor if submitted by CRO.

For drugs indicated in life-threatening, serious diseases or diseases of special relevance for India's healthcare needs, the toxicological and clinical data requirements may be abbreviated, deferred, or omitted, as deemed appropriate by the DCGI. The application dossier is usually complemented by a formal presentation to the DCGI by the sponsor or the appointed CRO.

Clinical trials can be conducted by any hospital in India and every application is considered on a case-by-case basis.

At the present time, the DCGI does not expect any extended commitment towards healthcare from clinical trial sponsors. Sponsors are not obliged to arrange for post-trial continuation of therapy when subjects show a benefit from the study drug. It is not a legal requirement for the sponsor to insure against trial-related injuries. However, ethics committees may insist on insurance or some form of provision for care of subjects in the event of a trial-related injury.

A clinical trial application, and application to import clinical trial supplies can be made only by a company with a legal basis in the country. This could be the Indian affiliate of an international company, the Indian partner of a foreign company, or an Indian company itself.

7.8.3. Approvals Process

The clinical trial application is first examined by the New Drug Division, followed by a detailed review by the IND Committee whose recommendations are presented to the DCGI. The DCGI may either grant approval or raise further queries.

For the purpose of granting permission, clinical trial applications are classified into categories A and B. Category A are those whose protocols have already been approved by one of the category A countries; namely, the

USA, the UK, Switzerland, Australia, Canada, Germany, South Africa, or Japan. The regulatory authorities of these countries are deemed to be more experienced, hence their approval signals that the protocol satisfies the requirements of a sophisticated clinical trial regulatory environment.

For a complete category A application, the first response from the DCGI office can be expected within 30 days for the clinical trial supplies import license and within 45 days for the application itself. The application status may be checked online; approval letters sent out by the DCGI office are displayed in the CDSCO website under the "Daily Dispatch" section.

Applications that do not fulfill category A criteria are of category B and require more time for processing. Once an application is deemed to be in cateogry B it cannot be transferred to category A. Table 7.1 describes the regulatory approval timelines in India. Any inadequacy found in the documents will lead to a query from the DCGI. The applicant's response to the query must navigate a queue and could result in delays of an additional 45 days.

Factors that expedite and support timely regulatory approval are a well-collated, complete application dossier supported by a justification of why India has been included in the study. A description of how the study drug might address an important healthcare issue in India could be of value in facilitating the approval process. The regulator obtains reassurance if the study is already initiated in other countries and has a number of subjects enrolled. Clearly, a category A classification is most beneficial; this may be achieved by a US Food and Drug Administration (FDA)-granted IND, regulatory authority, and ethics committee approval from one or more of the category A countries.

Table 7.1 Timelines for clinical trial regulatory approval in India

Regulatory body	Approval	Timeline
DCGI	Clinical trials: all phases	First response or approval within 45 working days
	Bioequivalence study	First response or approval within 28 working days
DCGI	Test license to import supplies	2 weeks in addition
Ethics committees	Local ethics committee approval by sites	6–8 weeks (in parallel)
Total (parallel processing)		8 weeks
Referral body for expert opinion	rDNA products, radiopharmaceuticals, stem cells, etc.	Additional 12–14 weeks

DCGI: Drugs Controller General of India; rDNA: recombinant DNA.

In situations where the DCGI office does not have the expertise to review the clinical trial application, the dossier is referred to external agencies such as the ICMR, DBT, and/or Genetic Engineering Approval Committee (GEAC) for their review and expert opinion. This is a long and unpredictable review process which often leads to substantial delays. Regarding permission for first-in-human, Phase I trials for new drugs discovered or developed in India, the timelines for DCGI approval are between five and eight months.

If the drug to be tested has a particular relevance for special populations such as pediatric, geriatric, pregnancy, or lactation, the sponsor may be asked to generate clinical data from testing in these special populations.

7.8.4. Single-window Clearance for Export NOC

Recently, a single approval process has been introduced for permission to export trial-related biological samples. Formerly, following DCGI approval, an applicant had to apply separately to the Directorate General of Foreign Trade (DGFT) for the export No Objection Certificate (NOC) for biological samples. This process has been simplified such that an applicant can apply for the export NOC simultaneously to DCGI, without the need for separate application to DGFT. As per revised timelines, an export NOC can now be expected within 10 working days, in contrast to the previous two- to four-week timeline for DGFT approval.

7.8.5. Handling of Protocol Amendments

Administrative, logistic, and minor protocol changes, and additional safety assessments do not require DCGI notification or permission. The addition of investigator sites, new appointment or a withdrawal of an investigator, amendments to the investigator's brochure or informed consent form require DCGI notification within 30 days but there is no requirement to wait for permission. The recruitment of additional patients, major protocol changes with respect to study design, dosing, treatment options, and inclusion or exclusion criteria require prior DCGI permission. All amendments must be approved by the concerned institutional ethics committee before their implementation.

7.8.6. Sponsor's Responsibilities

Sponsors are required to submit to the DCGI and ethics committee an annual status report on each clinical trial including ongoing, completed, and terminated studies. When a trial is terminated the reason should be stated. The sponsor is expected to be responsible for implementing and maintaining quality assurance to ensure compliance to GCP guidelines. The sponsor will have to submit status reports at prescribed intervals and inform the DCGI of reasons for premature termination of the study. The period for reporting serious adverse events (SAEs) to the DCGI and other investigators has been defined as within 14 calendar days.

7.8.7. Investigators' Responsibilities

All clinical trial investigators should possess appropriate qualifications, training, and experience and have access to diagnostic and treatment facilities relevant for the proposed trial protocol. A qualified physician who is a trial investigator or subinvestigator should be responsible for all trial-related medical decisions. During and following trial participation, the investigator should ensure that adequate medical care is provided to the participant for any adverse events. Investigators must report all serious and unexpected adverse events to the sponsor within 24 hours and to the approving ethics committee within seven working days of their occurrence.

Along with the sponsor and CRO, it is the investigators' responsibility to protect personal data collected during the conduct of the study as per the terms of the Indian Personal Data Protection Bill, 2006. Financial Disclosure is not mandatory for Indian company-sponsored trials. It is applicable only to international trials conducted in India. Schedule Y provides the format for the Letter of Undertaking similar to the FDA 1572 for this purpose.

7.8.8. Informed Consent

In all trials, freely given, informed written consent is required to be obtained from each study subject. The investigator must provide information about the study verbally as well as using a patient information sheet, in a language that is non-technical and understandable by the study subject. The subject's consent must be obtained in writing using an informed consent form. Ethics committee approval of the informed consent form is essential. The subject should initial all five consent clauses and make a full signature at the end of the form. There is a provision for mature minors and adolescents to sign the consent form. If the subject is illiterate, he or she is invited to nominate a literate relative to serve as the legally acceptable representative during the informed consent process. The subject's thumbprint impression is obtained on the informed consent form, accompanied by signatures of the legally acceptable representative, an independent witness, and the member of site staff overseeing the informed consent process.

7.8.9. Online Registration of Clinical Trials

Since June 15, 2009, online registration of clinical trials has been mandated by the DCGI. Any researcher who plans to conduct a trial involving human participants is expected to register the trial on the Clinical Trials Registry India (CTRI) before enrollment of the first participant. Registration is voluntary, but some fields are mandatory for registration to be complete; the trial is to receive a registration number and must fulfill WHO/International Committee of Medical Journal Editors (ICMJE) requirements. Incomplete entries are given a provisional registration number and are not eligible for publication in ICMJE-endorsed journals. Registration of trials in the CTRI is free. All registered trials will be made publicly available. The CTRI website may be accessed by anyone free of charge. The CTRI has been set up by the ICMR's National Institute of Medical Statistics (NIMS) and is funded by the Department of Science and Technology (DST) through the ICMR. It also receives financial and technical support through the WHO, WHO-South East Asia Regional Office (SEARO), and the WHO India country office. There are future plans to link the CTRI to the International Clinical Trials Registry Platform maintained by the WHO.

The DCGI has undertaken to conduct inspection of trial sites. The DCGI, in collaboration with the WHO and US FDA, has concluded initial training of 25 inspectors. A panel of three officials, comprising a drug inspector, a pharmacologist, and a clinical trials expert, is authorized to conduct surprise inspections at clinical trial sites. Inspectors will focus on infrastructure, skills, processes, and informed consent, with the authority to withdraw, suspend, or hold back clinical trial approval. Random selection of trials for inspection has been initiated by the DCGI.

7.8.10. Forthcoming Initiatives

The DCGI is planning to initiate a system of registration and accreditations of CROs, in an attempt to quality control their activities. A draft guideline on the CRO registration process has already been made with a proposal to incorporate this within Schedule Y of the Drugs and Cosmetics

Rules, 1945. There are plans to implement a process for the registration of clinical trial sites and ethics committees. The DCGI seeks to introduce laws for the penal punishment for clinical research fraud and misconduct in India. In due course, the DCGI will permit the conduct of Phase I trials in India on new drugs discovered outside India. Furthermore, there is interest in creating an environment for Phase 0 or microdosing studies in India.

The DCGI seeks to gear up its capabilities to keep pace with India's maturing clinical research environment and the growing international clinical trial activity by increasing the number of qualified staff. DCGI is collaborating with Health Canada, US FDA, and regulators in Brazil and South Africa to strengthen its capabilities in pharmacovigilance, regulating the approval of medical devices and biological drugs. The CDSCO has plans to implement an e-governance program which would enable companies to file, track, and review trial applications online.

7.9. ETHICS COMMITTEE APPROVAL

Most major clinical institutions and hospitals have ethics committees that comply with the ICH-GCP guidelines. India does not have a central ethics committee. Hence, all the approvals are given by individual ethics committees, which approve all study protocols prior to study commencement. Ethics committees customarily meet once a month. Approvals generally take up to 60 days but are occasionally granted in four weeks. The investigator submits the documents for ethics committee approval and may need to make a presentation on the study to the committee before its decision.

The following documents are required for ethics committee submission, in the English language:

- covering letter
- protocol and amendments (signature page signed and dated)
- case report form
- all relevant regulatory documents (copy of application to the DCGI if approval is not yet available)
- informed consent form (in English and in local language if required)
- patient information sheet (in English and in local language if required)
- investigator's brochure
- investigator agreements (as per ethics committee requirement)
- investigator CVs
- list of centers and investigators (as per ethics committee requirement)
- proposed financial agreement (as per ethics committee requirement)

- indemnity insurance
- any other patient information material to be used in the study, e.g. patient diaries, in English and local languages as per requirement
- the prescribed application form, if applicable.

Ethics committee approval should be notified to the DCGI before initiation of the clinical trial. The trial maybe initiated at site only after obtaining approval from that site's ethics committee. Sites without ethics committees can accept the approval granted to the protocol by the ethics committee of another site or an independent ethics committee, provided the approving ethics committee is willing to accept their responsibilities for the site without an ethics committee and that the protocol version is same at all trial sites. The ethics committee for pediatric trials has to include members knowledgeable about pediatric, ethical, clinical, and psychosocial issues.

Revised Schedule Y devotes significant attention to the roles and responsibilities of the ethics committee. It also describes the composition of the ethics committee as per the ICMR guidelines and provides formats for the approval letter of the ethics committee.

It is the responsibility of the ethics committee that reviews and accords its approval to a trial protocol to safeguard the rights, safety, and well-being of all trial subjects. The ethics committee should exercise particular care to protect the rights, safety, and well-being of all vulnerable subjects participating in the study, particularly those who are members of a group with a hierarchical structure such as prisoners or institutionalized communities, patients with incurable diseases, impoverished people, patients in emergency situations, ethnic minority groups, homeless persons, refugees, minors, or others incapable of personally giving consent. Ethics committees should document their standard operating procedures and maintain a record of their proceedings. They should make, at appropriate intervals, an ongoing review of the trials for which they have given approval. In case an ethics committee revokes its approval accorded to a trial protocol, it must record the reasons for doing so and at once communicate such a decision to the investigator as well as to the DCGI.

7.9.1. Ethical and Scientific Integrity Concerns

India has clinical trial subject populations with unmet medical needs, reverence for physicians who remain authority figures, and linguistic and cultural barriers for subjects to understand fully the implications of trial participation. ICH compliance in such an environment presents a challenge. Clinical trial conduct in India raises concerns about exploitative trial designs, informed consent, and post-trial access. Recognizing the importance of a robust ethical framework for clinical research in India, the ICMR has set up initiatives to audit the functioning of ethics committees, introduce a national ethics committee accreditation system, and offer training via the Independent Forum for Ethics Review Committees.

The DCGI has plans to monitor ethical practices of sites, investigators, and sponsors to ensure that there is compliance with the ICMR Ethical Guidelines for Biomedical Research on Human Subjects. This initiative is supported by the World Bank and is intended to strengthen the DCGI's control of the drug development process. Investigators and site personnel play a vital role in ensuring that all participants are provided with trial-related information in their language with adequate provision for consultation with family members. Best-in-class investigators in India fully recognize the importance of informed consent, more so keeping in mind the socioeconomic deprivation of their patients, and make every effort to comply. They do so not only for ethical reasons but also to ensure that clinical research in India continues to flourish, recognizing that unethical practices will be the death knell of this nascent sector. Medical social workers play an important role in ensuring that patients and their families understand the implications of clinical trial participation.

7.10. GOOD CLINICAL PRACTICE IN INDIA

Indian GCP guidelines to ensure uniform quality of clinical research were prepared by the CDSCO and endorsed by the Drug Technical Advisory Board (DTAB). The first clinical study conducted according to Indian GCP guidelines was performed in 1995. Indian GCP guidelines are in line with ICH-GCP. However, there are significant differences, as described below.

- Indian GCP insists that the investigator should be qualified as per the requirement of the Medical Council of India (MCI), thus disqualifying non-medical scientists or those with overseas medical degrees not prescribed by the MCI.
- Indian GCP mandates that the sponsor and the investigator should sign a copy of the standard operating procedures (SOPs). Besides, the investigator and his or her staff have to be aware of and comply with SOPs. In contrast, ICH-GCP expects the investigator to comply with the protocol and leaves the task of monitoring compliance to SOPs to monitors and auditors.

- As per ICH-GCP, when the trial is completed, the investigator has to provide the independent ethics committee (IEC) with a summary of the outcome of the trial. Indian GCP demands that the investigator should sign and forward their site's case report forms (CRF), interpretations, analyses, and reports to the sponsor and the ethics committee. In practice, many Indian investigators do not comply with this requirement and the CRFs are never sent to the ethics committee unless specifically requested.
- According to Indian GCP, the ethics committee has power to order discontinuation of a trial if it believes that the trial objectives have been achieved or unequivocal results obtained. As per ICH-GCP, this is the responsibility of an independent data monitoring committee established by the sponsor. In practice, for most global trials this committee is based outside India.
- Indian GCP guidelines define the content and format of the informed consent form, which also covers issues of biological samples. There is an expectation that subjects should be made to understand possible current and future uses of their biological samples, type of data that could be generated from the analysis of these samples, whether these data would be used for secondary purposes or would be shared with others, and the risks of discovery of biologically sensitive information. Also, subjects should be offered the right to prevent the use of their biological samples at any time during the conduct of the study. Not surprisingly, this clause makes consent more complex; hence, in practice, it is not always followed.
- Indian GCP guidelines prescribe mandatory clauses on compensation in the informed consent form. Free treatment is promised for research-related injury to be provided by the investigator and/or their affiliated institution. The sponsor agrees to compensate subjects for disability or death resulting from research-related injury for which it usually subscribes to insurance cover. Most ethics committees and clinical sites insist that sponsors comply with this guideline.
- Indian GCP mandates that when a subject is withdrawn from research for medical reasons related to the study, the subject should get the benefit for full participation.
- Indian GCP requires the monitor to inform the sponsor and ethics committee in case of any unwarranted deviation from the protocol or any transgression from the principles of GCP. As the monitor has no contact with the ethics committee, this requirement cannot be fulfilled.

Strangely, Indian GCP does not require the monitor to verify the informed consent from.

- Indian GCP states that the label on the investigational product should include the name and contact numbers of the investigator and the name of the site. This leads to practical difficulties as the standard practice is to have uniform labels for all sites and countries in a global trial.
- Indian GCP mandates that the sponsor should make arrangements for safe and secure custody of all study-related documents for a period of three years after the completion of the study or submission of the data to the regulatory authority, whichever is later. If the company does not obtain marketing approval within three years of completion of the trial and if there is a need for a postlaunch regulatory inspection, the records may not be available. Hence, it would be advisable, as in the case of ICH-GCP, to link the duration of records storage to marketing approval.
- Indian GCP guidelines state that the monitor should have adequate medical, pharmaceutical, and/or scientific experience.

7.11. FDA AUDIT AND INSPECTION: OUTCOMES AND FINDINGS

Over the past five years clinical data from India contributing towards pivotal global clinical trials have been accepted by the FDA and European Medicines Agency (EMA) on numerous occasions. The increasing volumes of data from India being submitted to the FDA prompted its first audit in India in 2005. Two clinical sites participating in a global study for an anti-infective were audited with no actionable findings. Since then, further FDA audits have taken place, and as yet none of the audits has passed the dreaded Official Action Indicated (OAI) verdict. Data submissions from India in the recent past have been part of 13 successful new drug application (NDA) submissions.

7.12. LANGUAGE

English is the language for state, business, and professional communication, and the language of clinical medicine and research. Medical education is delivered in English. All medical, nursing, and pharmacy professionals are fluent in English and use English for all medical notes and records. Similarly, all clinical trial documentation, including the clinical research forms, investigational brochures, drug labels, and regulatory and ethics committee submissions are in English.

Hindi is the national language and mother tongue of about 30 percent of the population. There are over 35 languages spoken throughout the country, each with their own script and strong literary tradition. Of these there are 14 official languages: Bengali, Telugu, Marathi, Tamil, Urdu, Gujarati, Malayalam, Kannada, Oriya, Punjabi, Assamese, Kashmiri, Sindhi, and Sanskrit.

In order to communicate with clinical trial subjects, it is necessary to translate patient-facing clinical trial literature, namely the informed consent form and patient information leaflet, into a selection of Indian languages. The choice of language and translations needed depend on the regions of India where the trial is being conducted; typically, an average of eight Indian languages is required for a multiregional trial in India. In this multilingual society, most site staff are conversant with the languages in which the informed consent document is provided to the site and able to explain verbally the informed consent document in a regional language with which the subject is comfortable.

Most widely used patient quality of life questionnaires are validated in about seven Indian languages, with ongoing efforts to extend this list. For example, QLQ-C30 is validated in Hindi, Marathi, Gujarati, and Telugu, and there is a validated Hindi version of mHAQ.

7.13. INDIA'S CLINICAL TRIAL SECTOR

India has been participating in global clinical development for less than a decade. Until a few years ago, generic pharmaceutical companies dominated the clinical development sector. The situation is changing rapidly, driven by a clinical research environment that has never been so conducive for international drug developers.

India is one of the fastest growing clinical research destinations, with a growth rate that is two and half times the overall market growth [12]. The number of registered international clinical trials that include India has increased by 30 percent each year for the past three consecutive years.

In 2005, India participated in 150 international clinical trials, just 1 percent of the total number of trials ongoing in the world then. In just two years, this increased by 75 percent to 260 international clinical trials, accounting for 1.2 percent of worldwide activity. In 2007, there were approximately 350 investigators in India engaged in clinical trials intended for FDA submission and 207 clinical sites registered to do so. In the second half of 2009, 373 clinical trials were registered with the Indian Clinical Trials

Registry (CTRI). This increase in clinical trial activity is reflected in the fact that the number of new investigators participating in FDA-registered clinical trials from India has increased over the past five years by 49 percent [13] and there has been a 116 percent increase in the number of clinical trial sites participating in international trials between the years 2002 and 2008.

The number of trials that include India registered on clintrials.gov has increased from 1191 in February 2010 to 1366 at the time of writing, a 15 percent increase in six months. However, this represents only 1.5 percent of all worldwide registered clinical trials as compared to 52 percent participation by the USA.

7.13.1. Players

The players within the Indian clinical development sector are local affiliates of multinational pharmaceutical companies, Indian pharmaceutical companies, Indian affiliates of global CROs, local CROs, and support service providers.

Traditionally, the Indian subsidiaries of multinational pharmaceutical companies have been sales and marketing units, with their involvement in clinical development restricted to the conduct of local postmarketing studies. This has changed greatly; today, Indian affiliates of multinational pharmaceutical companies have substantial clinical development capabilities and contribute increasing amounts of Indian Phase II and III clinical trial data for their global clinical development programs. They do this by utilizing in-house resources as well as engaging with CROs. The forerunners have included Pfizer, Lilly, and Aventis, which have substantially contributed to the early development of the sector by training investigators, ethics committee members, and clinical research professionals. As a consequence of clinical research studies placed at various hospitals, investments have been made in upgrading research infrastructure in India; for example, the purchase of bone densitometry diagnostic equipment following the conduct of a large osteoporosis clinical trial. They have also contributed to the maturation of the regulatory environment in India, including much needed revisions to regulations and the refinement of Indian GCP guidelines. Indian pharmaceutical companies such as Glenmark, Biocon, Ranbaxy, Dr Reddy's, and Sun Pharma, which are moving beyond their generic business, have begun to establish clinical development capabilities to meet the needs of their own innovative new molecules.

In 2006, the Indian CRO market was estimated to be valued at $265 million, a mere fraction of the total global CRO market of $14.3 billion. By

2010, based on a CAGR of 22.7 percent, the Indian market was expected to be $600 million. Among the global CROs, Quintiles, founded in India in 1997, is the market leader with the longest and broadest experience. Most global CROs now have a presence in India, notably, Parexel, PRA International, and ICON. Motivated by the relatively low barriers to entry and commercial hype, over 50 local CROs have set up operations. For the majority of local CROs, the provision of bioequivalence studies to domestic generic pharmaceutical companies forms a major part of their business; however, almost all have aspirations to develop their global clinical trials business. There are too many local CROs at the present time, and hence consolidation is inevitable. The better known local CROs include SIRO, Ecron AcuNova, Clinigene, ECCRO, and Veeda Research.

Pathology laboratories such as Dr Lal's, Metropolis, and SRL Ranbaxy offer central laboratory services for clinical trials. Their international College of American Pathologists accreditation, competitive prices and availability of reliable international courier services make these providers attractive not just for the Indian samples but also for those from worldwide sites. There is also a number of specialist service providers providing clinical trial supply management, translation, centralized electrocardiogram monitoring, and interactive voice response system (IVRS) services.

7.13.2. Training

Clinical research training to investigators, site staff, and professionals is being delivered in a number of ways. Study start-up activities include GCP training of investigators and site personnel. This is provided by sponsors and CROs, many of whom understand that this is a good way to build a pool of trained Indian investigators with whom they may continue to work. CROs are cognizant of the importance of trained clinical operation staff and continue to make substantial investment in this area. Global CROs utilize their global training resources and expertise to deliver both face-to-face and online training to their staff in India.

Over the past decade, training institutes have been set up that provide training to biomedical graduates seeking to develop a career in clinical research, including degree courses and diplomas. Several of these training institutes have collaborations with Indian and foreign universities, government, and industry. Examples of India's better known training institutes include the Academy for Clinical Excellence, the Indian Clinical Research Institute (ICRI) in association with the University of Cranfield, Clinical Research Education and Management Academy (CREMA), and Catalyst clinical services. These organizations and others have helped to meet the evolving training needs of participants in India's clinical research sector, be they from industry, government, regulators, or academia. Furthermore, they have facilitated interaction between senior members of stakeholder organizations, leading to collaborative improvements in this evolving sector [14].

Apart from formal and on-the-job training there is much conference and seminar activity which highlights best practices and current trends. The Drug Information Association (DIA) has had an office in Mumbai since 2007 and holds regular conferences in India. As a result of these activities, there is awareness among the Indian clinical research community of worldwide developments in global clinical research, interest among biomedical students to pursue a career in clinical research, and improving capabilities and quality standards.

7.14. ACCESS TO PATIENTS

The combination of the requirement for more patients per new drug approval and the increased numbers of new drugs in development has created increased competition for clinical trial participants in the world's favorite testing grounds, i.e. the USA and Western Europe. Internationally, recruitment and retention of clinical trial patients are a big challenge. The patient recruitment bottleneck is causing pharmaceutical companies losses. Delays in the completion of critical path studies push back the market launch of the new drug, with loss of revenues.

As clinical data from anywhere in the world are acceptable by global regulatory authorities, pharmaceutical sponsors have been accessing clinical trial participants from an increasing number of emerging countries. India represents one of the more recent additions to this group, with the potential to enroll large numbers of patients at a relatively low cost for global clinical trials. India has large numbers of treatment-naïve heterogeneous Caucasian patients with diseases of both the tropical and industrialized world.

So far, India has been underutilized for global clinical trials, and therefore there is little competition for patients. Furthermore, as increasing numbers of India's patients are using the Internet to access global healthcare databases and information sources, they are proactively seeking to participate in clinical trials so as to access new therapies. This is particularly true for serious diseases such as cancer with unsatisfactory available treatments.

Therefore, patient enrollment rates in India could be three to four times faster than in Western countries and up to seven times faster than in the

USA, for oncology studies. The concentration of potential clinical trial participants in metropolitan India further facilitates patient access. The 350 million individuals residing in India's cities and towns with an urban lifestyle and higher literacy levels form the population base best suited for international clinical trial participation. In addition, there is a well-established practice of patients from rural communities traveling to their nearest city or town to avail themselves of specialist healthcare. These patients often take up temporary lodgings in the city for the course of their treatment.

Many patients in India are still treatment naïve; they never have received drug treatment for their conditions. Furthermore, particularly in oncology, a large proportion of patients are undertreated and have not been through many courses of chemotherapy. This could make them good candidates for certain protocols. The large population exposed to infectious diseases is very suitable for vaccine trials. India's population is mainly Causasian, with an overall genetic makeup similar to that of the white populations in North America and Western Europe. The genetic diversity within the population and the practice of marriage within restricted communities have resulted in diverse but genetically well-preserved population groups. This could be very valuable for the conduct of pharmacogenetic clinical and genetic linkage studies, more so because extended families still live in proximity to one another.

In general, rich patients do not participate in clinical trials and the majority of participants belong to the lower socioeconomic groups. Participation in international clinical trials delivers valuable benefits to Indian patients. Study participation can offer access to high-quality healthcare and medicines that may otherwise not be affordable or available. There is evidence that patients taking part in a clinical trial enjoy better health outcomes than their counterparts, irrespective of whether they have been enrolled in the placebo or treatment arms of the trial. This is due to the increased attention they receive from the healthcare staff.

Access to biomedical innovation and free healthcare is an important motivator for clinical trial participation in all parts of the world. A survey of potential clinical trial participants in the USA revealed that about 50 percent of clinical trial participants claim that their primary motivation to participate in a clinical trial is to access free medication [15]. Thus, it is not surprising that in emerging countries access to free medication is an important motivator for clinical trial participation. Provided consent is free and informed, such a motivation remains wholly ethical and compliant with the principles of ICH-GCP.

7.14.1. Retention

Retention of subjects until study completion is important to ensure the study's statistical validity. In view of the socioeconomic setting in India, close attention to facilitating patient follow-up and study continuation is necessary. Strong patient–physician relationships encourage high levels of patient retention as patients in India tend to be in awe of the investigator and hence desirous of complying with their instructions. However, the challenges of their daily lives can make it difficult for them to adhere to visiting schedules. Site staff, typically the clinical research coordinator (CRC), able to build rapport with participants can effectively motivate them to comply with visiting and follow-up schedules. Patients in India could feel pressurized by the demands of the study protocol and feel unable to express their difficulties to the investigator. This gap between the investigator and patient needs to be bridged and is best done by site support staff. With these measures in place, subject retention rates in India could be among the highest in the world.

Study design plays an equally important role in retention. Studies with fewer blood draws and less frequent site visits often have better subject retention. Designing trial logistics that have been customized for the local setting can also be valuable.

7.15. ACCESS TO INVESTIGATORS

India has well-trained, motivated physicians, who can speak English, are computer literate, and are well suited to be investigators for global clinical trials. There is a large and growing number of GCP-trained physicians keen to participate in international clinical trials. Their motivation for this is more than financial, as they view participation in global clinical trials as a way to become involved in international clinical research, keep abreast with developments in their field, and access state-of-the-art treatments for their patients. Via clinical trial participation, hospitals in India are able to acquire sophisticated equipment and state-of-the-art treatments. Research funds enable the hiring of additional research staff. All of this provides training opportunities related to state-of-the-art clinical practice as well as learning clinical trial methodologies.

Indian physicians tend to be highly motivated and able and willing to adhere meticulously to study protocols. Clearly, not all Indian physicians are well suited to be investigators in global clinical trials; hence, relationships within the physician community enabling informed selection of investigators can be most valuable.

7.16. ACCESS TO CLINICAL SITES

The established practice of patients attending general and specialist hospitals for their healthcare makes India well suited for hospital-based clinical trials. As the primary healthcare system is less developed than the hospital system, patients attend hospitals for treatments that are delivered in the primary care setting in the Western world. Hospitals in India's cities and large towns, be they state managed, single specialty units, or part of corporate healthcare groups, have the required patient attendance, qualified physicians, and equipment to serve as high-performing clinical trial sites. As yet in India there is no process in place for site recognition, accreditation, or licensing.

7.16.1. Government/State Hospitals

These hospitals have large patient numbers and are staffed by academic physicians and their trainee doctors who are particularly keen on clinical trial participation. They are disadvantaged by limited site support resources, poor equipment, and a patient population that includes members of India's lowest socioeconomic groups. These patients tend to be illiterate, poorly nourished, and difficult to follow-up and retain in trials. Two of India's leading state hospitals are the All India Institute of Medical Sciences in New Delhi and the KEM Hospital in Mumbai. The latter is one of India's leading academic medical institutions, founded in 1926, admitting 180 undergraduate students each year, with 1.4 million patients attending outpatients clinics each year, 1400 inpatient beds, and 60,000 major operations each year. The institute has an annual budget of US$ 20 million, 400 academic staff, and 600 resident postgraduate doctors.

7.16.2. Corporate Hospitals

To meet the healthcare needs of India's 300 million-strong increasingly affluent middle class, corporate hospital groups are developing national, regional, and local hospitals. These newly established and well-furnished hospitals are keen to attract international clinical trials and to do so are investing in setting up clinical research secretariats to address the administrative and contractual requirements, as well as hotel-style beds to lodge clinical trial subjects and their families during their follow-up visits. The motivations for this are not only grants and additional resources but also the associated kudos that will attract the more sought-after physicians to be associated with these institutions. Although these hospitals serve fewer patients than the state-subsidized hospitals, their state-of-the-art equipment and more privileged patients make them particularly attractive for the conduct of international clinical trials. Examples of these corporate groups include Apollo and Fortis Hospitals Group.

7.16.3. Specialist Hospitals

These single-specialty hospitals attract referrals from large population bases, and hence could be well suited for complex protocols that seek patient subsets that are ordinarily difficult to access. Narayana Hrudayala, Bangalore, is spread over 10 hectares (25 acres), with 500 beds and the capacity to perform 25 heart surgery operations per day, and is South India's leading cardiology institute. The Tata Memorial Hospital in Mumbai, a specialist cancer hospital, is one example of a specialist oncology site with large patient numbers, many treatment episodes, state-of-the-art equipment, and the administrative support required for global oncology trials. Each year, 25,000 cancer patients visit, not only from India but also neighboring countries. One-thousand patients attend outpatients clinics each day and 10,000 major operations are performed each year. There are 441 inpatient beds, and 5000 radiotherapy and 5000 chemotherapy treatments are delivered each year. The institute has state-of-the-art facilities which include a spiral computed tomographic scanner, gamma cameras, a linear accelerator, bone marrow transplant facilities, a sophisticated blood bank, and laboratories. The clinical research secretariat coordinates clinical research activities that have been reviewed and approved by the scientific review committee and ethics committee.

Conducting clinical trials in the primary care setting in India is relatively challenging in the absence of a well-organized state healthcare service. However, it would be possible to recruit primary care physicians working in the private sector and/or state-managed primary care centers to conduct these trials. Follow-up over long periods of three years or more could be difficult in the absence of limited centralized medical records. In rural areas, healthcare camps, mobile health delivery units with trained staff and appropriate equipment that visit villages at regular intervals, have been effectively used to conduct prophylactic and preventive infectious disease clinical trials.

7.17. WORKING EFFECTIVELY WITH INDIAN CLINICAL TRIAL SITES

Clinical trial sites in India have the potential to be highly efficient, with relatively few poorly performing sites. High site productivity implies that study monitors have the opportunity to review large volumes of clinical data at each site visit, and this enhances project efficiency. However, efforts and investment must be made to ensure that sites have the support needed to perform at their full potential [16]. The investigator at a typical Indian clinical trial site faces a number of challenges, namely to balance resources between patient care and clinical research, to ensure that relatively low levels of patient education, literacy, and financial status do not compromise the principles of GCP, and to guard against allowing the sponsor's requirements for rapid patient enrollment to compromise quality standards [17].

There is recognition for the need for site support with dedicated, trained resources employed either by the site or by a site management organization (SMO). Their responsibilities include contract negotiations, ethical committee submissions, patient counseling, patient recruitment, patient follow-up, archival and maintenance of trial-related documents, SAE reporting, and ensuring protocol compliance. Site support staff and clinical research coordinators are biomedical graduates who typically view this role as the first step in their clinical research career. In contrast to other countries, relatively few Indian nurses take on the role of study nurse or site coordinator. A number of SMO businesses such as Excel Life Science, Neeman Medical, IRL Synexus, and ICube have been established to provide site support services. Communication between the SMO, CRO, sponsor, and site can be convoluted; as a result, CROs are beginning to place their own employees as site support staff.

Clinical trial agreements with the site can be bipartite or tripartite; the latter provides for an agreement between the sponsor or CRO and the investigator and the site. All agreements have a start date, an end date, and a confidentiality clause, and describe the obligations of each of the signatories, particularly those related to the management of SAEs and investigator payments.

7.17.1. Selection of Sites

The large number of potential sites and the substantial variation between their capabilities make careful site selection very important. Local knowledge is valuable to select judiciously clinical trial sites and investigators that will be capable of delivering quality clinical data.

Compared to the USA, there is limited competitor trial activity in India; however, this is rapidly changing at first tier sites that have developed experience and a track record. These sites are becoming increasingly busy with participation in numerous trials. Hence, some of these sites are experiencing capacity constraints and find that they are unable to meet sponsor expectations with respect to subject enrollment and data quality.

In order to ensure that there will be sufficient capacity within India's healthcare system to accommodate the increasing numbers of clinical trials being earmarked for India, it is essential for second tier clinical sites to develop capabilities. Often these sites are located within smaller cities, still large by most standards, with populations of around 2 million. Select second tier cities have well-developed healthcare facilities that also serve populations from surrounding rural and semirural areas. Initiatives to train potential investigators based in these places, allocation of funds to build infrastructure and resources at these sites, and commitment to grow the network of investigative sites will contribute to the development of necessary capabilities at the site level.

7.18. PHARMACOVIGILANCE

In 2005, the CDSCO launched India's National Pharmacovigilance Programme, sponsored by the WHO and funded by the World Bank. This seeks to monitor adverse drug reactions, institute periodic safety update reporting, exchange pharmacovigilance information with other international regulatory bodies, recommend label amendments, product withdrawals or suspension, and inform end users. Pharmaceutical companies are required to submit periodic safety update reports of all new drugs, every six months for the first two years following market launch in India, and then annually for the next two years.

As per Schedule Y of the Drugs and Cosmetic Rules, 1945, any unexpected SAE occurring during a clinical trial should be communicated by the sponsor or their representative to the DCGI and other participating investigators within 14 calendar days. There is no distinct requirement for expedited reporting. It further stipulates that the investigator should report the occurrence of all serious and unexpected adverse events to the sponsor within 24 hours and to the ethics committee that accorded approval of the study protocol within seven working days of their occurrence. This practice of reporting all SAEs to the DCGI is different from most other countries. SUSARs occurring at international sites must be reported to the DCGI. The SAE report should include patient details, information on the suspected drug(s), concomitant treatments, description of suspected adverse drug reaction(s), outcome, and investigator details. The report needs to be dated and signed by the reporting investigator.

7.19. DATA MANAGEMENT SERVICES

In recognition of India's leadership position within the IT and business process outsourcing space, offshoring global clinical data management to India is now considered to be a proven approach for sponsors seeking to reduce clinical development costs. The fundamentals to support efficient, high-quality, and cost-effective global clinical data management, namely, the skilled workforce and IT infrastructure, are available in India. Both sponsors and CROs have established data management capabilities in India and over the years have progressed from data entry services to programming, statistical analysis, and medical writing.

Numerous successful business models have been established to achieve this, including outsourcing in a piecemeal, project-by-project fashion, long-term functional outsourcing, and building capabilities within the local operating companies of multinational pharmaceutical sponsors. The leaders in the field are large IT and business process outsourcing firms such as Cognizant, Accenture, and TCS, who have secured multimillion-dollar, long-term contracts with some of the world's largest pharmaceutical sponsors and have set up units in India staffed with many hundred resources. Various business arrangements have been struck to manage these ambitious contracts that seek more than cost savings, as sponsors have been promised business process innovation and transformation as well.

Electronic data capture has been deployed at Indian clinical sites with positive results. Indeed, not having to overcome legacy issues arising from previous experience with paper studies has been distinctly advantageous.

In the Indian environment, electronic capture of patient-reported outcomes (PROs) is not widely practiced. However, there 550 million mobile phone users in India, with nearly 20 million new mobile accounts opened each month. This is transforming communication and working practices across socioeconomic groups. Even in India's urban slums where people live on less than $2 a day, almost everyone has a mobile phone. Hence, there is much potential to deploy short message service (SMS) texts for two-way communication with clinical trial participants. Challenges that need to be overcome include illiteracy, the large proportion of non-English-speaking users with limited texting systems in Indian languages, and the fact that many participants would be reluctant to give credence to a "faceless" message.

A pilot study on six patients where mobile phones were used for communicating chemotherapy-associated side-effects highlighted the problems related to the impersonal nature of communication, technical hitches including system downtime, and unsuitability if the patient was illiterate, too sick, or unfamiliar with technology [18]. Other ongoing pilot studies include the use of SMS texting in type 2 diabetes trials and to facilitate compliance in patients on first line antiretroviral treatment.

7.20. CLINICAL TRIAL SUPPLY MANAGEMENT IN INDIA

A number of specialist businesses provide clinical trial supply management services in India. These include international logistics companies such as World Courier and TNT, Indian divisions of global solution providers such as Bilcare and Thermo Fisher, and Indian companies such as Reliance Life Sciences. Their services include import to India, customs clearance, central temperature-controlled storage, packaging as per protocol specifications, labeling, randomization solutions, site distribution, cold chain transport, and package tracking. Additional services include current GMP sterile and non-sterile manufacturing and formulation development for clinical trial supplies; this requires permission from the respective state government. In addition, the return, reconciliation, and destruction of clinical supplies are handled as per regulatory requirements.

The investigational product should be labeled in exactly the same manner as proposed for the marketed product, but with the words "For clinical trials only" clearly shown. English is acceptable as the language for labels. Procurement of comparator drugs as well as additional trial supplies from anywhere in world is straightforward, facilitated through wholesalers and manufacturers.

7.21. COST OF CLINICAL TRIAL SERVICES

India offers the opportunity for up to 40 percent savings of clinical trial costs compared to North America and Western Europe. These cost savings accrue from lower labor costs for clinical operations personnel such as project managers, monitors, medical writers, data processors, and programmers. Furthermore, travel costs are lower because of the concentration of sites in metropolitan areas and enrollment of large numbers of patients per site, reducing the need to travel to many sites along with relatively cheap fares for domestic travel. Support services such as transport and storage of clinical trial supplies, printing, translation, and local courier services are cheaper than in the Western world.

The investigator and site fees are approximately one-half of those in the USA and 60 percent of those in Europe. In India, site budgets usually consist of a fixed and a variable component. The fixed component provides for the use of site infrastructure such as fax machines, printer, storage cabinet or refrigerator, and compensation for the site coordinator. An average site infrastructure cost per site would be a single payment of US $1500 and the site coordinator's salary per month would be about US $350. The variable component includes investigator fees, which are protocol dependent and calculated per completed subject. A portion of the site fees is utilized to purchase additional equipment and resources for use both within and beyond the related clinical trial. These low fees in India are a reflection of relatively low compensation structures for physicians, the desire of the physician community and hospitals to be attractive for global clinical trials, and the sector being in a relatively nascent stage. As the Indian clinical development sector matures the investigator fees will increase but they have a long way to go before they approach the levels in the USA and Western Europe. Treatment costs are half those in the USA, with lower costs for medication, investigations, and hospitalization. Thus, although there is an expectation for the trial sponsor to cover the costs of hospitalization, comparator and supplementary treatments, and investigations, the amounts involved are low by Western standards.

A number of forces, in particular the escalating salary expectations of clinical research staff, are increasing the costs of clinical trial conduct in India; nevertheless, the opportunity for cost savings will remain for the foreseeable future. When computing the costs of clinical trial conduct in India it is important to recognize the often hidden costs of training and additional support. This often requires global travel for trainers and trainees, as most global organizations seeking to build capabilities in India realize the need for this investment in training. Business travel costs in India for the international traveler are comparable to those in Western countries and need to be considered when calculating the cost of conducting business in India.

7.22. CLINICAL TRIAL STAFF

India's expanding clinical trials sector has created many career opportunities. Clinical research associates (CRAs) usually have a life science degree, in biology, pharmacy, or biochemistry. Qualified nurses are also well suited for a career in clinical research; however, in India few nurses do so, in contrast to other parts of the world. A qualification, either a diploma or degree, in clinical research is available from one of the many recently set up training institutions and does assist with obtaining the first job in this field. The attributes that employers seek in their CRAs are communication skills, strong integrity, an eye for detail, and the willingness to undertake extensive domestic travel. CRAs with experience in the oncology, neurology, and cardiology therapeutic areas are particularly sought after. A few years ago, recruiting and retaining staff was a challenge reflecting the resource gap created by a fast growing but nascent sector. However, as the sector has begun to mature with the availability of increasing numbers of trained and experienced resources, this staffing challenge has ameliorated. India's strength remains it large number of English-speaking biomedical graduates with a strong work ethic and desire for self-improvement. Therefore, employers willing to invest in mentoring their staff have overcome the problem of staff churn.

7.23. CONDUCTING TRIALS IN INDIA

In order to utilize the clinical trial opportunities in India most effectively, it is essential to confirm protocol feasibility and factor in realistic timelines for regulatory approval and study start. Study feasibility should comprise ascertaining the disease profile in India, standard of care, and healthcare practices. Including India in a global clinical trial remains a considered decision for most sponsors. India should be viewed as one of the high-recruiting countries, able to contribute between 30 and 80 percent of trial participants. The exact proportion depends on the protocol, study objectives, and commercial imperatives. However, the effort of including India for fewer than 30 percent of patients would need careful evaluation. On the other hand, pivotal data submissions to international regulatory authorities with over 80 percent of subjects from India would be inadvisable unless there were compelling reasons to do so.

Questions have been raised about the acceptability and extrapolation of Indian clinical data because of the perceived differences in diet, lifestyle, attitude to pain, and genetic variation.

As the Indian population is Caucasian there are very few differences in the genetic makeup that could influence drug metabolism. Quality of life questionnaires that include attitude to pain may need to be modified to accommodate any variations in sociocultural attitude to pain. It is important to be aware of the factors that could affect the extrapolation of Indian clinical trial data so that potential issues may be overcome by careful study design. For certain protocols it could well be advisable to avoid including India.

Sponsor engagement, ideally in the form of site visits either at the stage of investigator selection or early in the study cycle, positively influences site performance and recruitment. Demonstration of a sponsor's interest in the environment in which subjects are recruited can substantially influence study performance. If geographical challenges make site visits difficult, investigator interaction via electronic, telephone, and written communication is essential.

Although it is satisfying that some sponsors no longer perceive data quality to be a concern in India [19], a broader sampling would highlight the variability in data quality depending on site and CRO selection. Implementation of global SOPs designed for established countries to emerging clinical trial environments remains difficult. CROs with SOPs adapted to local environments, an understanding of industry leading practices, along with local knowledge and relationships are best able consistently to deliver quality data.

The advantage of working with local CROs is that they offer cheaper prices, thereby enabling the sponsor to maximize the cost savings that India offers. This is because local CROs do not have the expensive overheads of global structures, processes, and systems. Global CROs offer the advantage of a one-stop shop as they are able to execute trials in a number of countries, including India, with uniform data quality. This is of particular value if sponsors lack the internal resources to manage multiple CRO providers. As sponsors recognize the value of utilizing specialist local CROs, they are opting for a combination of global and local clinical trial service providers and investing in the in-house resources necessary to manage multiple vendors.

7.24. IMPORTANCE OF INDIA'S CLINICAL TRIALS SECTOR

The inclusion of India in increasing numbers of international clinical trials brings important benefits for Indian patients. Trial participants have the opportunity to access cutting-edge biomedical innovation which could be life saving. Indian hospitals receive cash, equipment, and additional staff for participating in clinical trials, which benefit all patients served by that hospital. Exposure of the Indian healthcare system to the discipline of international clinical research will enhance the practice of evidence-based medicine, thorough record-keeping, and better patient communication.

Thus, the inclusion of India in global clinical trials will not result in the exploitation of Indian patients but, on the contrary, enable pharmaceutical sponsors to fulfill some of their corporate social responsibilities. The Indian government realizes the value of encouraging India's clinical research industry and has implemented a number of initiatives to encourage it, including abolishing import duties on clinical trial supplies.

The industry acknowledges that up to 40 percent cost and up to 70 percent time savings can be achieved by conducting Phase II–III clinical trials within emerging countries such as India. This, coupled with the need to make the most of every research and development dollar, will further heighten the interest of sponsors in placing their clinical trials in India. The drive to commercialize new products and capture market share in the fast growing pharmaceutical markets of countries such as China, India, and Brazil is becoming stronger. Given that the conduct of Phase II–III clinical trials in a country facilitates the introduction, adoption, and commercialization of a new product in that market, multinational pharmaceutical companies seeking to build market share in emerging countries must consider including them in clinical trials.

Today's global environment presents tremendous opportunities for India's clinical trial sector. Organizations that are able to deliver for international sponsors the promise of India with quality assurance will make a valuable contribution towards developing more affordable and better quality medicines.

REFERENCES

[1] Kearney AT. Make Your Move: Taking Clinical Trials to the Best Location. A.T. Kearney, http://www.atkearney.com/index.php/Publications/make-your-move.html; 2009.
[2] Saxena R. The Middle Class in India. Deutsche Bank Research, http://www.dbresearch.de/PROD/DBR_INTERNET_DE-PROD/PROD0000000000253735.pdf; 2010.
[3] Census of India 2001.
[4] PricewaterhouseCoopers. Emerging Market Report: Health in India, http://www.pwc.com/en_GX/gx/healthcare/pdf/emerging-market-report-hc-in-india.pdf; 2007.
[5] Verghese C. Cancer Prevention and Control in India. National Cancer Registry Programme 2001:48–59.
[6] Mohandas KM, Jagannath P. Epidemiology of digestive tract cancers in India. Projected burden in the new millennium and the need for primary prevention. Indian Journal of Gastoenterology 2000;19:74–8.
[7] Yeole BB, Sankaranarayanan RL, Sunny R. Survival from head and neck cancer in Mumbai (Bombay), India. Cancer 2000;89:437–44.
[8] Krishnaswami S. Prevalence of coronary heart disease in India. Indian Heart Journal 2002;54:103.

[9] Handbook of Medical Education. New Delhi: Association of Indian Universities; 2004.
[10] Survey IMS. Global Pharmaceutical Markets – Current and Future Trends, 2003.
[11] Varawalla N. Conducting clinical trials in Asia. Applied Clinical Trials 2006:108–13.
[12] FICCI and E&Y Knowledge Paper 2009.
[13] Tufts CSDD. Analysis of FDA's Bioresearch Monitoring Information System File, 2006.
[14] Varawalla N. India's growing clinical research sector: opportunity for global companies. IDrugs 2007;10:391–4.
[15] Center Watch Survey. What motivates participation in clinical research? 2004.
[16] Varawalla N. Investigative sites unlock the door to success in India. Applied Clinical Trials 2007:48–54.
[17] Bhatt A. Clinical trials in India: pangs of globalization. Editorial. Indian Journal of Pharmacology 2004;36:207–8.
[18] Marimuthu P. Projection of cancer incidence in five cities and cancer mortality in India. Indian Journal of Cancer 2008;45:4–7.
[19] Henderson L. Asia Pac grows with China focus. Applied Clinical Trials 2010:52–4.

Chinese Regulatory Framework

Jenny Zhang
Tigermed Consulting, Shanghai, PR China

Contents

8.1. OVERVIEW OF DRUG ADMINISTRATION IN CHINA

8.1.1. Regulatory Affairs Profile

China has established a quite streamlined drug regulatory system from a nearly zero base during the past 30 years, along with its national policy of reform and opening up [1]. The State Food and Drug Administration (SFDA) and its affiliates play a key role in this system; they make decisions on approvals, additional requirements, or exemptions. However, there are still many defects and loopholes in both process and regulations, such as provisions and guidelines being vague, or a lack of explanations, and these flaws are part of the reason that regulation time takes longer than other

Global Clinical Trials
ISBN 978-0-12-381537-8, Doi:10.1016/B978-0-12-381537-8.10008-1

countries. Knowing its deficiencies, the Chinese regulatory system is continuing to evolve, and trying to keep up with the standards of the European Union (EU), the USA, and Japan.

With protection of intellectual property, quality standards embodied by good practices, efforts on regulatory clarity, and more and more emphasis on innovation, China is striving hard to make progress on drug regulation to ensure safe and effective usage in medication [2]. Following the national strategy of innovation, it is becoming inclined to encourage "real" new drugs and is being more and more stringent on generic drugs, and a more science-driven regulatory system with enough clarity is expected in China.

8.1.2. Regulatory Agencies

In 2003, the SFDA was established to replace the State Drug Administration (SDA), with expanded functions. The SFDA was modeled after the US Food and Drug Administration (FDA), and operates under supervision of the Ministry of Health.

With its provincial, municipal, and county branches, the SFDA is the governing body that regulates all drugs, medical devices, food, health food, and cosmetics, and controls all registrations, inspections, sales, research, and advertisings for these products. The SFDA is also in charge of all new drug registration approvals.

Among its affiliated units, the National Institute for the Control of Pharmaceutical and Biological Products (NICPBP), Center for Drug Evaluation (CDE), Certification Committee for Drugs (CCD), Center for Drug Re-evaluation (CDR), and Chinese Pharmacopoeia Commission (CPC) play very important roles in the registration and manufacturing regulation process.

8.1.3. Laws and Regulations

8.1.3.1. Laws

The basic regulations for the government's administration of the pharmaceutical industry in China are outlined in two fundamental laws. These laws regulate all pharmaceutical areas, including drug manufacturers, drug distributors, pharmaceutical use in medical institutions, new drug registrations, drug packaging, pricing, advertising, and postmarketing surveillance.

First, the Drug Administration Law of the People's Republic of China [3] was revised at the 20th Session of the Standing Committee of the 9th National People's Congress on February 28, 2001, and became effective on December 1, 2001. This law controls all pharmaceutical areas from research and development (R&D), through manufacturing, to marketing and circulation of drugs. Since its first edition in 1985, it has formed the basis of the whole Chinese drug regulation system.

Second, the Regulation for the Implementation of the Drug Administration Law of the People's Republic of China [4] was approved by the State Council and became effective on September 15, 2002. It is a corresponding law for the maneuverable operation of the Drug Administration Law.

8.1.3.2. Regulations and Guidance Related to Product Registration

Drug Registration Regulation

The Drug Registration Regulation in China has undergone four revisions since its first version in 1985. It has gone through versions 1999, 2002, 2005, and the latest version (Order 28), which was enacted on October 1, 2007. The regulation has been changed significantly over its 20-year development. The newly revised regulation emphasizes safety, efficacy, and quality control of drugs, and ensures that the registration processes and procedures are judicial, fair, and open. The current regulatory system is built on a joint accountability system, staff briefing and challenge system, and responsibility assigning system. The most important improvement is that all regulatory processing activities should be under public surveillance. Compared to the previous versions, the newly revised regulation encourages innovation and sets a lot of limitations for generic drug applications.

Special Evaluation and Approval Procedure

The introduction and implementation of the Special Evaluation and Approval Procedure [5], issued on January 1, 2009, outlines detailed regulations on the special review and approval process for innovative new drugs. Other than raising good manufacturing practice (GMP) standards in more facilities and more stringent intellectual property protection, the Special Evaluation and Approval Procedure shows the government's commitment to encouraging the research and development of new drugs and to treating difficult and life-threatening diseases by exercising approval via a special process designated for innovative new drugs, as outlined in SFDA order 28. For drugs meeting the following criteria, the applicant can win more support from the government. The threshold for applying this special process is to meet at least one of the following criteria:

- new drug material and its preparation, active ingredients and its preparation extracted from plant, animal and minerals, which have not been marketed in China

- chemical drug substance and its preparations, and/or biological product that have not been marketed domestically and outside China
- new drugs for the treatment of acquired immunodeficiency syndrome (AIDS), cancer, and orphan diseases which have significant efficacy
- new drugs for the treatment of diseases that have no effective therapy.

The Special Evaluation and Approval Procedure means that applications can be evaluated in 10 days fewer than normal (90 working days) and facilitates communication with CDE evaluation staff.

Rules for Registration of Drug Technology Transfer

New drugs are granted 20 years' protection under Chinese patent law, five-year market exclusivity and six years of protection for the data from local clinical trials.

In 2009, a new regulation on Rules for Registration of Drug Technology Transfer [6] was published for public opinion. The objective of this regulation is to facilitate the licensing of new drug and technology and the transfer of manufacturing processes. The government is expecting to optimize and upgrade the Chinese pharmaceutical industry and product structure. According to this regulation, it is permitted to transfer the production process of an imported drug to a domestic manufacturer in China. Although the details of execution of this process have not yet been made available, this regulation opens a new pathway for the development of imported drugs.

Technical Guidance Related to Drug R&D

To improve and standardize drug R&D activities in China, the SFDA has promulgated 80 technical guidelines, covering guidance on chemical drugs, traditional Chinese medicine, biological products, integrated disciplines, and general principles. Although it is a long way from establishing a comprehensive guidance system like the FDA, which has more than 500 technical guidelines, the SFDA is continuing to work on it. The introduction of International Conference on Harmonisation (ICH) guidelines and promoting international mutual recognition are also on the SFDA's to-do list.

8.2. DRUG REGISTRATION

8.2.1. Drug Category and Classification

In Drug Registration Regulation SFDA Order 28 [7], a drug should fall into one of three categories: small molecule, biological product, or traditional Chinese medicine. Under these three categories, there are various classes. There are six classes of new chemical drugs, 15 classes of biological drugs, and nine classes of traditional Chinese medicine. The provisions define these classes mostly by demonstrating and distinguishing how a drug and its preparation process have been marketed, especially with regard to China. For example, class 1 chemical drugs refer to new drugs that have never been marketed in any country, while class 3 chemical drugs refer to new drugs that have only been marketed outside China.

8.2.2. Drug Registration Application Category

Besides the drug category and classification, according to Article 11 of SFDA Order 28, drug registration applications are also divided into three different types, namely domestic new drug application, domestic generic drug application, and imported drug application. These three types of application are described below.

8.2.2.1. Domestic New Drug Applications

A new drug application means a registration application for a drug that has not been marketed in China, and the drug should be produced on the Chinese mainland. Applications for a change in dosage form, route of administration, or an additional new indication of marketed drugs shall be regulated as a new drug application procedure. Classes that fall into this category include class 1–5 for new chemical drugs, class 1–14 for new biological drugs, and class 1–8 for new traditional Chinese medicine.

8.2.2.2. Domestic Generic Drug Applications

A generic drug application means a registration application of drugs for which SFDA has already established the official standards. A generic drug application can be used for drugs "that already have a national standard in China", which are class 6 of chemical drugs, class 15 of biological drugs, and class 9 of traditional Chinese medicine.

8.2.2.3. Imported Drug Registration

Chinese regulations take the manufacturer as the main body of an application; therefore, if the manufacturer is located outside the Chinese mainland, the drug will be under the administration of imported drug registration. According to SFDA Order 28, the precondition for approval of imported drug registration is that the drug should have obtained marketing authorization approval in another country or region.

- For an imported drug application, if a Certificate of Pharmaceutical Product (CPP) is not available at the beginning of the clinical trial application (CTA) submission, it will be under the administration of new drug application category 1 for requirement on dossier, clinical trial, and timeline.
- For an imported drug under the original brand, if a CPP is available at the beginning of the CTA submission, it will be under the administration of new drug application category 3 for requirement on dossier, clinical trial, and timeline.
- For an imported generic drug, even if the CPP is available at the beginning of the CTA submission, the SFDA will only regulate it as a generic drug application on requirement of dossier, clinical trial, and timeline.

8.2.3. Drug Registration Requirement

8.2.3.1. Application Documents

The application documents of drug registration application consist of four sections:

- summary materials
- pharmaceutical research materials
- pharmacological and toxicological research materials
- clinical research materials.

For each section, the SFDA has issued a detailed documents list for chemical drugs, biological drugs, and traditional Chinese medicine. Different drug classes have different requirements for the above documents. Registration document requirements in China all follow ICH guidelines, meaning that requirements are fairly similar for China, the USA, and the EU.

Overall, during the registration process, pharmaceutical companies should not encounter problems with different data requirements in China compared with other foreign countries. However, problems often arise when an applicant is reluctant to submit sensitive, proprietary information, or to divulge information about the manufacturing process or quality control of raw materials.

8.2.3.2. Specification Verification

Drug registration testing is mandatory for all kinds of drug registration, with differences in when and how many sample batches will be requested. For example, for imported drug registration, the applicant should submit a sample of three batches automatically to NICPBP for testing at the CTA stage. For a domestic new drug application, for a chemical entity, three batches will be sampled by the provinicial FDA only at the beginning of the production application stage. For biological products, three batches will be sampled by the provinicial FDA both at the beginning of the CTA and at the production application stage.

For drug registration testing, the NICPBP will first develop a Chinese specification based on the proposed in-house specification and Chinese pharmacopeia, and then carry out quality control testing according to the Chinese specification draft. When the product has been granted marketing authorization by the SFDA, the Chinese specification draft will become the official drug registration specification and will be used as the specification for quality control of commercial goods.

In addition, a local quality control testing report is requested by the CDE so that it can draw its final conclusion on the evaluation. Failure to submit the quality control testing report to the CDE will lead to delay in the evaluation timeline.

8.2.3.3. Requirement for Local Registration Trial

Drug registration in China has become more stringently regulated over the past two decades. Local drug clinical trials are mandatory to achieve product registration.

SFDA Order 28 has detailed the clinical requirements for different drug categories, drug classifications, and application types. In addition, clinical trials can only be conducted in China after obtaining CTA approval from the SFDA, and are only permitted to be carried out at hospitals that have been accredited by the SFDA. Effective from March 1, 2005, no registered clinical trials in China can be conducted in non-accredited clinical research centers.

For imported drugs, only data from local clinical trials can be counted in the local registration. Even if a product is approved elsewhere in the world, the SFDA is still likely to require the foreign manufacturer to conduct at least one study in China before approval. Results from overseas clinical trials can only be regarded as referential clinical data.

Requirements for clinical trials in different situations of imported drugs are shown in Table 8.1.

8.2.4. Drug Registration Process

In China, the drug registration process is different from that in the EU and North America, and it also differs from that in other Asia Pacific countries. The process includes two applications, before and after the clinical trial.

Table 8.1 Clinical trial requirements for imported drugs

Regulatory status	**Clinical requirement**
New drug never marketed in any country at CTA stage	Phase I, Phase I and Phase III trials The sample size should meet the statistical power
Drug marketed outside China but not yet marketed domestically	Pharmacokinetics study and abbreviated Phase III trial with 200 cases Or abbreviated Phase III trial with 100 cases for treatment arm only
Drug product with changed dose form, but no change of administration route, and the original preparation already approved in China	Bioequivalence study only
Drug substance or product following a national standard	Clinical trial waived Or bioequivalence study Or clinical trial with 100 cases for treatment arm

The first step in this process is to submit an application to the SFDA for a CTA. The SFDA will conduct a preliminary review of the submission package and then transfer the dossier to the CDE. Reviewers with a background in pharmaceuticals, pharmacology, and clinical studies will run a technical review, while local sample testing will be conducted in parallel. Few CTAs pass through the CDE review in one single round. However, most applications will receive a written supplement notice(s) requesting additional information for further assessment. In such cases, the CDE will allow a four-month period for the applicant to gather and submit additional requested information to the CDE. This entire CTA step usually takes at least 125 working days.

The second step in drug registration is production application (or imported drug license application), which involves submitting a clinical report and other relevant dossiers to obtain an imported drug license. The process is basically the same as the CTA step. This second step will take approximately 145 days.

In general, CTA is analogous to an investigational new drug (IND) application in the USA, and production application (or imported drug license application) is comparable to a new drug application (NDA) in the USA.

8.2.4.1. Different CTA Approaches

As mentioned above, in China, it is not possible to obtain marketing authorization without a local registration clinical trial. In addition, no clinical trial can be initiated without the SFDA's approval on CTA. A brief summary of all kinds of CTA falling into different drug registration application categories is given below.

CTA Under The Name of Imported Drug Registration

If the sponsor starts the CTA under the name of imported drug registration with CPP, the requirement of a clinical trial is plain and straightforward. Usually, a local bridging study (data on 100 pairs of subjects and/or a pharmacokinetics study) is sufficient to move forward to the marketing authorization stage. This is the most common scenario for products already marketed in other countries.

If the sponsor starts the CTA under the name of imported drug registration without CPP, an entire clinical trial program from Phase I to Phase III is required to move forward to the marketing authorization stage. This is an aggressive scenario for imported drugs not marketed anywhere in the world.

CTA Under the Name of Domestic New Drug Application

This CTA pathway follows the requirements of a domestic new drugs application, and on-site inspection from the provincial FDA is a must. It is the most common scenario for a product developed by a domestic company, and local benefits and sometimes support from local government are possible.

Since the controlling system of drugs in China is based on the manufacturer's location, if manufacture of the final product is transferred to a subsidiary or joint venture, or to a third party in China, CTA from this entity is regarded as the first stage of domestic NDA, regardless of whether the subsidiary or joint venture is 100 percent owned by a foreign company. Though it is a full local clinical program independent of global development, it provides an alternative R&D development path.

International Multicenter CTA

In Article 44 of Order 28, if the drug has already been registered abroad or entered into Phase II or Phase III clinical trials in another country, and the foreign applicant would like to conduct an international multicenter clinical trial in China, the Chinese Drug Registration Regulation provides a channel known as international multicenter clinical trial application (Int'l CTA). The only difference between Int'l CTA and imported drug registration is that from the regulatory perspective, the purpose of an Int'l CTA is to conduct clinical trials in China only. Upon completion of the study, the applicant shall submit the final study report to the SFDA, and then this regulatory procedure is over automatically. However, Int'l CTAs are forbidden for first-in-human or vaccine studies.

8.2.5. Drug Registration Strategy for Imported Drugs

Though differ types of CTA provide different ways to conduct clinical trials in China, for any sponsor, the ultimate goal is to receive marketing authorization. Because of the long regulatory process and duration, it leads to a long registration lag compared with other countries. Meanwhile, China's growing pharmaceutical market has provided many new opportunities for drug companies to conduct clinical trials in China. The greatest incentive for these opportunities is the enormous potential for cost savings in clinical study as well as a large population pool, which translates into faster enrollment.

The Chinese SFDA does not currently engage in a simultaneous IND submission process, so for applicants who are seeking a shortcut within the existing regulations to reduce the registration lag, the only way is to add China to their global development and start CTA under a different name as soon as possible after the overseas clinical program enters Phase II. Optimally, it is possible to obtain an imported drug license within one year of obtaining marketing authorization, one year after approval in the original country.

REFERENCES

[1] Hu YL. Drug safety regulation in China: institutional transition and current challenges (1949–2005). Chinese Journal of Health Policy 2009;2:45–61.
[2] Guise J, Carson B. Biogeneric regulatory policies in China and India: a comparison study. Drug Information Journal 2010;44:55–67.
[3] Chairman of P.R. China. Drug Administration Law of the People's Republic of China. Order 45.
[4] Implementation Regulation of the Drug Administration Law of the People's Republic of China.
[5] Notification on Special Evaluation and Approval Procedure, SFDA No. 17, 2009.
[6] Notification on Rules for Registration of Drug Technology Transfer, SFDA No. 518, 2009.
[7] Drug Registration Administration Provision (SFDA Order 28).

Clinical Trials in China

James (Dachao) Fan* and Suzanne Gagnon**
*Medical and Safety Services, ICON Clinical Research, Singapore
**Medical, Safety and Scientific Services, ICON Clinical Research, USA

Contents

Global Clinical Trials
ISBN 978-0-12-381537-8, Doi:10.1016/B978-0-12-381537-8.10009-3

9.1. INTRODUCTION

9.1.1. Demography

China is the most populous country in the world. In 2009, the population of mainland China was 1334.74 million, not including the populations of Taiwan and the Special Administrative Regions of Hong Kong and Macao [1]. Since the implementation of the one-child policy in 1979, China has experienced an unprecedented demographic transition, with decreasing birth rates and an increasingly elderly population (Table 9.1). In addition, the economic reform and development since 1978 have resulted in accelerated urbanization. The proportion of the country's urban population increased from 17.9 percent in 1978 to 46.6 percent in 2009, although there remains a large rural population of 713 million [1]. Fifty-six ethnic groups are recognized in China. The Han Chinese are the largest ethnic group, comprising 91.6 percent of the population; the remaining 55 ethnic groups are referred to as ethnic minorities [1]. Chinese is spoken by the Han ethnic group, but most of the 55 minorities have their own languages. Standard Chinese or Mandarin (Putonghua) is the official national language.

The sixth national population census was conducted in 2010, beginning November 1, and is expected to provide valuable information for the government's future development plans.

9.1.2. Economy

Since the economic reform in 1978, China has undergone rapid and steady economic development, with an annual growth rate averaging 8–9 percent. The country's economy depends largely on investment and

Table 9.1 Demography of China

Population size (2009) [1]	1334.74 million
Population pyramid (2008)[1]	Younger population (under 15 years) 17.3% Productive age (15−64 years) 73.1 % Aged population (65 years and over) 9.5%
Life expectancy (2008)[2]	Female 76 years, male 72 years, average 74 years
GDP (2008)[3]	US $4,326,200 million
Per capita GDP (2008)[3]	US $2940
Languages	Standard Chinese or Mandarin (official), Chinese dialects and minority languages

[1]Chinese Health Statistical Digest 2010: http://www.moh.gov.cn/publicfiles//business/htmlfiles/zwgkzt/ptjty/digest2010/index.html
[2]World Health Statistics 2010: http://www.who.int/whosis/whostat/EN_WHS10_Part2.pdf
[3]International Statistics 2009: http://www.stats.gov.cn/tjsj/qtsj/gjsj/2009/

export. In response to the global financial crisis in 2008, the Chinese government announced a four-trillion yuan (US $586 billion) stimulus package to boost domestic demand and to offset falling exports. The stimulus package will be spent on upgrading existing and construction of new infrastructure, raising rural incomes, social welfare, and disaster reconstruction. Owing to the stimulus package and other measures such as tax cuts and consumer subsidies, the growth rate for 2009 was 8.7 percent, exceeding the official target of 8 percent for the year. Total gross domestic product (GDP) was 33.54 trillion yuan (US $4.91 trillion). However, the country continues to face challenges in restructuring the economy in order to achieve balanced development and sustainable economic growth [4].

9.1.3. Infrastructure of Healthcare in China

The Ministry of Health under the State Council oversees the health service system in China. The country has significantly increased its investment in health in recent years. The total expenditure on health was 4.96 percent of the GDP in 2009; a per capita expenditure of 1192.2 yuan and a 0.13 percent increase over 2008 [5]. Of the total health expenditure in 2008, 59.6 percent came from public funds and 40.4 percent was private expenditure investments [5]. China had 1.62 physicians and 2.96 hospital beds per 1000 population in 2009 [5], for a total of 20,291 hospitals in the country, of which 13,364 were general hospitals, 2728 were traditional Chinese medicine (TCM) hospitals, 245 were TCM–Western medicine hospitals, and 3716 were specialized hospitals [1]. According to the standards set by the Ministry of Health, hospitals in China are classified into levels 1, 2, and 3 by their function and scale, and under each level there are grades A, B, and C, indicating technical competency, service quality, and management efficiency of the hospital. Level 3A is the top class and represents the best of general or comprehensive hospitals at national, provincial, or city level. In 2009, there were 1233 level 3 hospitals (765 level 3A), 6523 level 2 hospitals, 5110 level 1 hospitals, and 7425 unclassified hospitals [1].

The Chinese healthcare system has been criticized for soaring medical fees, having unaffordable medical services and low medical insurance coverage. These problems have called for a reform in public healthcare. The large rural population has been a challenge for China's medical insurance system. In 2003, the New Cooperative Medical System (NCMS) was introduced and piloted in some rural areas. By 2009, the NCMS covered 94 percent of the rural population [5].

The medical insurance system in China has been gradually reshaped and is now comprised of a Basic Medical Insurance of Employees, Basic Medical Insurance of Residents, NCMS, and Medical Relief, covering the working population in urban areas, unemployed residents in urban areas, rural populations, and poor populations in both urban and rural areas, respectively. On April 7, 2009, the government announced an investment of 850 billion yuan (US $123 billion) over the three-year period between 2009 and 2011 for a new round of healthcare reform [6]. Five key measures of this reform are:

- to speed up the construction of the medical insurance system for both rural and urban populations
- to establish a national essential medicine system, in which essential drugs are manufactured and distributed under government control and covered by medical insurance
- to improve the grassroots medical service system for less developed areas
- to promote gradually equality in basic public health services in both rural and urban areas
- to pilot reform throughout public hospitals to improve their service quality [6].

It is hoped that through these measures medical service with improved quality can be made accessible and affordable for ordinary people.

9.1.4. Disease Profile in China

In 2009, the main causes of death in both urban and rural areas of China were malignant neoplasms, cerebrovascular disease, heart disease, and diseases of the respiratory system (Figures 9.1 and 9.2). The death rate for malignant neoplasms was 167.57 per 100,000 population in the cities and 159.15 per 100,000 population in counties [1].

9.2. REGULATORY CLIMATE

The State Food and Drug Administration (SFDA) of China was founded in 2003, expanding out of the former State Drug Administration (SDA), which was established in 1998. The SFDA is in charge of the registration and supervision of drugs and medical devices in China, following the Drug Administration Law of the People's Republic of China (2001), and the Regulations for Implementation of the Drug Administration Law of the People's Republic of China (2002). To regulate the drug registration

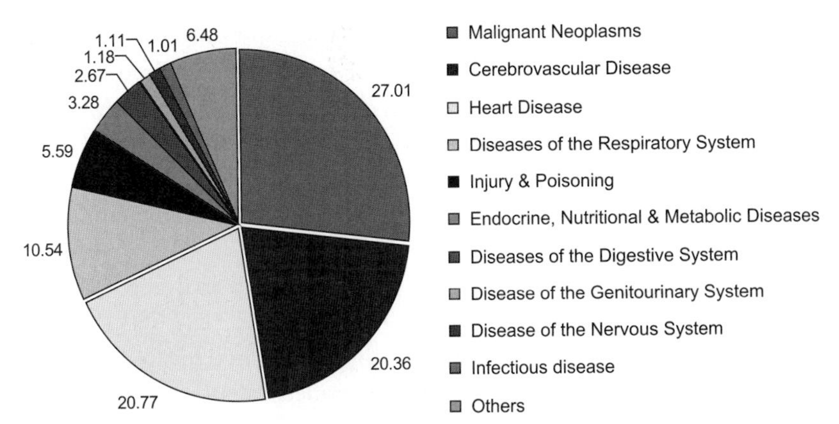

Figure 9.1 Percentages of the 10 main diseases causing death in China's cities in 2009 [1]. (Please refer to color plate section)

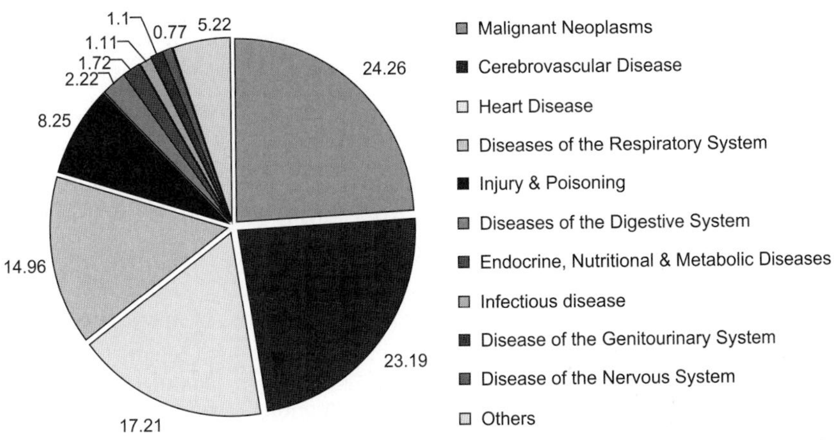

Figure 9.2 Percentages of the 10 main diseases causing death in China's counties in 2009 [1].

process, the SFDA formulated the Provisions for Drug Registration in 2002 (revised in 2005 and 2007). The latest version of Provisions for Drug Registration was issued on October 1, 2007, and the evaluation timelines have been shortened considerably.

9.2.1. Center for Drug Evaluation

The Center for Drug Evaluation (CDE) is responsible for the evaluation of drugs, traditional Chinese medicines and biologic products. Clinical trial approval time is of primary concern to sponsors. Between January 2004 and

April 2008 CDE received 248 Clinical Trial Applications (CTAs) for international, multicenter trials [7]. Agency review time ranged from 4 to 12 months (median, 7 months) (Table 9.2). Total time from CTA submission to approval, including supplementary document review, ranged from 4 to 20 months (mean, 8 months; median, 7 months). The lengthy review time has clearly been seen as a negative for sponsors considering conducting multicenter trials in China.

In an effort to speed up the approval process, CDE analyzed 168 CTAs submitted during that time period. Of 168 CTAs, 138 (82.1%) were approved, 23 (13.7%) were withdrawn by the sponsor and seven (4.2)% were not approved. Reasons for disapproval included safety concerns, CMT testing not meeting the regulatory requirements, and the CTA not providing sufficient information for review. Reasons for lengthy time to approval included insufficient safety data provided in the original application, inappropriate protocol design, updated information provided by the sponsor after original submission requiring re-review of the dossier by the Agency, and lack of a face-to-face meeting of the sponsor with CDE, resulting in inadequate or miscommunication and the need for resubmission of documents. In order to improve the process, CDE has introduced periodic joint consultation meetings attended by experts, sponsors and CDE reviewers. Under the 2007 version of Provisions for Drug Registration, the evaluation timeline has been shortened from 120 days to 90 days.

Table 9.2 Center for Drug Evaluation evaluation times of multicenter trials between 2004 and 2008

Agency review time (months)	No. of applications	Total time to approval (months)*	No. of applications
4	19	4	2
5	21	5	20
6	32	6	15
7	45	7	49
8	17	8	36
9	11	9	10
10	9	10	6
11	9	11	10
12	5	12	14
		13	4
		16	1
		20	1

*Total time to approval: Time from CTA submission to approval, including review of supplemental documents submitted.

9.2.2. Good Clinical Practice and Training

Good clinical practice (GCP) guidelines were introduced into China in 1998 by the Ministry of Health and revised in 1999 by the SDA, the predecessor of the SFDA. The current version came into effect in 2003 [8]. The SFDA has promoted GCP by releasing strict usage rules in 2003 and 2004 and introducing compulsory GCP training in April 2004. In most technical aspects, the Chinese GCP regulations are similar to those in other countries. However, there are some differences. In China, clinical trials must be approved by the SFDA and can only be conducted in SFDA-certified clinical trial institutions. Clinical trial institutions cannot allow the same department to conduct concurrent trials using the same type of drug under different sponsors, and there is a limit to the number of different drug types that may be investigated at one time.

9.2.3. Human Specimen Export Regulations

The export of human specimens is subjected to strict regulation in China. For human specimens which contain genetic materials, an export permit must be applied for through the Ministry of Health. International collaborative projects involving human genetic resources are reviewed by the Human Genetic Resources Administration of China (HGRAC), an office under the Ministry of Science and Technology. In international multicenter trials, the Chinese collaborator must apply for permission from the HGRAC for the export of whole blood and tissue samples and the success rate is extremely low in real practice. The export permit of human serum and plasma samples had been administered by Ministry of Health, but the process was transferred to the HGRAC with effect from August 20, 2010. Whereas a Ministry of Health export permit is usually issued within 10–30 working days, the processing time with the HGRAC may be as long as four to five months. First time applications to the HGRAC must include an application for the project approval, which is subjected to a specialist committee appraisal held on a quarterly basis. The HGRAC will approve or disapprove the project within 20–30 working days. After the project approval, the first export permit can be applied for at any time and the HGRAC will issue the permit after one working day. Export permits of subsequent shipments of the same project can also be obtained within one working day after the submission of application documents to the HGRAC.

9.3. CURRENT STATUS AND STATISTICS ON CLINICAL TRIALS

9.3.1. Total Investigational New Drugs in China

The number of investigational new drugs (INDs) in China between 2006 and 2008 is shown in Table 9.3.

Table 9.3 Number of investigational new drugs (INDs) in China by year [9]

Year	Number of INDs
2006	1426
2007	758
2008	581

9.3.2. Multicenter Trials in China

The number of multicenter trials conducted in China between 2002 and 2008 (Figure 9.3) increased substantially from 2005 to 2007. The annual number of applications increased from fewer than 20 before 2005 to about 70 thereafter and appeared to stabilize. However, in 2008, the number of multicenter trials decreased significantly because of the slowing global pharmaceutical research and development (R&D) market and the SFDA's focus on historical problems in the pharmaceutical industry, such as centralized evaluation.

Among multicenter trial applications submitted to CDE in China, a total of 248 applications from 41 registered applicants was submitted between 2004 and the first half of 2008; 31 of the applicants were multinational companies

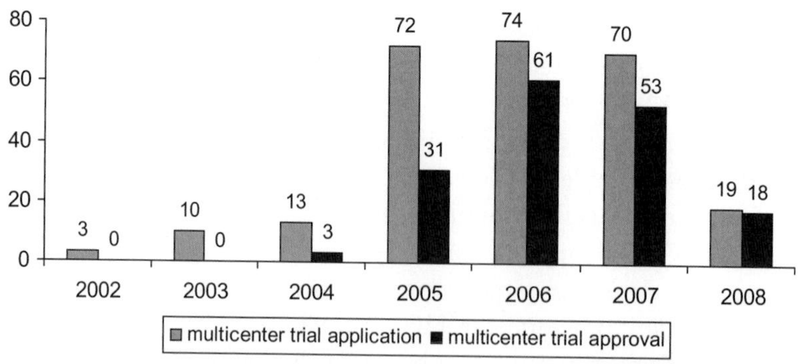

Figure 9.3 Number of multicenter trials in China from 2002 to the first half of 2008 (*Source:* adapted from China Prescription Drug, 2009 [10]). (Please refer to color plate section)

(83.9% from Europe and America, 16.1% from Asia) and 10 were contract research organizations (CROs) [7]. The majority (87.9%) of multicenter trial applications were chemical compounds; only a small number were biologicals. Ranking chemical compounds by therapeutic indications from most to fewest applications received, the order is as follows: oncology, psychiatry and neurology, hypertension, hyperlipidemia, diabetes, viral infection, hepatitis, obesity, respiratory system diseases, surgical, gynecology, and radiology. Biological applications involved mainly tumor-related indications (approximately 60%), endocrine disease, cardiovascular disease, and kidney disease.

9.4. OVERVIEW OF INVESTIGATORS AND SITES IN CHINA

9.4.1. Study Investigator Profile

Investigators employed by clinical trials institutions in China are usually experienced physicians who are highly respected in their fields. Only physicians in SFDA-certified sites are qualified to conduct sponsored clinical trials in their therapeutic areas. This is required in order to ensure uniform quality and GCP compliance. Although medical terminology and education are in Chinese, many therapeutic area heads have had some clinical training experience in hospitals in Europe or the USA and therefore understand English. As in other countries, language barriers may exist among lower level staff if the documents are only in English, so translation is required. Because of the large number of domestic trials and increasing number of global trials currently underway, investigators are familiar with clinical trial processes and Good Clinical Practice (GCP).

9.4.2. Study Site Characteristics

As required by China's drug administration laws and regulations, only SFDA-certified drug clinical trial institutions are qualified to conduct clinical trials in specifically approved therapeutic areas. When the first Drug Clinical Trial Institution Qualification Announcement was issued on February 3, 2005, a total of 17 medical institutions was verified [11]. Up to January 2010, the SFDA had issued 22 qualification announcements and 275 clinical trial institutions were certified [11], all of which are level 3 hospitals and the majority are 3A hospitals. The clinical trial institutions are mostly located in Beijing, Shanghai, Guangzhou, and other provincial capital cities. In addition, three hospitals in Hong Kong (Prince of Wales Hospital, Queen Mary Hospital, and Hong Kong Eye Hospital) have been certified to participate in SFDA-approved clinical trials. The number and type of certified clinical trial institutions satisfy most trial-related needs in China. However, too few institutions specialize in therapeutic areas such as acquired immunodeficiency syndrome (AIDS), vaccines, pediatrics, and orphan diseases.

In order to ensure uniform quality and GCP compliance, the SFDA issued a new regulation in 2009 requiring all clinical trial institutions to be re-evaluated for certification every three years. In 2008, a survey conducted to assess the levels of satisfaction among clinical research associates (CRAs) revealed that over 60 percent are satisfied with the service provided by clinical trial institutions [12]. In particular, CRAs cited satisfaction with the infrastructure and equipment, documentation, and management style. Notably, CRAs considered that the infrastructure and equipment were of an exceptionally high standard. Inadequacies were noted in the areas of work efficiency, coordination, quality control, and servicing attitudes.

9.4.2.1. Study Subject Recruitment

In China, advertisements for clinical trial subject recruitment are generally posted on hospital and community center bulletin boards, in outpatient reception areas, or published by Internet media and online patient communities. Almost none of the recruitment advertisements are submitted to the institutional review board (IRB) for prior approval [13]. Often, the advertisements do not specify that the investigation is a "clinical trial" or "clinical research", and tend to suggest that the drug or device under investigation has good clinical efficacy and safety, or is even better than other drugs or existing devices. Phrases such as "free treatment", "high reimbursement for participation", and "SFDA approved" are deliberately used to attract attention. However, China's IRBs seldom check recruitment advertisements when approving the trial protocol. Typically, an investigator fails to plan for subject recruitment and issues advertisements in a hurry, when it is realized that time is running out to recruit sufficient patient numbers. Thus, patient recruitment methods need to be improved.

Certified clinical trial institutions in China are generally well-known hospitals located in the larger cities, with one unique feature: the fact that many of the patients are from outside the local area. Patients may be from far-away cities or rural areas, coming in the hope of obtaining better treatment. Feasibility studies should take this phenomenon into account and record the number of local patients, in addition to the total number of patients. Non-local patients may participate in trials, but with a low rate of compliance.

9.4.3. Costs of Clinical Trials

An advantage of conducting clinical trials in China is the associated lower operational cost (Table 9.4), estimated to be about half of that in the USA or Europe [14]. Procedural test fees involving patients in China are on average 39 percent of the cost in the USA, while professional fees such as wages and facility fees are 50 percent of those in the USA [15]. Most contracts are negotiated directly with the investigators, and budgets are usually on a cost-per-visit basis. The various procedural fees of specific hospitals are publicly available and can be used as a guideline. Pharmacy and hospital fees need to be included in the hospital contract. Professional fees per visit are dictated by physicians, who may require additional coordinators to be hired on a per-visit basis or on a monthly salary. Most investigator and hospital contracts include overhead charges of 10 percent that apply to procedural and professional fees.

9.4.4. Quality Control at Trial Sites

In many clinical trial institutions, clinical trial monitoring is the responsibility of monitors sent by the sponsor. Some monitors are only concerned about the progress of the trial instead of its quality, and lack the ability or technical expertise to follow up on any problems that may arise upon inspection. To ensure that trials meet quality standards, some of the

Table 9.4 Comparisons of procedural and professional fees regarding clinical trials in China and the USA [15]

Line item description	China	USA	% Difference
Brief office or other outpatient examination: includes vital signs, blood pressure, and weight/height. Typically, five minutes are spent performing or supervising these services during the visit	$20	$37	−45%
Urine pregnancy, human chorionic gonadotropin (hCG, beta-hCG), qualitative	$10	$35	−71%
Blood count: complete (CBC), automated (Hgb, Hct, RBC, WBC, and platelet count), automated differential WBC count	$13	$51	−75%
Electrocardiogram, routine ECG (EKG) with at least 12 leads: includes tracing, interpretation, and report	$50	$119	−58%
Physician per visit (complex, e.g. initial visit, final visit)	$128	$175	−27%
Study coordinator per visit (complex, e.g. initial visit, final visit)	$64	$125	−49%

more established clinical trial institutions in China have their own quality control systems, often known as "tertiary quality control" [16]. There are three tiers of quality control: the first is the departmental quality control delegate, who should have proper GCP training and previous experience in clinical trials. The departmental quality control delegate is responsible for ensuring implementation of standard operating procedures (SOPs) and for checking the original clinical data against the case report form (CRF). The second tier is the institutional quality control delegate, who oversees the entire trial process and continuously inspects the trial in collaboration with the IRB. The last tier of tertiary quality control is represented by the monitors designated by the sponsor; they regularly monitor all aspects of the trial and the job functions related to the first and second tiers. It is important that sponsor and clinical trial institutions make a concerted effort to ensure the high quality of a trial.

9.4.5. Institutional Review Board

The recent, substantial increase in local and international trials in China has led to confusion concerning the IRB review process. In some cases, all IRBs from all participating institutions will review the necessary trial-related documents, while at other times only the IRB at the leading investigator's institution in a multicenter trial will review documents. There is ongoing debate as to whether a multicenter trial should be reviewed only by a central IRB or separately by all participating IRBs. According to a recent survey of clinical trial practices [17], the majority of respondents (65.1%) do not use a central IRB, citing uncertainty about the quality of the central IRB review process, its inability to protect subjects properly, having poor communication with the IRB, and being unable to obtain a sufficiently localized informed consent form. The survey results indicate that the central IRB concept is not widely accepted in China. Those respondents with experience using a central IRB are attracted by the advantages of shorter inspection times and less influence on the research progress. Since 53.6 percent of IRB review processes take up to four weeks [17], individual IRB reviews can significantly affect the progress of an ongoing trial that has to revise its protocol or informed consent form (ICF). However, whereas the US Food and Drug Administration (FDA) has issued specific guidance on the use of the centralized IRB [18], China's SFDA does not specify an ethical review process for multicenter trials [8] and many clinical trial organizations lack SOPs for applying to central IRB processes.

China therefore needs to develop a more comprehensive system for the review process. In 2009, the SFDA issued draft Guidelines for Ethics Committees on Drug Clinical Trial Ethical Review and is currently soliciting public feedback [19]. These guidelines recommend that the central IRB be responsible for reviewing the scientific and ethical aspects of a study, with local IRBs conforming with all central IRB decisions and for reviewing feasibility aspects of the study to be conducted at a local institution. Normally, local IRBs do not provide commentary on the protocol design, but have the right to refuse to conduct the study. After the study has obtained approval and is carried out, local IRBs should coordinate with the central IRB to undertake a continuous review process that records all drug-related events such as any serious adverse reactions. It is hoped that a final version of the guidelines will be issued soon to guide IRB functions in China.

9.4.6. Clinical Research Professionals

According to a recent survey of 229 CRAs from 31 cities in China [20], 73 percent of respondents were in their twenties. Almost half of all respondents reported that 30–60 percent of their working hours were spent traveling for monitoring purposes and almost a quarter (23%) spent over 60 percent of their working hours traveling. The high travel demand is less of a barrier to CRA recruitment among younger age groups. Most CRAs in China have a medical degree and some have a pharmacy degree or a nursing background. Besides routine monitoring duties, CRAs in China often adapt and take on numerous other job responsibilities such as writing protocols, designing the ICF and CRF, and writing final reports (Table 9.5). These are clearly strengths of CRAs in China. However, the majority work in small CRA teams and lack extensive clinical trial experience (Figure 9.4) and experience in managing large numbers of trials or sites and small CRA teams. Another potential weakness regarding CRAs in China relates to the limited number of global trials conducted in China. Further training on monitoring activities, SOP regulations, and the drug development process is needed to improve standards among CRAs in China, and that clearly depends on how enthusiastic pharmaceutical companies are about expanding their activities in China.

9.4.7. Differences Between Local Trials and Global Trials

As the number of clinical trials being conducted in China increases, they can be classified as either local or global. “Local” trials are conducted for

Table 9.5 Job responsibilities among clinical research associate survey respondents [20]

	Yes	No
Protocol writing	152 (67%)	75 (33%)
ICF design	159 (70%)	68 (30%)
CRF design	170 (74.9%)	57 (25.1%)
Final report writing	109 (48%)	118 (52%)
Study grant negotiation	176 (77.5%)	51 (22.5%)
Investigator selection	156 (68.7%)	71 (31.3%)
CT agreement	174 (76.7%)	53 (23.3%)
IRB submission	162 (71.4%)	65 (28.6%)
Protocol training to CRC	110 (48.5%)	117 (51.5%)
Initiation of a clinical trial	211 (93%)	16 (7%)
Some data verification	196 (86.3%)	31 (13.7%)
Site monitoring report	203 (89.4%)	24 (10.6%)
Collection and maintenance of study documents	202 (89%)	25 (11%)
Investigational product handling and accountability	201 (88.5%)	26 (11.5%)
Audits of clinical trials	94 (41.4%)	133 (58.6%)
SAE reporting	202 (89%)	25 (11%)

ICF: informed consent form; CRF: case report form; CT: clinical trial; IRB: institutional review board; CRC: clinical research coordinator; SAE: serious adverse event.

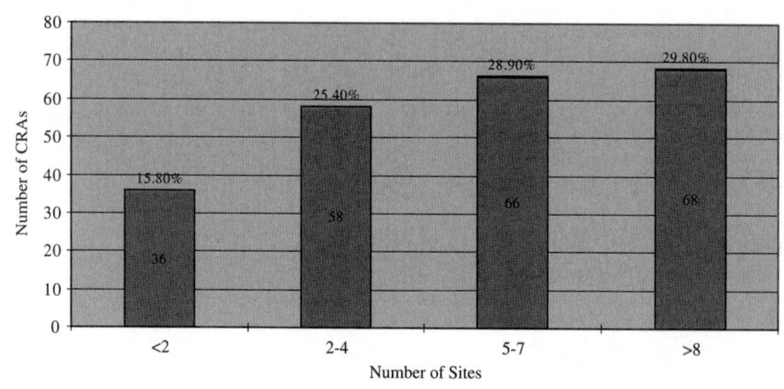

Figure 9.4 Number of sites for each clinical research associate (CRA) survey respondents [20]

local registration of medicinal products that are new to China but are already marketed in the USA and/or Europe. "Global" trials involve new medicinal products requiring marketing registration by the FDA and/or the European Medicines Agency (EMA), and which are not yet marketed in the USA and/or Europe. Since the regulatory requirements for new drugs differ among China, the USA, and Europe, a survey was conducted to determine the views of clinical research professionals and to assess any differences in the quality standard between local and global trials in China [21]. The survey was posted on a clinical trial website (www.druggcp.net) and interested clinical research professionals were asked to complete the survey online. The survey was available online for six months from October 21, 2009, to April 21, 2010, and the data were analyzed in April 2010. A total of 273 clinical research professionals responded to the survey.

When asked the question, "Do you agree that local trials in general should have the same standard of quality as global trials?" the majority of the 233 respondents to this question (90%) agreed (i.e. "yes" and "maybe") that both local and global trials should have the same standard of quality (Figure 9.5). Only 6 percent of respondents disagreed with this statement. In response to an optional follow-up question, the main reason for disagreeing was that local regulatory requirements and practice guidelines are lower for local trials than global trials because these medicinal products are already marketed overseas and therefore have proven safety and efficacy. Therefore, funding for local trials is not so critical, since they only need to meet local regulations to obtain marketing authorization.

In response to the question, "Do you agree that in reality, local trials have the same standard of quality as global trials?" most respondents (71%) stated that the quality standard of local trials is lower than that of global trials. Fourteen percent stated that it is the same and only 3 percent thought that the standard was higher in local than in global trials (Figure 9.6). When aspects of local trials where quality is lower were analyzed from a list of options provided, most respondents chose study monitoring (61%), lack

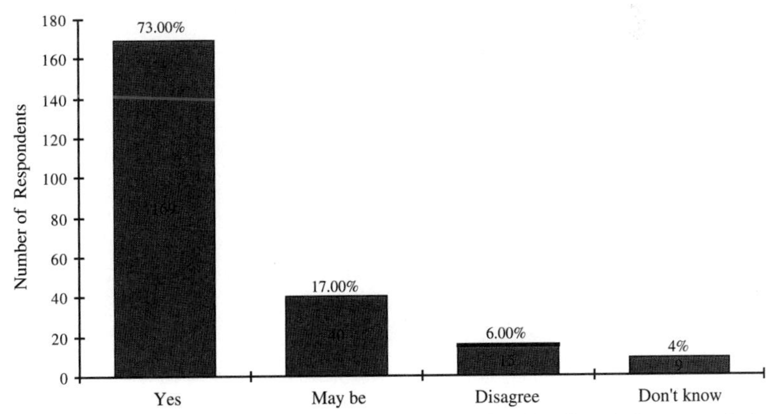

Figure 9.5 Answer to the question: "Do you agree that local trials in general should have the same standard of quality as global trials?"

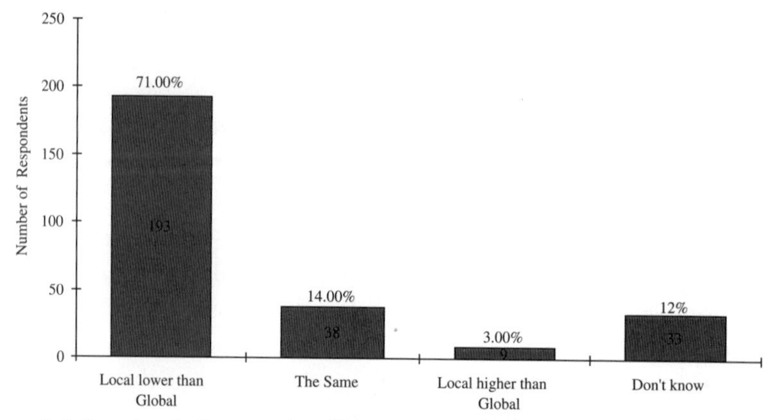

Figure 9.6 Answers to the question: "Do you agree that in reality, local trials have the same standard of quality as global trials?"

of insurance or an indemnity policy (55%), relatively simple study design (54%), and data quality/data management (54%) as the main aspects contributing to lower quality (Figure 9.7). When reasons for the perceived lower quality of local trials were further explored, from a list of options provided, the main reasons chosen by respondents were low trial funding (72%), small study grants (65%), lack of site/investigator's interest in local

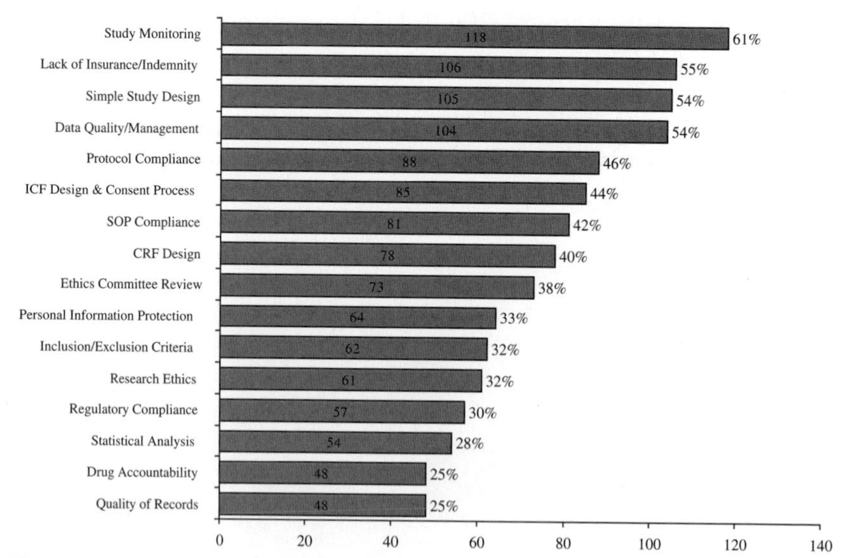

Figure 9.7 Aspects of clinical trials related to lower quality. ICF: informed consent form; SOP: standard operating procedure; CRF: case report form

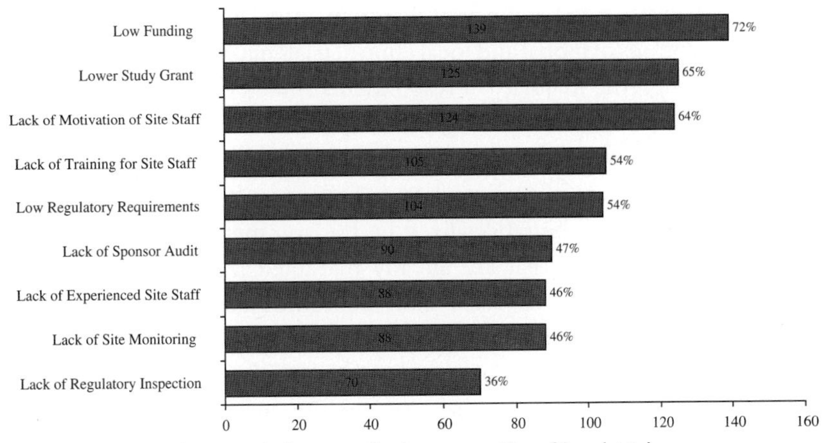

Figure 9.8 Reasons for lower quality of local trials

trials (64%), lack of training for site staff (54%), and low regulatory requirements (54%) (Figure 9.8). No additional reasons to those listed were provided in response to a follow-up optional question.

9.4.7.1. Motivation, Recruitment, and Retention

About one-third of respondents (37%) believed that patients are more willing to participate in a global trial than a local trial. The reasons given included the advantages provided by global trials such as insurance, a "free" import new drug, a "free" laboratory test, higher quality medication and monetary incentives (e.g. transportation expenses). However, a similar proportion (38%) of respondents felt that there was no difference between global and local trials with respect to the patients' motivation to participate. When asked about the difference in the recruitment rate for global compared with local trials, more respondents (42%) believed that the recruitment rate was better in global trials but 31 percent thought that there was no difference. The reasons given for the better recruitment in global trials included the use of larger hospitals with bigger patient populations, the payment of transportation expenses, and access to free import new drug.

In terms of the difference in patient retention and compliance between global and local trials, an equal proportion of respondents (39%) stated that they are better in global than in local trials, or the same (38%). Reasons given for better retention and compliance in global trials included superior study design and patient follow-up and greater investigator motivation owing to increased study funding.

The survey provided some useful insight into clinical research professionals' perceptions of global and local clinical trials in China. The key finding was that although the majority of respondents agreed that local trials should have the same quality standard as global trials, most thought that, in reality, local trials have a lower quality standard than global trials. The survey suggests that the main difference in quality relates to lower regulatory requirements for local trials, which may be rationalized by the argument that local trials are conducted to test drugs that are already marketed overseas and therefore have already had their efficacy and safety demonstrated in clinical trials.

Despite GCP standards having been introduced in China more than 10 years ago, the results of this survey indicate that local and global clinical trials are still conducted to different quality standards. To improve the quality of local trials, positive measures should be taken in a number of areas including study monitoring, insurance, data quality/management, and increased investment in trial funding and training site staff. In addition, the regulatory requirements for local trials should be consistent with those for global trials.

9.5. CONTRACT RESEARCH ORGANIZATIONS

The first CRO in China was established in 1996 by the MDS Pharma Service. Subsequently, some of the larger international CROs have entered the Chinese market, such as Quintiles, Covance, ICON, Kendle International, PPD and Parexel-Apex international Co. The international CROs have operations in China that provide the management infrastructure to conduct global trials for multinational pharmaceutical companies. There are also local CROs in China whose main customers are national pharmaceutical companies, local affiliates of multinational pharmaceutical/medical device companies, and academic institutes. Most local CROs are rarely involved in global trials. The main factors determining the outsourcing to local CROs are cost and need for medical personnel. Many local pharmaceutical companies do not have a medical department and therefore may not be able to conduct local trials without the support of a CRO. Local affiliates of multinational pharmaceutical companies do have medical departments and may have strong clinical teams; therefore, they may conduct the global clinical trials themselves, using local CROs as needed for assistance in conducting local trials for the new drug registration in China.

Currently, there are about 300 CROs in China, more than 100 of which are located in Beijing. Differences exist between foreign and domestic

CROs. Foreign CROs usually conduct clinical trials for foreign pharmaceutical companies, because the cost of their services remains a barrier for domestic pharmaceutical companies. Domestic CROs perform mainly late phase studies on products already marketed elsewhere, whereas international CROs perform more Phase II and III registration trials. Advantages of domestic CROs include their low service fees, and better relationships with the SFDA and local hospitals. However, domestic CROs often operate on a smaller scale than foreign CROs.

As more local CROs have been established in recent years, the competition for local trials has intensified, driving down costs. It could be argued that the low costs and quality standards of local trials create a business opportunity for local CROs. For the Chinese operations of international CROs conducting global trials, this low-cost strategy poses the challenge of how to adjust their pricing policy to meet the demands of the local market while maintaining the quality standards required.

9.6. CHALLENGES AND OPPORTUNITIES

9.6.1. Challenges

9.6.1.1. Language

Documents submitted for SFDA and IRB approval should be in Chinese, including the protocol, investigator's brochure, and ICF. Multinational clinical trial sponsors should factor in the time and cost related to document translation, and be prepared to deal with any language-related misunderstandings.

9.6.1.2. Regulation

Despite China's publication of many regulatory laws and guidelines, the regulatory system is still evolving (e.g. a site inspection system is not yet fully established). CDE's lengthy regulatory review process, seen as an impediment to performing clinical research in China, is being analyzed in an attempt to shorten timelines and make China a more attractive place to perform clinical trials.

9.6.1.3. Independent Review Board

China has no independent IRB. Instead, IRBs are attached to clinical trial institutions; challenges include potential conflicts of interest with the composition of members.

9.6.2. Opportunities

9.6.2.1. Patient Recruitment

The large patient pool in China's population is obviously a major advantage. China has a broad disease profile. With an increasingly aging population, the number of people with cancer, diabetes, and cardiovascular diseases is also increasing. It is estimated that there are 40 million diabetes patients in China, with about 1.2 million new cases each year [22]. Many patients are treatment naïve and want to participate in clinical trials because of the free treatment and their preference for imported drugs over domestic drugs. There is also less competition for the same patient pool for similar trials among different sponsors. These factors can speed up the patient recruitment process and, despite a lengthy start-up time, quick enrollment of patients may ultimately make up for time lost elsewhere in the development process.

REFERENCES

[1] Ministry of Health of the People's Republic of China. Chinese Health Statistical Digest 2010, http://www.moh.gov.cn/publicfiles//business/htmlfiles/zwgkzt/ptjty/digest2010/index.html

[2] World Health Organization. World Health Statistics 2010, http://www.who.int/whosis/whostat/EN_WHS10_Part2.pdf

[3] National Bureau of Statistics of China. International Statistics 2009, http://www.stats.gov.cn/tjsj/qtsj/gjsj/2009/

[4] China's GDP grows 8.7% in 2009. China Daily, January 21, http://www.chinadaily.com.cn/bizchina/2010-01/21/content_9354887.htm, 2010 (accessed July 15, 2010).

[5] Ministry of Health of the People's Republic of China. Briefing of the Nation's Health Service Development in 2009, http://www.moh.gov.cn/publicfiles/business/htmlfiles/mohwsbwstjxxzx/s8208/201001/45652.htm

[6] Chinese Central Government's Official Web Portal. China unveils action plan for universal access to basic health care, April 7, 2009, http://english.gov.cn/2009-04/07/content_1279122.htm

[7] Yang ZM, Wen BS, Shi JF, Yang JH, Feng Y. Research in profile of application and evaluation for global clinical trials in China. Chinese Journal of Pharmaceuticals 2009;40:155–7.

[8] State Food and Drug Administration, People's Republic of China. Good Clinical Practice, http://www.sda.gov.cn/WS01/CL0053/24473.html

[9] State Food and Drug Administration, People's Republic of China. Drug Administration Statistics, http://www.sfda.gov.cn/WS01/CL0010/

[10] Yu L. Review of current status and looking forward into the future of clinical trials in China. China Prescription Drug 2009;93:22–4.

[11] State Food and Drug Administration, People's Republic of China. Drug Clinical Trial Institution Qualification Announcement, http://www.sfda.gov.cn/WS01/CL0069/ (accessed July 15, 2010).

[12] Wu J, Zhang W. An investigation of the satisfaction of clinical research associates about Drug Clinical Trial Institution service. Medicine and Society 2009;22:49–50.

[13] Zeng S. Investigation on the clinical trial recruitment advertising management status and strategy in China. China Prescription Drug 2010;94:12–3.

[14] China beckons to clinical trial sponsors, http://www.nature.com/drugdisc/news/articles/nbt0705-768.html (accessed July 27, 2010).

[15] Jankosky J, Jiang Y, Farwell T. Grant budgeting and negotiating in India and China. Applied Clinical Trials 2007;16:56–62.

[16] Zhao, Y., Zhang, Y., Wang, S. Practice and thought for "tertiary quality control" management mode of clinical trials. Chinese Journal of New Drugs and Clinical Remedies 2209;28:472–474.

[17] Huang J, Liu H, Hu J. Inspection of ethical issues in multicenter clinical trials in China: a survey of current status. Academic Journal of Second Military Medical University 2009;30:1182–5.

[18] US Food and Drug Administration. Guidance for Industry. Using a centralized IRB review process in multicenter clinical trials, http://www.fda.gov/RegulatoryInformation/Guidances/ucm127004.htm

[19] State Food and Drug Administration, People's Republic of China. Announcement about soliciting feedback on Guidelines for Ethics Committees on Drug Clinical Trial Ethical Review, http://www.sfda.gov.cn/WS01/CL0237/40546.htm (accessed July 15, 2010).

[20] Fan J. A route to success: CRAs in China. GCP Journal 2008;15:22–4.

[21] Fan, J. Quality standard of local and global clinical trials in China: a survey of clinical-trial professionals. The Bulletin, Pharmaceutical Contract Management Group (September, No. 11), 2010:9–11.

[22] Diabetes becoming "major problem". China Daily, February 20, http://www.chinadaily.com.cn/china/2008-02/20/content_6468137.htm 2008 (accessed July 15, 2010).

CHAPTER 10

Clinical Trials in Taiwan

James (Dachao) Fan* and Suzanne Gagnon**
*Medical and Safety Services, ICON Clinical Research, Singapore
**Medical, Safety and Scientific Services, ICON Clinical Research, USA

Contents

Global Clinical Trials
ISBN 978-0-12-381537-8, Doi:10.1016/B978-0-12-381537-8.10010-X

10.1. INTRODUCTION

10.1.1. Demography

The population in Taiwan registered by Census was 23.1 million as of May 2010. According to the official statistics, 98 percent of Taiwan's population is made up of Han Chinese, and the remaining 2 percent are Taiwanese aborigines [1]. About 7 percent of the population follows either one or a mixture of 26 recognized religions. About 68 percent of the religious population practices a mixture of Buddhism, Taoism, and Confucianism, while a minority (4%) follows Christianity [2]. The demography of Taiwan (2009) is as shown in Table 10.1. The official language of Taiwan is Mandarin, although the majority of the people speak in Taiwanese Minnan and Hakka dialects [2].

10.1.2. Economy

Taiwan's quick industrialization and rapid growth during the latter half of the twentieth century is called the "Taiwan miracle" or "Taiwan economic miracle". Developing along with Singapore, South Korea, and Hong Kong, Taiwan is one of the industrialized developed countries; hence these four countries are collectively known as the "Four Asian tigers". Through decades of hard work and sound economic management, Taiwan has transformed itself from an underdeveloped, agricultural island to an economic power which is a leading producer of high-technology goods. Taiwan's economic growth from 2002 to 2007 ranged from 3.7 percent to 6.2 percent per year. Economic growth slowed down to 0.73 percent from 2007 to 2008 and to −1.87 percent from 2008 to 2009 in the recent global financial crisis. In the year 2009, the gross domestic product (GDP) was US $379,408 billion and the per capita GDP was US $16,422. As the economy recovers, the projected growth rate for 2010 is 6.14 percent [3].

Table 10.1 Demography of Taiwan (2009)

Population size[1]	23.1 million
Population pyramid[2]	Younger population (under 15 years) 17%
	Productive age (15−64 years) 72.6%
	Aged population (65 years and over) 10.4%
Life expectancy[1]	Female 82.01 years, male 75.49 years, average 78.5 years
GDP[2]	US $379,408 billion
Per capita GDP[2]	US $16,422
Languages	Mandarin (official), Taiwanese Minnan and Hakka dialects

[1]Department of Health, Executive Yuan: http://www.doh.gov.tw
[2]Directorate-General of Budget, Accounting and Statistics, Executive Yuan: http://www.dgbas.gov.tw

10.1.3. Healthcare Infrastructure in Taiwan

Healthcare in Taiwan is administrated by the Department of Health of the Executive Yuan. In 2009, Taiwan had nearly 1.6 physicians and 5.7 hospital beds per 1000 population [1]. There was a total of 515 hospitals in the country. Per capita health expenditure totaled about US $1015 in 2009 [1]. Health expenditure constituted 6.39 percent of the GDP in 2009 and 59.2 percent of expenditure was from public funds [1].

The insurance system of current healthcare in Taiwan, known as the National Health Insurance (NHI), was instituted in 1995. NHI is a single-payer compulsory social insurance plan which centralizes the disbursement of healthcare funds. The system promises equal access to healthcare for all citizens, and the population coverage had reached over 99 percent by the end of 2009 [3]. NHI is mainly financed through premiums which are based on the payroll tax, supplemented with out-of-pocket payments and direct government funding. When conducting a clinical trial, it is imperative to classify which parts could be covered by the NHI and which parts should be covered by the sponsor. Usually, sponsors assess insurance coverage for a clinical trial prior to its initiation. Currently, two insurance companies in Taiwan (The First Insurance Co. Ltd and Fubon Insurance Co. Ltd) can provide this kind of business service.

10.2. REGULATORY CLIMATE

10.2.1. Taiwan Food and Drug Administration

The Taiwan Food and Drug Administration (TFDA) was established on January 1, 2010, to streamline product regulation, consumer protection, crisis management of product safety, and international affairs. A functional Office of Evaluation was created by merging the reviewers from the Center for Drug Evaluation (CDE) and the TFDA. The TFDA is now leading an Asia Pacific Economic Cooperation (APEC) Best Regulatory Practice project under APEC's Life Science Innovation Forum. Under this project, good review practices for medical drugs and devices and regulatory dialog through an APEC Pharmaceutical Evaluation Scheme will be conducted.

10.2.2. Center for Drug Evaluation

The Center for Drug Evaluation (CDE) was established by the Department of Health of Taiwan on July 13, 1998, as a non-governmental, non-profit organization.

Its original mandate included "review and benefit–risk assessment of new drug applications (NDAs), evaluation and approval of clinical trial protocols for investigational new drugs and medical devices, implementation of bridging study, Taiwan's critical path program for biomedical product development, and regulatory consultation service for the industry" [4].

After 12 years' evolution, the principal responsibilities of the CDE now are focused on assisting the TFDA in the set-up of the review system and enhancing its ability to review NDAs, supporting the Department of Health in establishing a mechanism for drafting guidelines, and providing a consultation service to domestic biopharmaceutical industries, in order to upgrade Taiwan's ability to review the NDAs of products developed domestically and those that have not been approved in any developed countries. The sponsors of the pharmaceutical industry are encouraged to consult with the CDE throughout the research and development (R&D) process. The CDE provides free scientific advice according to the specific need of each individual sponsor. The sponsor may request consultative services via the Internet or by telephone. In 2008, approximately 200 consultations were requested and face-to-face meetings were granted to 40 percent of them. The average investigational new drug (IND) review time, including all sponsor time, was 40.6 calendar days. Of 205 INDs reviewed, 76 percent were multinational clinical trials. According to a 2009 article published in the Drug Information Journal, "Taiwan has become one of the most preferred sites for clinical trials in Asia" [5].

10.2.3. Good Clinical Practice and Training

Taiwan's good clinical practice (GCP) guideline was officially announced in 1996 and was revised in 2002 according to the International Conference on Harmonisation (ICH) E6 guidance. In 1999 and 2000, the Department of Health in Taiwan announced the Guidance for Clinical Trials of Drugs in a variety of medical subspecialties. In 2001 and 2002, the Department of Health further announced the Guidance for Clinical Trials of Drugs in Special Populations such as geriatrics, pediatrics, and populations with impaired hepatic or renal function. During the same period, the Guidance for Bridging Studies – Ethnic Factors in the Acceptability of Foreign Clinical Data was published, as was the guidance for the Content and Format of Clinical Trial Reports.

With the introduction of GCP to Taiwan, it has become important to ensure that the clinicians and principal investigators complete their training in clinical trials methodology and are qualified in research capabilities as per GCP requirements. There is a shortage of skilled workers in the clinical research monitoring segment because of the booming demand, even though many sponsors and contract research organizations (CROs) have set up intensive internal training courses to nurture their staff. To meet the market demand and in response to the increasing need for training in the clinical research field, the Taiwan government invited academies and medical centers to collaborate with CROs. By joining hands, GCP and clinical trial research training are currently able to be executed at three levels, namely government (e.g. CDE), university, and CRO and professional organizations [e.g. Foundation of Medical Professionals Alliance in Taiwan (FMPAT)].

The training curriculum includes pharmaceutical regulatory guidelines, ICH-GCP guidelines, ethical and practical considerations, clinical trials applications, development of Chinese herbal medicines, biological products, bridging studies, pharmacokinetics/pharmacodynamics, and biostatistics. Included are clinical practice training modules specifically designated for investigational site personnel such as physicians, pharmacists, and nurses. The Research Medical Council has accredited some of the training courses for Continuing Medical Education (CME) credits and Continuing Nursing Education (CNE) hours. Many conferences are held to promote GCP and clinical trial research, including the APEC Network of Pharmaceutical Regulatory Sciences annual conference, which has had a major focus in this area since 2000.

10.2.4. Good Clinical Practice Inspections

The quality of clinical trials performed in Taiwan is equivalent to those executed in Western Europe, the USA, and Japan. To assure the quality and credibility of clinical trials, study sites are often inspected by the Department of Health. The Department of Health performs GCP inspections on all registration trials done in Taiwan. The inspection results are categorized as: acceptable; acceptable upon further clarification; reinspection required; case dismissed; pending supplement from trial companies (sponsors); or failed. The results of inspections from 2001–2010 are shown in Table 10.2. They

Table 10.2 Clinical trial inspections in Taiwan

	2001	2002	2003	2004	2005	2006	2007	2008	2009	2010
No. of trials inspected	31	37	47	36	34	38	23	29	24	35
Failed	1	4	4	5	2	2	0	5	3	4
Failure rate (%)	3%	11%	9%	14%	6%	5%	0%	17%	13%	11%

Source: CDE database.

suggest that most of those trials met the regulatory requirements in Taiwan. In 2008, five trials had non-conformity with quality requirements, with the main findings being protocol deviations. No safety issues were identified during the inspections. Three Food and Drug Administration (FDA) inspections, one European Medicines Agency (EMA) inspection, and one Pharmaceuticals and Medical Devices Agency (PMDA) inspection took place in Taiwan, and the outcome revealed a remarkable achievement in meeting the FDA requirements in recent years; the inspected study drugs are marketed in the USA now.

10.3. CURRENT STATUS AND STATISTICS ON CLINICAL TRIALS

The status of phases of clinical trials reviewed by the Department of Health between January 2004 and December 2010 is shown in Table 10.3.

Table 10.3 Phases of clinical trials in Taiwan (2004 to 2010)

	2004		2005		2006		2007		2008		2009		2010	
Phase	**P**	**S**	**P**	**S**	**P**	**S**	**P**	**S**	**P**	**S**	**P**	**S**	**P**	**S**
I	8	12	14	26	12	20	10	18	11	14	18	19	19	21
II	22	57	33	78	32	98	46	158	46	120	60	167	49	137
III	85	237	69	242	86	300	106	391	132	527	95	407	119	527
IV	4	10	4	5	3	4	6	14	16	21	14	19	21	30
Total	119	316	120	351	133	422	168	581	205	682	187	612	208	715

P: Number of protocols; S: number of sites.
Source: CDE database.

Table 10.4 Study sites of clinical trials in Taiwan (2004 to 2010)

	2004		2005		2006		2007		2008		2009		2010	
	P	**S**	**P**	**S**	**P**	**S**	**P**	**S**	**P**	**S**	**P**	**S**	**P**	**S**
Taiwan single site	32	32	24	24	11	11	21	21	34	34	38	40	47	55
Taiwan multicenter	25	88	10	43	22	74	20	81	16	49	9	35	17	60
Multinational, multicenter	62	196	86	284	100	337	127	479	155	599	140	537	144	600
Total	119	316	120	351	133	422	168	581	205	682	187	612	208	715
% of multinational	52.1		71.1		75		75.6		75.6		74.9		69	

P: Number of protocols; S: number of sites.
Source: CDE database.

The total number of protocols increased by 175 percent (208/119) from 2004 to 2010, especially in early Phase I and II clinical trials. The proportion of multinational trials in Taiwan was 52.1 percent (62/119) in 2004; this increased to 75 percent (100/133) in 2006 and has stayed around 75 percent in the following years through 2009 (Table 10.4). Thus, Taiwan has demonstrated its ability to conduct quality early phase clinical trials and to participate in multinational clinical trials, as evidenced by the increasing number of studies being performed in Taiwan.

10.4. OVERVIEW OF INVESTIGATORS AND SITES IN TAIWAN

10.4.1. Study Investigator Profile

Even though lectures are given in the Chinese language, most Taiwan physicians are familiar with English literature since the medical universities use English textbooks. Clinical trial documents written in English are used at sites and accepted by the regulatory authority. Investigators are capable of performing both local and global trials following GCP guidelines. For domestic clinical trials, investigators work closely with the CROs and may be involved in the study design and protocol development. They may work with data management and the statistician(s). They fully understand the clinical trial process and have a clear concept of the drug's development and regulatory processes and requirements. For global trials conducted in Taiwan, some of Taiwan's key opinion leaders (KOLs) may be involved in the early preparation stage, such as providing consultation for the protocol design and endpoint selection, particularly in areas of extensive expertise such as hepatocellular carcinoma.

10.4.2. Study Site Characteristics

Taiwan adheres closely to ICH-GCP guidelines. As per government requirements, only about 130 teaching hospitals are qualified (authorized) to perform clinical trials of new drugs and investigators must complete a required 30-hour GCP training as one of the qualifications to participate in clinical trials.

10.4.3. Cost of Clinical Trials

Performing a multinational global clinical trial in Taiwan may cost about 50–70 percent of the US cost. The cost for domestic clinical trials is usually approximately one-third that of global trials. The main sponsors for domestic clinical trials in Taiwan are local pharmaceutical companies, local affiliates of global biopharmaceutical/medical device companies, and academic institutes. The trials are mainly categorized as "local trials" for local registration, academic research, government-sponsored new drug development projects, and investigator-initiated study (IIS).

Since Taiwan is a small country, round-trip travel to any of the hospitals can be completed within one day. Most major sites for clinical trials are located in the big cities of the west coast and clinical research associates (CRAs) are usually centrally located in Taipei. Because of geographical proximity, the cost of travel to monitor studies is not a main concern when utilizing sites in Taiwan.

10.4.4. General Clinical Research Centers

In 1999, the Department of Health began subsidizing the setting up of general clinical research centers (GCRCs) in 20 regional teaching hospitals and medical centers. Their purposes are to integrate hospital and laboratory resources, develop training programs, set up research management units, and improve the quality of clinical research. The GCRCs also provide a forum for education and career development in clinical research for healthcare professionals. Among GCRCs, in 2006, the Department of Health sponsored four additional medical centers identified as National Centers of Excellence in Clinical Trials and Research (CECTRs) in order to build up Taiwan's infrastructure for early phase clinical trials, focusing on Phase I, Phase II and pharmacokinetics studies of new drugs, as well as Chinese herbal medicine. When a site is needed for a Phase I trial, any of the CECTRs can provide this service.

The four original CECTRs are located in the National Taiwan University Hospital, the National Cheng Kung University Hospital, the Tri-Service General Hospital, and the Taipei Medical University Wan Fang Hospital. China Medical University Hospital and Chang-Gung Memorial Hospital have been newly added in 2010.

10.4.5. Institutional Review Board, Joint IRB and Taiwan Association of IRBs

In addition to the local institutional review board (IRB) at each study site, the Joint Institutional Review Board (JIRB) was set up in March 1997 to provide an efficient and qualified IRB review for multicenter trials. Some of the objectives of the JIRB are to shorten the IRB approval time for clinical trials, to avoid repetitive reviews by local IRBs, and to improve communication between sponsor and sites. At present, there are 93 hospitals participating in JIRB. Seventeen are medical centers; the others are regional hospitals and specialized clinics. JIRB and CDE applications for clinical trials can be submitted in parallel. The submission dossier may be either in English or in Chinese. The review process takes an average of 40.6 calendar days.

The Taiwan Association of Institutional Review Boards (TAIRB) was established in 2009 to serve as a platform for IRB-related issues. Currently, there are over 20 IRBs in Taiwan accredited by the Forum for Ethical Review Committees in Asia and the Western Pacific (FERCAP).

10.5. CONTRACT RESEARCH ORGANIZATIONS

Many entrepreneurs like Quintiles, Covance, PPD, and ICON, have already queued up to set up their clinical research organizations in Taiwan. They currently offer a full complement of services for all phases of clinical development including clinical data management, regulatory affairs in IND and NDA applications, feasibility studies, and clinical outcomes/observational studies. The global CROs located in Taiwan acquire some of the projects from the headquarters of the multinational pharmaceutical companies as multinational trials. Operations are fully GCP and ICH compliant. The demand for a skilled workforce in the CRO segment is very strong as the number of global/regional CROs increased from five to eight in 2007. Two Japanese CROs, CMIC and EPS, opened branch offices in Taiwan in 2007 and one new local CRO, Formosa, was established. Taiwan's largest CRO, Apex, was acquired by Parexel in 2007, now Parexel-Apex has extensive coverage in the Asia Pacific area. Regional operators include PPC, StatPlus, Formosa, EPS, CMIC, Qualitix, and VCRO. There are also niche CROs, mainly for bioequivalence, and absorption, distribution, metabolism, and excretion (ADME), toxicology and pharmacology, such as Mithra, MDS Pharma, Rosetta, QPS, Level, MedGaea and the Center of Toxicology and Preclinical Sciences of Development Center for Biotechnology (DCB).

In additional to performing clinical trials, CROs may act as training centers in collaboration with academic institutes, schools of medicine, and the government.

10.5.1. Local Contract Research Organizations

The main customers of local CROs in Taiwan are local pharmaceutical companies, local affiliates of big biopharmaceutical/medical device companies, and academic institutes. The types of trials performed are mainly trials

Table 10.5 Local contract research organizations (CROs) in Taiwan and their services

CRO	Date	BA/BE	PhaseI–IV	RA	Lab	DM	Stat
Mithra Bioindustry Co. Ltd	1988	√	√	√	√	√	√
Protech Pharmaservices Corp., PPC	1997	√	√	√	√	√	√
Genovate Biotechnology Co.; Qualitix Clinical Research Co. Ltd	1996		√	√		√	√
Virginia Contract Research Organization Co. Ltd	1997	√	√	√	√	√	√
StatPlus, Inc.	1998		√	√		√	√
Parexel Apex International	1999		√	√		√	√
Adv. Pharma Inc.	2001	√			√		
Rosetta Pharmamate Co. Ltd	2002	√			√	√	√
Quest Pharmaceutical Services, Taiwan	2004	√		√	√	√	√
Formosa Biomedical Technology Corp.	2006		√	√		√	√

BA/BE: bioassay/bioequivalence; RA: Regulatory Affairs department; Lab: laboratory; DM: data management; Stat: statistics.
Source: With permission from Chin Ching-Yao [6].

for local registration, academic research, government-sponsored new drug development projects, Investigator Initiated Study (IIS), and listing trials (Table 10.5). Occasionally, clinical trials are contracted by regional pharmaceutical companies performing studies in Taiwan.

The primary factors determining outsourcing to local CROs are cost sensitivity, number of medical staff, and experience. Most of the local pharmaceutical companies in Taiwan do not have medical departments. Some local affiliates of the global pharmaceutical companies have medical departments as well as CRAs and regulatory affairs staff, but they also utilize local CRO support. Many local CROs can provide the full spectrum of services from protocol design to final reporting writing. Frequent communication with CDE throughout the development process facilitates implementation and helps the local CROs to maintain quality and compliance with regulatory requirements.

10.6. CHALLENGES AND OPPORTUNITIES

10.6.1. Challenges

Some of the challenges associated with performing clinical trials in Taiwan relate to a relatively immature local business model compared to the international practices and standards required for global clinical research. Business partnerships between local pharmaceutical companies and local CROs are not well established. There is currently a competitive demand for experienced clinical research staff, with the local affiliates of international companies offering incentives that attract clinical research staff away from domestic companies. The ability to participate in global clinical trials is attractive to clinical project managers and CRAs, moving research experience from local to international pharmaceutical companies and CROs, and making it difficult for domestic companies to retain staff. As in many parts of the world, owing to competitive demand, CRA turnover in some cases may be as high as 20–30 percent annually.

Another consideration in performing clinical trials in Taiwan is the requirement for export licenses to export blood samples, which should be considered in the clinical development plan.

Challenges may also be related to contracting with investigators in Taiwan. Each study site has its own Chinese version of the contract. Each IRB requires the sponsor to translate the English version of the contract into Chinese and follow the hospital-specific contract. Study grant allocation is usually under the administration of the hospital-specific contract. Some IRBs also require independent review and approval of study protocols, despite JIRB approval. Since there may be several study sites involved in a clinical trial, the requirement for individual site contracts is a time-consuming process.

10.6.2. Opportunities

10.6.2.1. Government

The Taiwan government has put a great deal of effort into establishing regulatory and clinical trial environments conducive to supporting clinical research, beginning in 1996 when ICH-GCP was introduced. The quality of clinical trials performed in Taiwan is equivalent to those executed in more advanced nations, such as in Western Europe, the USA, and Japan. With the full support of the government, the requirements for performing clinical research have given impetus for excellence not only to domestic clinical research activities, but also to continuous improvement of the domestic pharmaceutical industry. The visible commitment to improving clinical research training and infrastructure suggests that Taiwan can provide a solid foundation to support clinical product development.

10.6.2.2. Development Center for Biotechnology

The Development Center for Biotechnology (DCB) of Taiwan is a non-profit organization founded in 1984 for R&D in biotechnology. Co-sponsored by government grants and private donations, DCB's purpose is to advance Taiwan's biotechnology industry by building infrastructure, developing key technologies, and training talented workforces in coordination with government, industry, and academic institutions. The Industry & Technology Intelligence Services (ITIS) Program of DCB provides real-time information and analysis of the industry through publication of monthly review newsletters, a Yearbook of Biotechnology, and special reports which focus on trends, strategies, and analyses of the clinical trial industry in Asia.

10.6.2.3. National Health Research Institute

The National Health Research Institute (NHRI) of Taiwan is another important non-profit R&D foundation established by the government in 1996. The focus is on mission-oriented medical research and basic biomedical science, as well as early research on specific diseases. Promising research results are transferred to translational medicine and the domestic pharmaceutical industry for further new drug investigation.

10.6.2.4. Clinical Trial Talent Pool

Most of the local CROs that were set up 10 years ago still exist. With 10 years of experience performing clinical trials in Taiwan, there is now an experienced clinical trial talent pool, although many of the most experienced clinical research staff have moved to global pharmaceutical companies and CROs doing business in Taiwan.

Because of the well-established hospital infrastructure, partnership with the government and required training, investigators are capable of performing both local and global trials in accordance with GCP-ICH guidelines. Some experienced KOLs are already participating in the early preparation stages of global trials, including consultation for protocol design, endpoint selection, and liaison with Taiwan's regulatory authorities.

10.6.2.5. Taiwan Biotechnology Takeoff Diamond Action Plan

In an effort to ensure a successful pathway for the biotechnology industry, the Taiwan government announced the Taiwan Biotechnology Takeoff Diamond Action Plan in 2009. This strategic plan includes four component goals: strengthening the industrial value chain and preclinical development in the commercialization process, establishing a biotechnology venture capital fund, promoting an integrated incubation mechanism, and establishing the TFDA. The TFDA has been formally functioning since January 2010.

10.6.2.6. Regional Cooperation

China–Taiwan regulatory cooperation has been proposed for simultaneous clinical trial evaluation and new drug development. The Cross Strait Biotechnology and Medical Device Industry Partnership and Dialogue held in May 2010 in Taipei opened up dialogue on many issues related to the biopharmaceutical industry. Geographically close, China and Taiwan have the same official language. Their cultures and economies are highly interchangeable. Mutual benefit and collaboration can be achieved through continuous international cooperation and harmonization on regulation and review processes. It is reasonable to expect that clinical trial data obtained following the best international practices will be mutually accepted in the near future. The APEC LSIF Best Regulatory Practice Project from Taiwan is a high-priority two-year project recommended by APEC Life Sciences Innovation Forum Regulatory Harmonization Steering Committee (LSIF RHSC) to invite regulators from regional developing countries to work together.

ACKNOWLEDGMENTS

We wish to thank Ching-Yao Chin, Industrial Consultant of Industry & Technology Intelligence Services (ITIS), Development Center for Biotechnology (DCB) of Taiwan; Herng-Der Chern, CDE Executive Director, and Jung-Chun Wang, CDE Senior Specialist; Meir-Chyun Tzou, TFDA Division Chief of Drug & New Biotechnology Products; and Shirley Pan, TFDA Section Chief of Clinical Trial; who provided their expertise and opinions with permission to publish the figures included in this chapter.

REFERENCES

[1] Department of Health, Executive Yuan, Republic of China. http://www.doh.gov.tw
[2] Department of Civil Affairs, Ministry of the Interior, Republic of China (Taiwan). Taiwan Yearbook. http://www.moi.gov.tw
[3] Directorate-General of Budget, Accounting and Statistics, Executive Yuan, Republic of China. http://www.dgbas.gov.tw
[4] Tarn Y-H, Chern H-D. A proposal for a parallel submission process for new drug applications and health technology assessment in Taiwan: a win–win solution. Drug Information Journal 2009;43:319–23.

[5] Chern H-D, Gau C-S, Hsu Chen H-M, Liao CCh. An experimental model of regulatory science in Asia: Center for Drug Evaluation in Taiwan. Drug Information Journal 2009;43:301–4.
[6] Chin, C.-Y. CRO Industry in Taiwan, Yearbook of Biotechnology Industry 2010, ITIS, DCB, 2010.

CHAPTER 11

Clinical Trials in the Philippines

Romillie E. Cruz[*] and Suzanne Gagnon[**]
[*]Medical Safety Services, ICON Clinical Research, Singapore
[**]Medical, Safety and Scientific Services, ICON Clinical Research, USA

Contents

11.1. INTRODUCTION

11.1.1. Demography

The Republic of the Philippines is an archipelago made up of 7107 islands. It is divided into three main geographical regions: Luzon, Visayas, and Mindanao. Manila is its capital city. As of July 2009, the population was estimated to be 97.9 million, which ranks the Philippines as the 12th most

Global Clinical Trials
ISBN 978-0-12-381537-8, Doi:10.1016/B978-0-12-381537-8.10011-1

populous country in the world. It has a population density of 5.617 people per square kilometer.

The demography of the Philippines is shown in Table 11.1. Less than 5 percent of the population is over the age of 65 years. The infant mortality rate is relatively high, at more than 20 deaths per 1000 live births. Perinatal conditions are a significant cause of mortality. The growth rate is 1.957 percent. The birth rate is 26.01 births per 1000 population and the death rate is 5.1 deaths per 1000 population, according to a July 2010 estimate.

11.1.2. Economy

The Philippines weathered the 2008/09 global recession due to relatively low dependence on exports and high levels of domestic consumption. This was

Table 11.1 Demography of the Philippines

Age breakdown	
0–14 years	35.2%
15–64 years	60.6%
≥ 65 years	4.1%
Median age (years)	
Male	22.2
Female	23.2
Total	22.7
Gender ratio (M/F)	
At birth	1.05/1
0–14 years	1.04/1 (17,606,352 M/16,911,376 F)
15–64 years	1.04/1 (29,679,327 M/29,737,919 F)
≥ 65 years	0.76/1 (1,744,248 M/2,297,381 F)
Total population	1/1
Infant mortality rate (deaths/1000 live births)*	
Male	23.17
Female	17.83
Total	20.56
Life expectancy at birth for the total population (years)**	
Male	68.17
Female	74.15
Total	71.09

M: male; F: female.
Country comparison to the world: *102; **133.

fueled by large remittances from four- to five-million overseas Filipino workers. Economic growth in the Philippines has averaged 5 percent per year since 2001 but despite this growth, poverty has worsened because of the high population growth rate and unequal distribution of income. The Philippines must maintain the momentum of reform to catch up with regional competitors, improve employment opportunities, boost trade, and alleviate poverty.

The Philippines maintains a close bilateral relationship with the USA. Aside from historical factors, ties are strong for a number of reasons, including important commercial relationships and close military associations (the US army provides support for the Philippine government's campaign against Muslim separatists on Mindanao). In recent years, the country has also established relations with China, the new dominant power in the region.

11.1.3. Disease Profile

The most common causes of morbidity and mortality in the Philippines are listed in Table 11.2. Infectious diseases, including those for which vaccinations are available, are a significant cause of morbidity.

11.1.4. Infrastructure of the Philippine Healthcare System

Although medical treatment has improved and services have expanded in recent years, widespread poverty and lack of access to family planning detract from the general health of the Philippine people. The Philippines has a social security system including Medicare with wide coverage of regularly employed urban workers. It offers a partial shield against disaster, but is

Table 11.2 Mortality and morbidity in the Philippines

Top 10 causes of mortality	Top 10 causes of morbidity
1. Heart disease	1. Pneumonia
2. Vascular system diseases	2. Diarrhea
3. Cancer	3. Bronchitis/bronchiolitis
4. Accidents	4. Influenza
5. Pneumonia	5. Hypertension
6. Tuberculosis	6. TB, respiratory
7. Signs and symptoms, abnormal laboratory clinical findings, NEC	7. Heart disease
8. Chronic lower respiratory diseases	8. Malaria
9. Diabetes	9. Chickenpox
10. Perinatal conditions	10. Measles

NEC: necrotizing enterocolitis.
Source: Department of Health, Republic of the Philippines [1].

limited both by the generally low level of incomes, which reduces benefits, and by the exclusion of most workers in agriculture. The Philippines has about 95,000 hospital beds or about 1 per 800 people. There are approximately 1700 hospitals, of which 60 percent are private.

The Philippines has both private and public healthcare institutions. Despite some misconceptions, most of the government hospitals provide quality healthcare in the same way that private hospitals do. The main differences between public and private hospitals are the facilities and technologies offered. Most of the public hospitals are not equipped to the same standard as the private ones. However, some of the best physicians work in the government hospitals.

The Department of Health is the principal health agency in the Philippines. It is responsible for ensuring access to basic public health services to all Filipinos through the provision of quality healthcare and regulation of providers of health goods and services.

The Philippine Council for Health Research and Development (PCHRD) is the primary oversight organization for health research activities in the Philippines. PCHRD is an agency of the Department of Science and Technology and is responsible for coordinating and monitoring research activities in the country. PCHRD organized the National Ethics Committee (NEC) in an effort to ensure that all health research and development proposals conformed to ethical standards. PCHRD also promoted the establishment of institutional ethics review committees (IERCs) within institutions that conduct biomedical and behavioral research.

PCHRD and the Department of Health jointly established the Philippine National Health Research System (PNHRS) in 2003. PNHRS launched a working group on ethics, which evolved into the Philippine Health Research Ethics Board (PHREB). The board was formally constituted in early 2006 as a national policy-making body on health research ethics. PNHRS envisions creating regional ethics boards that will function similarly to PHREB and will be supervised by PHREB.

11.2. REGULATORY CLIMATE

11.2.1. Philippine Food and Drug Administration

On August 18, 2009, Republic Act No. 9711, otherwise known as The Food and Drug Administration (FDA) Act of 2009 was signed and the Philippines FDA replaced the Bureau of Food and Drugs (BFAD) as the country's food and drug safety monitor. This strengthened and rationalized the Agency's regulatory capacity. The previous structure of BFAD included Divisions in charge of Product and Services (PSD), Laboratory Services (LSD), Administration (AD), Legal Information and Compliance (LICD), Policy Planning and Advocacy, (PPAD) and other regulatory divisions classifying reviews according to cosmetics, food, or drugs. The new Philippine FDA will have its strength in the creation of a separate center for every major product category including a Center for Drug Regulation and Research. The new organizational structure aims to focus its attention on the regulation of clinical trials.

This new law empowers the FDA to ban, seize, recall, and withdraw health products from the market that may have caused death or other serious adverse events to consumers.

Clinical trials can be conducted in the Philippines provided prior permission from the Philippine FDA (formerly BFAD) is granted. The approval process is outlined in Table 11.3.

11.2.2. Submission Requirements

Application for protocol approval is on a per-phase, per-product basis. Applicants are required to disclose fully all pertinent documentation and information regarding the product, the subjects and disease process to be evaluated, the study endpoints, the clinical trial sites, existing resources and infrastructure at the proposed trial sites, and other field site information such as location, personnel, resources, equipment, and facilities.

Application to the Philippine FDA for permission to conduct a clinical research study follows a series of steps and timelines. For FDA approval, the preparation time is about three weeks and approval is given within three to

Table 11.3 Approval process

	Source of approval	Timeline	Fee*
Phase I—IV	Philippine FDA	3—5 months	Php 2500.00 for initial submission
Export permit	Bureau of Quarantine	1—2 weeks	Php 500.00 per shipment
Amendment (protocol and ICF)	Philippine FDA	< 3 months	Php 1000.00 per amendment
Additional sites — submitted as amendment	Philippine FDA	< 3 months	Php 1000.00 per amendment

ICF: informed consent form; FDA: Food and Drug Administration; Php: Philippine pesos.
*Price quoted as of July 2010.

five months. Basic documents required are the informed consent form (ICF), protocol, protocol amendment, if applicable, and investigational brochure. This approval is a prerequisite for obtaining an import license for the study.

11.2.2.1. Imports and Customs

All sponsors of clinical trials of developmental or investigational products need to apply for an import permit or Permit for Clinical Investigational Use (PCIU) before beginning clinical trials. For the import license, approval time is one month. All clinical trials of investigational, new, or established biological products and drugs require clinical trial protocol approval by the Philippine FDA.

Import license applications are usually submitted after an approval to conduct a clinical trial is granted. Sponsors and contract research organizations (CROs) are required at a minimum to submit pro forma invoices, good manufacturing practice (GMP) certificates, and certificates of analyses specifying the exact number of investigational products that will be shipped.

Previous procedures allowed companies to submit applications summarizing the accumulated shipments that will be made throughout the duration of the clinical trial. This allowed the use of one permit for all shipments that would occur in the study.

Current regulations only allow one permit per shipment. Companies now have to state the exact number of investigational products to be imported for a single shipment. Subsequent shipments need new permits. Release of investigational products from customs requires the payment of certain duties and taxes.

For importation of laptops and other electronic devices, a National Telecommunications Committee (NTC) approval/license must be acquired for a quick release of equipment. The permit will not exempt payment of customs duties and taxes.

11.2.2.2. Exports and Bureau of Quarantine

Export permits are given for the shipment of laboratory samples out of the country. Currently, each shipment requires one permit. Approvals are usually released within one to two weeks from time of submission. The export license is submitted to the Bureau of Quarantine together with the protocol and FDA approval.

Pharmaceutical companies and CROs are required to pay an amount (currently Php 500.00) for each date of shipment. The address of the central laboratory, list of clinical trial sites, and a copy of the protocol must be submitted as an attachment to the application letter. Applicants must ensure strict adherence to the codes of good clinical practice (GCP), good laboratory practice (GLP), and GMP. The specifications, preparation, and composition of batches or lots of developmental biological products or drugs should be the same as the batches or lots to be registered and commercially produced in the future.

The Philippine FDA reserves the right to consult appropriate clinical pharmacologists and consultants in evaluating the clinical trial protocols. At any time, the FDA may terminate any clinical trials that fail to comply with the codes of GCP, GLP, and GMP, or after careful evaluation of an incident report, deviation report, adverse events following immunization (AEFI) report, or information and findings from other national regulatory authorities (NRAs) and international bodies, such as the World Health Organization (WHO).

Most of the documents are submitted in two languages, English and Tagalog. However, in some sites located in the provinces, translation to native dialect is necessary. Examples of documents translated into Tagalog or certain dialects include the ICF, patient's diary, information card, recruitment materials, and advertising brochures.

11.2.3. Regulatory Reporting Requirements

There is no written requirement addressing timelines for serious adverse event (SAE) and suspected unexpected serious adverse reaction (SUSAR) reporting. However, in practice, the Philippines follows US FDA global reporting timelines.

For both marketed drugs and drugs in clinical trials, all related local SAEs, whether expected or unexpected, should be reported to the Philippine FDA. Death and life-threatening SAEs should be submitted within seven days and other SAEs within 15 days.

For international SAEs and SUSARs, the usual practice of the company or the sponsor is followed. This varies from one company to another, so some can be submitted as quarterly line listings, annual listings, by batch sending, or even individual SAE/SUSAR reports.

The Philippine FDA should be notified of the end of the study and the final study report should be submitted once the study has been completed. No specific timeframes are set; however, it is expected that submission is done as soon as possible.

11.2.4. Good Clinical Practice

GCP training in the Philippines is conducted by the University of the Philippines Manila–National Institutes of Health (UPM-NIH). GCP training by UPM-NIH is open to everyone for a fee. International pharmaceutical companies may hire lecturers from UPM-NIH for their own in-house training. UPM-NIH is a recognized training center for health research ethics and GCP. GCP training is also offered by the University of the Philippines (UP) College of Pharmacy and by CROs.

There are no GCP inspections, but the institutional review boards (IRBs) and the FDA do monitor clinical trial sites. IRBs specifically focus on the welfare of trial participants and the FDA focuses on safety and compliance with regulations. IRBs can suspend a clinical site, and FDA can recommend trial termination, if necessary. In the Philippines, it is usually the sponsor that terminates the trial.

Members of the IRB are also members/board directors of the Strategic Initiative for Developing Capacity in Ethical Review (SIDCER) and the Forum for Ethical Review Committees in Asia and the Western Pacific (FERCAP).

Steps are currently being taken to begin to unify the processes for the monitoring of clinical trials nationwide, particularly the review of protocols and the standard operating procedures (SOPs) of the IRBs.

11.3. CURRENT STATUS AND STATISTICS ON CLINICAL TRIALS

As of 2010, 403 clinical trials have been, or are being, conducted in the Philippines. These studies are both local and international and range from Phase I to Phase IV clinical trials. Over the past five years, the number of global clinical trials has increased. The trials that were being conducted in

Table 11.4 Current clinical trials in the Philippines (as of July 2010)

Phase	No. of trials
I/II	4
II	21
II/III	3
III	113

Source: Citeline Trial Trove [2].

July 2010 are listed in Table 11.4. The Philippines is becoming a favorite among international pharmaceutical companies and CROs owing to the abundance of qualified professionals and the low cost of performing clinical research.

The most commonly studied therapeutic areas include type 2 diabetes mellitus, thrombotic disorders, breast cancer, non-small-cell lung cancer, renal disease, coronary artery disease, respiratory infections, schizophrenia/depression, vaccines and hepatic conditions.

11.4. OVERVIEW OF INVESTIGATORS AND SITES IN THE PHILIPPINES

11.4.1. Study Investigator Profile

Investigators in the Philippines are capable of performing both local and global trials following GCP guidelines. All investigators are trained in International Conference on Harmonisation (ICH)-GCP, with training performed by UPM-NIH or as pharmaceutical company-sponsored in-house training. Lectures are given in English.

Investigators contracted to conduct clinical trials in the Philippines are mostly consultant (academic) physicians who are board certified. They may include therapeutic area experts and/or key opinion leaders (KOLs). For local trials, KOLs may be invited to participate in the protocol development. For global trials, most investigators' participation is limited to conducting the study.

Study coordinators generally have completed tertiary-level education and possess a diploma in biomedical sciences or related courses. They are usually nurses or pharmacists, but at some sites may be resident physicians who are learning to conduct research. Coordinators are also ICH-GCP trained and fluent in English, and have experience in various electronic case report forms (eCRFs)/electronic data capture (EDC) systems, interactive voice and web response systems (IVRS/IWRS), and specimen handling. They are encouraged to maintain updated training on the conduct of clinical trials and ICH-GCP provided by sponsors, UPM-NIH, and reputable medical institutions.

11.4.2. Study Site Characteristics

Site selection depends on a number of factors, all of which are reviewed and analyzed during feasibility studies to determine whether a site is capable of conducting the trial successfully. Several sites have established specific research centers within hospitals. Some of these may be the institution's research center and some are set up by individual investigators.

The majority of the trials are conducted in large healthcare institutions and government hospitals since they have greater access to the patient populations required and the investigators are more likely to have experience conducting clinical studies. Sites with large numbers of potential patients and greater recruitment capability, presence of an IRB, and necessary facilities and equipment (e.g. laboratory, computed tomography scanner) are preferred. It is also very important that there are investigators and support staff with established clinical trial backgrounds and experience present at the site. Major clinical research sites in the Philippines include the cities of Manila, Cebu, Davao, and Iloilo.

11.4.3. Site Contracts

Contracts for clinical studies may be between the sponsor and investigator (two-party agreement) or they may include the institution where the study is being performed (three-party agreement). This is determined during budget negotiations. Historically, most trials in the Philippines involved a two-party contract, signed only by the sponsor or its representative, and the investigator. This has evolved, however, and involvement of the institution is now included in many trials. In a three-party agreement, the institution where the trial will be conducted by the investigator is paid an institutional fee. This covers the use of facilities, laboratory, and infrastructure, and can range from a fixed amount to about 10 percent of the total study budget. Some sponsors prefer a three-party agreement to ensure that the institution is aware that a certain clinical trial is being conducted.

11.4.4. Cost of Clinical Trials

Conducting global trials in Philippines may cost about 80 percent of the US cost. The total cost of a trial includes the CRO service cost, investigator payment, and pass-through costs. Lower investigator payments, airfares, and salaries of clinical monitors decrease the total cost of global clinical trials conducted in the Philippines compared to the USA. Most clinical research conducted in the Philippines is performed as part of global trials. The cost of local trials varies depending on the indication, the number of sites, and the staff needed to conduct the trials. Local trials can be pharmaceutical company sponsored, institution sponsored, or supported by UPM-NIH.

11.4.5. Institutional Review Board Approval

Every clinical trial needs to obtain IRB [also called institutional ethics review board or ethics review board (ERB)] approval. Several different IRBs are used in the Philippines. Unlike other countries which have joint IRBs, there is no central ethics committee for different hospitals. Most sites have their own IRB. Sites without an IRB can obtain approval from the NEC, which reviews protocols for areas with no IRBs. This review function has recently been transferred to the NIH-ERB.

Written approval is necessary for all studies, including amendments and addition of sites. The timeline runs between 30 and 60 days with fees ranging from Php 20,000 to Php 40,000 for Phase I–IV trials, excluding amendments. Application for permission to conduct the clinical trial can be submitted to IRBs simultaneously with the FDA application. The fee for IRB submission can be variable, depending on additional requirements such as institutional fees or section fees, which can amount to approximately 10 percent of the study budget.

11.5. CHALLENGES AND OPPORTUNITIES

11.5.1. Challenges

Some of the challenges associated with performing clinical trials in the Philippines relate to its relatively immature system. Regulations are in their infancy and will need to be improved as the number of clinical trials conducted in the country increases. Some investigational sites are not fully equipped.

The establishment of the Philippine FDA in 2009, as the country's food and drug safety monitor, will provide the agency with greater regulatory powers than those of its predecessor, BFAD. There will be created a separate center for every major product category, including a center for drug regulation and research. The new organizational structure aims to focus its attention on the regulation of clinical trials and speed up their review and approval. This is still being newly implemented.

Currently, there are still no specific regulations or guidelines for the review and approval of clinical trials or the safety reporting process. Timelines for approval range from four to six months from the time of submission. There is no consistency in the safety reporting timeline. Some studies follow the US FDA timelines, whereas some follow the sponsor pharmaceutical company's manner of reporting, which may be by line listing, batch sending, or individual event reporting. Standardizing reporting requirements will be important as clinical research increases in the Philippines.

11.5.2. Opportunities

The Philippines is a fertile land for conducting clinical trials. With its large population and well-established network of hospitals and physicians, it is an attractive location for drug development programs. Most of the time, physicians performing the studies are also members of academia and are very interested in conducting high-quality clinical research.

There has been a significant increase in clinical trial activities over the past five years, especially in diabetes, psychiatry, oncology, and respiratory medicine. Some hospitals have established clinical trials offices to service the specific needs of pharmaceutical-sponsored clinical trials. The Philippines is actively training and building new research teams all over the country, and trials are being conducted beyond the major cities of Manila, Cebu, Davao, and Iloilo, in provincial areas such as Cavite, Baguio, Bataan, Batangas, Pampanga, and Lucena. These areas have physicians with good medical backgrounds and solid practice experience, and can provide a growing talent pool for clinical research and additional patients outside the hospital setting.

REFERENCES

[1] Department of Health, Republic of the Philippines. Mortality: ten leading (10) causes, number and rate/100,000 population, Philippines 5-year average (2000–2004) and 2005.

[2] Citeline Trial Trove July 2010.

CHAPTER 12

Clinical Trials in the Middle East and North Africa

Mira Serhal
American University of Beirut, Beirut, Lebanon

Contents

12.1. INTRODUCTION

The increasing demand for new patient populations has led pharmaceutical companies to several emerging markets in recent years, from Eastern Europe to China and India. During this quest for new regions, the Middle East and North Africa (MENA) has been quietly developing a strong infrastructure, powered by highly qualified human resources and supportive governments. Today, the region stands ready to embrace a new era in clinical research activity that expands from regularly conducted Phase IV and bioequivalence studies, to the more interventional Phase II and III

Global Clinical Trials
ISBN 978-0-12-381537-8, Doi:10.1016/B978-0-12-381537-8.10012-3

clinical trials. With a disease profile that very much resembles the industrial world, the MENA is also ready to receive trials from a diverse pool of therapeutic indications.

12.1.1. MENA Background

The MENA is constituted of 21 Arab states: Algeria, Bahrain, Djibouti, Egypt, Jordan, Iraq, Kuwait, Lebanon, Libya, Mauritania, Morocco, Oman, Palestine, Qatar, Saudi Arabia, Somalia, Sudan, Syria, Tunisia, United Arab Emirates (UAE), and Yemen. Stretching from the Red Sea and over the Mediterranean, to the Atlantic Ocean the MENA comprises one of the most heterogeneous regions in terms of economic growth, clinical research activity, and disease prevalence. The MENA population is currently estimated at 417 million throughout 21 countries. Generally characterized by a young demographic profile, where 36 percent of the population is below 15 years old, the MENA is also noted for active population growth rates and average life expectancy of over 70 years [1]. Almost 20 percent of the MENA population resides in Egypt, whose population counts a total of 80 million, of whom 20 million live in Cairo. The population density is much less in other major cities of the region. Saudi Arabia currently has a population of 27 million, of whom 4 million live in Riyadh. In general, the major cities of each country have a higher population density than rural areas.

12.1.2. Epidemiology and Geography

The demographic profile of the region shows similar endemicity of a number of diseases.

Diabetes mellitus (DM) is officially classified by international associations as an epidemic in the MENA region, which contains five of the 10 countries with the highest DM prevalence worldwide. The high DM prevalence is concentrated in the Gulf Cooperation Council (GCC) countries of the MENA, where 18.7 percent of UAE's population are DM patients, followed by Saudi Arabia (16.8%), Bahrain (15.4%), Kuwait (14.6%), and Oman (13.4%). DM prevalence in most other MENA countries falls around the 10 percent range (e.g. Jordan, Egypt, Iraq). Genetic factors, together with lifestyle, constitute the underlying cause of this exploding epidemic. According to the International Diabetes Federation, 24.5 million diabetic patients currently reside in the MENA, and this population is expected to double by 2025.

Cardiovascular diseases (CVDs) are another highly prevalent disease in the MENA countries, where multiple CVD risk factors (DM, tobacco, diet, obesity, and physical inactivity) contribute to the increasing prevalence. The prevalence of hypertension, for example, averages 29 percent and affects 125 million individuals. Of all MENA countries, the prevalence of hypertension is highest in Iran (43.6%), followed by Kuwait (38.6%) and Iraq (37.6%). Hypercholesterolemia and hyperlipidemia are also highly prevalent; however, there are no official statistics on the prevalence figures of other CVD factors.

Hepatitis is particularly recognized for its high prevalence in Egypt, where prevalence is estimated at 6 million chronic hepatitis C virus (HCV) patients [1]. In terms of hepatitis B virus (HBV), countries of the region vary between intermediate endemicity in Tunisia and Morocco, and high endemicity in Saudi Arabia, Oman, Jordan, and Yemen [2]. Hepatitis A virus (HAV) rates have declined significantly in the MENA, but endemicity in Saudi Arabia and Egypt remains in its intermediate stage [3].

Hemoglobin disorders account for another highly prevalent disease in the MENA. Beta-thalassemia is prevalent in Mediterranean countries of the MENA (Syria, Lebanon, Jordan, Egypt, Tunisia, and Morocco). Sickle cell anemia is also common in Saudi Arabia. In addition to these, hemophilia appears to be prevalent in Iran, Syria, Lebanon, and Egypt.

Cancer has also achieved significant prevalence rates in the MENA region. Though present rates remain less than industrial countries, cancer prevalence is increasing in the MENA region. Cancer malignancy affects 60–216 per 100,000 of the population, with the highest prevalence in Lebanon [4]. Lung cancer and liver cancer rates are particularly on the rise. Stomach and hepatic cancers also show significant prevalence rates.

Traveler's diarrhea constitutes a common infection that affects many visitors to the MENA, particularly tourists traveling to Egypt all year round, pilgrims to Saudi Arabia in the Hajj season, and immigrants returning to Lebanon in the summer time.

The high prevalence in consanguineous marriages within the MENA leads to an increased prevalence of associated diseases and genetically related metabolic disorders.

The official language in all MENA countries is Arabic. In clinical research, the informed consent form (and any other patient-related document, such as patient diaries) therefore needs to be translated into Arabic, and one translation would be sufficient to submit to all MENA countries. In French-speaking countries (Algeria, Morocco, and Tunisia), an additional

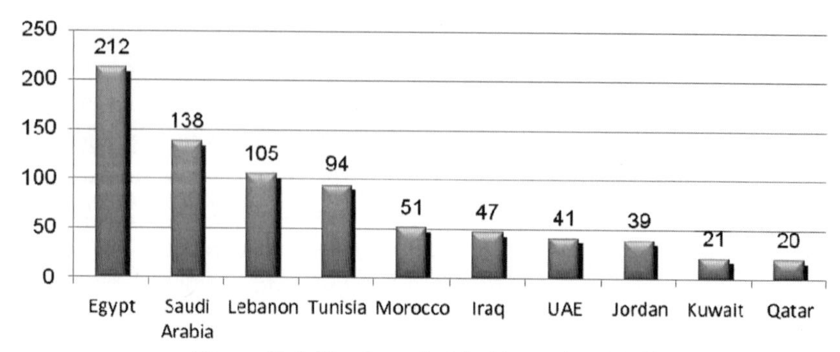

Figure 12.1 Number of trials (November 2010)

translation into French is required, which makes a total of two language translations in a total of 21 countries.

12.2. CLINICAL TRIAL STATUS

The number of clinical trials conducted in the MENA has been increasing in the past few years. The official documentation of clinical trials conducted in the MENA region is published at www.clinicaltrials.gov, as shown in Figure 12.1. In addition to these trials, it is estimated that another 500 investigator-initiated trials (IITs) are conducted throughout the MENA.

Until a few years ago, the MENA region was mainly a location for observational trials, usually for marketing purposes. Recently, however, a trend towards Phase II and III trials has emerged. This is partially due to the presence of treatment-naïve patients, but also due to largely attractive costs. In fact, the North African region is believed to be 30 percent less expensive than the European market, while the remainder of the Middle East is 15–20 percent less expensive. The increasing competition in this virgin market, evidenced by the inauguration of a number of local and international contract research organizations (CROs), is expected to sink the costs further, making the MENA region a more competitive market in the near future.

12.2.1. Contract Research Organizations

The MENA region has been occupied by small local CROs for several years, offering services mostly to the pharmaceutical industry within the region, with scarce experience among international pharmaceuticals and biotechnology companies. Most of the local CROs are family businesses, operating as sole proprietorship or limited liability companies. In addition, these CROs have been initiated by people who gained experience in the industry, either having worked for other CROs in Europe and the USA, or having worked in clinical research roles for multinational pharmaceuticals in the MENA. In either case, present CROs offer local clinical research activities in their home country, three of which currently provide regional coverage. The rate of regional coverage is expected to grow as more local CROs seek to expand to several MENA countries, in order to gain competitive advantage and increase market share.

In terms of experience, recent trends of increasing research activity have naturally led to the development of infrastructure and experience of several local CROs, while also attracting a number of international CROs. Today, four international CROs and 14 local CROs are known to have offices in the MENA region. Other international CROs with no official presence in the region work in the MENA by subcontracting the local CROs. The services performed by present CROs are mainly focused on site management, study management, clinical monitoring, and logistic support, in terms of drug import and distribution.

It is expected that more local CROs will continue to emerge in the MENA, as more experienced professionals create spin-offs from the present CROs. International CROs shall unhurriedly continue to subcontract the local players, while they evaluate the market demand for this region and the best strategy for expanding, either by employing their own staff or by acquiring one of the local CROs.

12.2.2. Logistics

In line with emerging CROs, ancillary services in the MENA are also on the rise. In particular, a number of multinational transport companies engage in the shipment of blood samples to central laboratories, and are known to provide prominent logistic support as per international standards and sponsor requirements. Custom clearance and drug release are also offered by such international couriers, with custom duties considered low in comparison to other regions of the world. Once the study drug has been imported and released, specialized, monitored drug depots are also available for drug storage and distribution, providing the logistic support necessary for the smooth conduct of clinical trials.

Local clinical trial insurance is normally provided for regional Phase IV trials. For earlier stages, a sponsor's global insurance is normally used to cover the MENA countries included in the study, and is generally accepted during the regulatory and institutional approval processes. In such cases, local insurance brokers may adapt the global insurance policy to local requirements where required (e.g. Egypt, Turkey, Jordan).

Interactive voice response systems (IVRS) and electronic data capture (EDC) services are not provided locally, also owing to low demand, where MENA sites usually comprise a part of international trials. In such cases, IVRS and EDC services are normally provided by global CROs and used locally by clinical trial staff.

12.2.3. Clinical Trial Staff

The modest volume of clinical research in the MENA implies a limited volume of experienced local human resources in the field of clinical research. In other words, clinical research staff are not widely available in the MENA. Therefore, new clinical research associates (CRAs) and study nurses need to be trained in order to acquire the necessary skills and good clinical practice (GCP) compliance necessary to conduct international trials. Clinical trial staff are initially highly qualified in the MENA, with CRAs having degrees in pharmacy, life sciences, nursing, and even medicine (as in the case of Egypt). In addition to such qualifications, CRAs receive International Conference on Harmonisation (ICH)-GCP training, and many are certified by international organizations, such as the Association of Clinical Research Professionals (ACRP), the Institute of Clinical Research (ICR), and others. CRA experience varies among countries and therapeutic areas, with four years currently being the average CRA experience. Study nurses are also well trained at university medical hospitals, and are familiar with clinical trial regulations. A growing number of university hospitals with well-established clinical research units currently provide formal training to their nurses. Site management organizations do not yet exist in the MENA, but are expected to emerge as the clinical research market grows.

12.2.4. Site

Clinical research conduct is generally restricted to governmental university hospitals in the MENA, especially in Phase II and III clinical trials. At present, there is no certification process under which sites may obtain licensure to conduct clinical trials. Health authorities in different countries are regularly revising current regulations, and may authorize the participation of private medical institutions in clinical research practice.

Investigator certification is not a requirement either; the investigator's qualifications to conduct a clinical trial are currently evaluated as part of the institutional review and approval process of a new protocol.

Site and investigator contracts may be combined or signed separately, depending on site requirements and sponsor preference. Both negotiation models are normally acceptable.

12.3. INFRASTRUCTURE AND SYSTEM OF HEALTHCARE

The healthcare infrastructure varies widely from one MENA region to another.

GCC countries are recognized for huge facilities, with the latest technology and equipment [e.g. positron emission tomography (PET), magnetic resonance imaging (MRI), computed tomography (CT), advanced laboratory equipment]. The major university hospitals in these countries offer high-quality medical services through well-educated local physicians, while also employing foreign physicians from the USA and Europe, in an attempt to facilitate patient access to medical care. The current physician to patient ratio is 18/10,000 in most GCC countries, with the exception of Qatar and Bahrain, which have achieved a 27/10,000 ratio [1]. Many of the GCC institutions have also acquired Joint Commission International (JCI) accreditation, and continue to pursue recognition through other international accrediting bodies, given the governmental support for better healthcare systems.

Levant countries (Lebanon, Syria, Jordan) have smaller facilities, but also include advanced healthcare technology for the conduct of clinical research. These countries are recognized for reputable healthcare services in the region, given that practicing physicians often receive their specialized education in the USA or Western Europe, where they participate in clinical trials as (sub)investigators and acquire clinical research experience in compliance with GCP standards. The physician to patient ratio is the highest in this part of the region, with Lebanon having 29 physicians per 10,000 patients, followed by Jordan with 25 per 10,000 patients.

In North Africa, large healthcare facilities that are capable of conducting clinical research are restricted to the main cities of Egypt, Tunisia, Algeria, and Morocco, where qualified physicians bring their education and expertise from France and the surrounding European Union countries back to these once French territories. Of all North African countries, Egypt has the highest physician ratio (25 per 10,000), while the French-speaking areas have a much lower count (estimated at 9 per 10,000). These numbers are even lower in Libya, Yemen, and Sudan, where very scarce clinical research activity is reported.

The cost of medical care in GCC countries is generally higher than Levant or North African countries. Compared to industrial countries, however, all costs for medical procedures are considered intermediate, being at least 25 percent lower.

12.3.1. Regulatory Authority

Each MENA country has its own regulatory structure, and the regulations among MENA countries are totally independent. In terms of compliance, all countries generally comply with ICH-GCP standards as a golden rule. The extent to which governmental oversight is implemented varies from one country to the other. At present, GCP training is not mandatory by any regulatory authority, but is a topic of discussion within a number of regulatory bodies. In the interest of ensuring the quality of their clinical trials, a number of pharmaceutical companies and CROs currently sponsor GCP training for investigators and clinical research staff. Timelines also vary from one MENA country to another. Table 12.1 summarizes the timelines and process in the MENA.

12.3.1.1. Egypt

The Ministry of Health in Egypt is the official regulatory authority responsible for approval of clinical trials in Egypt. According to the latest updates of the Ministry of Health guidelines in Egypt, a sequential process is employed, where the approval of at least one independent ethics committee (IEC) is required prior to submission to the Ministry.

Ethics Committee Submission

Each ethics committee may request additional documents (other than the ones listed below), and it is the responsibility of the CRA to ensure that all required documents are collected from the sponsor and from the investigative site and submitted to the IEC.

The submission deadline is generally one to two weeks before the targeted meeting date.

The required documents include, but are not limited to the following:

- study protocol: final version plus amendments (if applicable)
- investigator's brochure
- subject informed consent form (Arabic; English is not required for submission)

Table 12.1 Timelines and process in the Middle East and North Africa (MENA)

Country	Time to IEC approval*	Time to MoH approval**	MoH/IEC submissions†	Time to import license‡
Lebanon	2–6 weeks	Not required	N/A	1–3 weeks
Syria	4 weeks	1 month	Sequential	2 weeks
KSA	4–6 weeks	N/A	Sequential	Institutional
Morocco	4–6 weeks	N/A	Sequential	2 weeks
Egypt	2–3 months	2 months	Sequential	2–4 weeks
Jordan	Institutional at JFDA-registered hospitals	2 months	Sequential	Included in MoH approval
Iran	4–6 weeks	1 month	Parallel	2 weeks
UAE	4–6 weeks	1–2 months	Parallel	2 weeks
Turkey	8 weeks	8 weeks	Sequential	10 days

KSA: Kingdom of Saudi Arabia; JFDA: Jordanian Food and Drug Administration; N/A: not applicable.
* Average time required to obtain the approval for a clinical trial from the Independent Ethics Committee (IEC) of a given site per each country.
** Average time required to obtain approval from the Ministry of Health (MoH) of each MENA country, for a given clinical trial.
†Order of IEC and MoH submissions; in case of sequential submissions, the IEC approval is required before submission to MoH. Parallel submissions may progress simultaneously.
‡Average timeline required to obtain an import license for the investigational medicinal product (IMP).
Source: www.clinicaltrials.gov

- curricula vitae (CVs) of participating investigators
- any other patient-related materials (e.g. questionnaire, diary cards)
- insurance policy.

Approval timelines differ for each ethics committee, but are generally two to six weeks.

Regulatory Process

After obtaining ethics committee approvals from all participating sites in Egypt, a Ministry of Health submission package is prepared by the CRA.

Submission to Ministry of Health

The list of documents provided below is the minimum requirement for initiation of the review process. Additional documents may be requested by the committee on a case-by-case basis. A review will not be initiated until all the requested documents are provided at the time of submission unless an exemption has been made by prior communication with the chairperson of the committee.

Submission Procedures:

Procedure for Obtaining the Necessary Documents Required for the Proposal to the Clinical Research Scientific and Ethical Committee

- The researcher (companies, government departments, universities and scientific institutes) applies for the necessary documents and the conditions required for the research studies through an authorized person (with a certified authorization letter from the researcher).
- The authorized person will fill in the form (Application Form for Protocol Submission/Review Code No. FQP-RHD-007/01), completing all fields in full details and arranging them in the same order as in the folder provided.

Required documents: The person authorized by the researcher will provide the necessary documents for the clinical Research Scientific and Ethical Committee, including the following:

- a letter from the entity that made the research requesting a scientific and ethical review by the clinical research scientific and ethical committee (original certified letter plus 10 copies)
- study protocol with the last amendments (10 copies)
- research summary in Arabic and English (10 copies)
- investigator's brochure (10 copies)
- informed consent in Arabic and English (10 copies)
- case report form (CRF) (10 copies)
- insurance policy for trials within the Arab Republic of Egypt detailing the name of the study and the number of persons subject to trial (valid for at least six months from the date of application) (10 copies)
- approval of sites where the study will be conducted (certified copy) (10 copies)
- approval of the ethics committees of all the sites where the study will be conducted (if available) (10 copies)
- CVs of the researchers (signed and dated) (10 copies)
- the quantities of drugs expected to be used in the study as well as the scientific name and the effective ingredient all detailed in a separate certified letter submitted by the researcher
- any additional documents related to the study or annexed to the protocol (ID of persons subject to trial, advertisements for the recruitment of subjects, etc.)
- a check for the amount of 10.000 LE in Arabic language addressed to the Fund for the Service Improvement, the Central Directorate for Research and Health Development – Research item.

- CD with all the necessary documents required (including Application Form for Protocol Submission/Review) (four copies).

Procedure for Receipt of the Necessary Documents and its Time Schedule

The official employee in charge has the right not to receive the study documents in the case they were incomplete or the recorded data were invalid, and a grace period of 15 days is given to complete the documents, otherwise the study registration would be considered annulled.

The official employee in charge will deliver the form FQP-RHD-007/14 stamped with the directorate seal to the researcher. The research receipt entails the number of copies and the name of the study in addition to the date of receipt, and is delivered to the researcher only upon the completion of all documents.

Sixty working days are calculated from date of delivering the research receipt form for presentation to the Scientific Committee and an opinion is given in accordance with Ministerial Decree No. 95 for the year 2005 as regards each of the new research and the research being amended. As for the research, which is renewed every year, counting starts 30 working days (30 days) before the actual renewal date in order to begin the study renewal process and a maximum of 15 working days (15 days) after the deadline for the completion of the renewal process.

Procedure of Following Up the Committee Decisions

- In the case of decision to approve:
  The following conditions are the included with the approval according to the type of the study:
  - The Commission has the right to withdraw the approval if not provided with a report on the research every three months.
  - The approval is valid for one year from its date.
  - In case of any serious side-effects, the committee must be informed within 24 hours.
  - No research results must be published before their being first presented to the Committee.
  - Any human samples from persons subject to study are not to travel outside the Arab Republic of Egypt except after obtaining the approval of national security, the security office of the Ministry of Foreign Affairs (which is an independent body from the Ministry of Health).
- In the event of decision to approve with amendments:
  - The proposed amendments are attached to committee decision and handed to the researcher who is supposed to respond to the Committee within 30 days from the date of receipt of the decision.

    - A copy is sent to the company by mail, fax, and e-mail, and recorded in form FQP-RHD-007/12.
- In the event of the decision to postpone:
    - The reasons for the delay are attached to the decision and handed to the researcher, who is supposed to respond to the Committee within 60 days from date of receipt of the decision so as to take a decision on the study or reject it.
- In the event of the decision to refuse the study:
    - The reasons for refusing the study are attached to the committee decision and delivered to the researcher by mail, fax, and e-mail, and recorded in the registration form Mail FQP-RHD-007/12.
- In the event of termination of the study:
    - The reasons for terminating the study are attached to the committee decision and delivered to the researcher by mail, fax, and e-mail, and recorded in the registration form Mail FQP-RHD-007/12.
    - In case of terminating the study by the researcher, the Scientific and Ethical Committee will be informed within 15 days from the date of termination and the reasons completely attached with the decision.
- In the case of a decision to suspend the study:
    - The reasons for suspension are attached to the decision and the researcher is informed by mail, fax, and e-mail, and the decision recorded in the registration form Mail FQP-RHD-007/12.
    - In the event of the study being suspended by the researcher the Scientific and Ethical Committee will be informed within 15 days from the date of suspension and reasons completely attached with the decision; otherwise the study would be considered annulled and will be closed.

Submission of Amendments:

- a letter from the researcher to the Research Scientific and Ethical Committee requesting an amendment (original certified copy and 10 copies)
- a comparison table showing the amendment and the difference between the original and the amended forms with demonstration of modified points (10 copies)
- all documents that were actually subject to amendment (10 copies)
- a CD with all the documents that were subject to amendment (four copies)

- a check with the amount of 10.000 LE in Arabic language addressed to the Fund for the Service Improvement, the Central Directorate for Research and Health Development – Research item.

Submission of Quarterly Reports: A quarterly report (every three months) should be submitted to the Ministry of Health stating study progress. In case of failure to submit progress reports, the study is to be suspended and the problem endorsed to the Scientific Committee to decide on these studies. A letter would be sent to the researcher inviting him or her to send a reply within a week from that date.

Submission of Serious Adverse Events: Serious adverse events (SAEs) should be notified within 24 hours of being diagnosed on fax no. +2 279 47 369 or +2 279 50 678 and by e-mail: rhd@mohp.gov.eg, or to National REC as an initial report, followed by follow-up reports.

A provisional report on the side-effects from all studies is to be documented and submitted (hard and electronic copies) by the researcher within 24 hours of their appearance; this is to be followed by patient follow-up reports. Follow-up reports can be sent together with the quarterly reports.

Annual Renewal for Study Approval:

- a letter from the researcher to the Research Scientific and Ethical Committee requesting a follow-up approval (original certified copy and four copies)
- the last copy of the study (and its amendments, if any) (two copies)
- renewal of the insurance policy as well as approvals for the sites where the study is to be conducted
- a copy of the last periodic follow-up report showing the number of patients subject to trial and the quantities of drugs used in the study as well as a statement on the location of the study and the resulting side-effects (four copies)
- a check for the amount of 10.000 LE in Arabic language addressed to the Fund for the Service Improvement, the Central Directorate for Research and Health Development – Research item
- a CD with all the documents that were subject to amendment (including the Application Form for Protocol Submission/Review).

Submission Dates and Location of Office: Study protocol submission packages are only accepted on Sundays and Mondays of each week from 9 am until 3 pm. Submission packages are accepted by Coordinator of Research Administration of Egyptian Ministry of Health, Floor 7, Room 125.

Translations: The following documents must be translated into Arabic for Ministry of Health submission:

- protocol summary
- informed consent form
- any patient related forms.

Labeling of an Investigational New Drug

Labels can be in either English or Arabic, if the labels will not be seen by the patient. They should be provided by the sponsor. If the labels will be seen by the patient, then they must be provided in Arabic.

The reference for any detail on the label for the investigational product is the approved study protocol. It must include:

- sponsor's name, address, and telephone number
- dosage form, route of administration, quantity of dosage units (e.g. volume or number of tablets), and name of the active ingredient or in case of blinded trials code of the medication, in open trials strength/ potency
- batch/lot number or the code that is used to identify the contents and packaging operation
- protocol number
- investigator's name
- caution statement: "For clinical trial use only"
- dosage and administration (reference should be made to the protocol or another document prepared according to the protocol)
- manufacturer's address (the telephone number of the manufacturer is not required)
- storage conditions
- expiry date
- manufacturing date
- caution statement: "Keep out of reach and sight of children" for the products that will be taken home by the patient.

Patient ID number and visit number are not requirements (unless the medication is for compassionate use); however, to keep multinational labels harmonized, these can appear on the label and be left blank by the investigator.

Insurance Certificate

An insurance policy is required for trials within the Arab Republic of Egypt, detailing the name of the study and the number of persons subject to trial, and valid for at least six months from the date of application.

Contracts

The CRA will be responsible for completing the contract negotiations between the site and the sponsor.

A tripartite contract is usually required among the principal investigator, the institution, and the sponsor.

Specific regulations at Alexandria Clinical Research Center (CRC) for contracting:

- The CRC fee for each project is 40 percent of the investigator fees in cases where the CRC is the source of study and 30 percent if the principal investigator is the source of study.
- The fees of clinical research coordinators, pharmacists and nurses (affiliated to CRC) should be transferred directly to the CRC's account.

Drug Importation

An import license is issued within two to four weeks after Ministry of Health approval for the calculated amount needed for study conduct.

Required Documents for Import License Submission:

- cover letter directed to:
  - Ministry of Health
  - Central Administration for Pharmaceutical Affairs
  - Planning and Importation Approvals Department
- copy of Ministry of Health study approval letter
- study protocol (one copy)
- investigator's brochure (one copy)
- marketing authorization (if applicable)
- quotation on amount of study drug to be dispensed throughout study duration, based on study drug protocol dose and number of patients to be recruited
- pro forma invoice
- certificate of analysis [investigational medicinal product (IMP)/placebo]/certificate of origin
- list of participating sites in Egypt
- approval of sites where the study will be conducted.

Drug Conservation

To be performed according to provided guidelines of study drug storage and according to good manufacturing practice (GMP).

Sample Export

Any human samples from persons subject to study are not to travel outside the Arab Republic of Egypt except after obtaining the approval of national security. Further documents shall be required for obtaining an export license for exporting biological samples to be defined soon by the Ministry of Health.

12.3.1.2. Saudi Arabia

The Saudi Food and Drug Administration (SFDA) is the official regulatory authority for clinical trials in Saudi Arabia. Established since 2003, the SFDA employs many of the US FDA regulations with respect to clinical trials. As such, the SFDA requires IEC approval prior to receiving application for Ministry of Health approval.

Submission to Local IEC: Generally Required Documents

- study protocol and any approved amendments (English)
- an updated investigator's brochure (English)
- informed consent form (ICF) (Arabic and English)
- updated CVs of participating investigators and subinvestigators (English)
- CRF (English)
- written information to be provided to subjects (if applicable)
- information about payments and compensation available to subjects.

Approval differs between hospitals and depends on each hospital's local ethics committee.

Timeline: The usual time to obtain local IEC approval from each site is approximately four weeks.

Submission to the SFDA

- The online application form has to be completed and printed out, to be delivered by mail or by appointment meeting with the necessary documents required to the SFDA.
- A cover sheet for the application should contain the following information:
  - the name and contact details (e.g. address and telephone number) of the sponsor, the date of the application, and the name of the investigational new drug (IND)

  - identification of the phase or phases of the clinical investigation to be conducted
  - a commitment not to begin clinical investigations until the IND application is approved by the SFDA
  - a commitment that an institutional review board (IRB) will be responsible for the initial and continuing review and approval of each of the studies in the proposed clinical investigation and that the investigator will report to the IRB proposed changes in the research
  - a commitment to conduct the investigation in accordance with all other applicable regulatory requirements
  - the name and title of the person responsible for monitoring the conduct and progress of the clinical investigations
  - the name(s) and title(s) of the person(s) responsible for review and evaluation of information relevant to the safety of the drug
  - if a sponsor has transferred any obligations for the conduct of any clinical study to a CRO, a statement containing the name and address of the CRO, identification of the clinical study, and a listing of the obligations transferred; if all obligations have been transferred, a general statement of this transfer including a list of the specific obligations transferred must be submitted
  - the signature of the sponsor or the sponsor's authorized representative.
- The application form contains the following parts:
  - trial identification
  - sponsor responsible for the request and source of funding
  - information on IMP(s) being used in the trial: medicinal product being tested or used as a comparator
  - information on placebo
  - authorized site responsible in the community for the release for the IMP
  - general information on the trial
  - population of the trial
  - proposed clinical trial site concerned by this request
  - authority concerned by the request
  - signature and printed name of the applicant in the Gulf state.
- updated version of the protocol (English)
- updated investigator's brochure (English)
- consent form (Arabic and English), in addition to patient information and patient card, if any (Arabic and English)

- CRF (English)
- insurance certificate (fronting policy for the global insurer)
- updated CVs for the principal investigator and the subinvestigator (English) at the investigative sites
- site local IRB approval letter (Arabic or English). The submission could be done in parallel with the IEC submission but the approval will not be granted before submission of the IEC approval
- a copy of the agreements between the sponsor, the investigators/site, and CRO (Arabic or English)
- financial disclosures
- GMP certificate for the test product (to show that the test products have been manufactured under cGMP conditions) and its certificate of analysis and dissolution profile tables and charts (English)
- drug label (English), with a sentence added in Arabic (For clinical trial use only); labels to be translated into Arabic if the medication is to be given to the subject at home
- certificate of analysis
- samples of imported drugs (IMP and comparator/standard of care if applicable)
- submission fees: 15,000 SR per site (US $4000).

Regulatory Approval Process

- The protocol should be submitted to the SFDA any time prior to meetings.
- There are two clinical trials unit committees:
  - the Clinical Trials Committee that meets once a week
  - the Clinical Trials Evaluation Committee that meets regularly
  - the authorized parties for submission: Sponsor: Saudi Medical Director or Saudi country manager
  - principal investigator or institution
  - licensed CRO.

Timelines:

- Validation of the submission file: 10 days
- if the drug is already on the market: 30 days
- for an IND: 120 days.

Labeling of an Investigational New Drug

- Drug label (English) with a sentence added in Arabic (For clinical trial use only); labels to be translated into Arabic if the medication is to be given to the subject at home.

Insurance Certificate

International insurance is demanded.

Contracts

Negotiated and executed only after IRB/IEC approvals.

Drug Importation

There is no need for an import license to import the drug. The SFDA grants approval on the protocol and drug importation at the same time.

Drug Conservation

As per protocol requirement.

Safety Reporting

- In cases where a SUSAR occurred at a site, either the investigator or the sponsor should submit to the SFDA, but the information should not be submitted twice.
- Method of submission of safety reports to the SFDA: Cases can be submitted manually or by e-mail using Council for International Organizations of Medical Sciences (CIOMS) forms. If the study was blinded and/or the product is not approved by the SFDA then reports should be submitted to Clinical Trials Unit. Unblinded approved products can be submitted to the National Pharmacovigilance Center or Clinical Trials Unit.
- Frequency of submission: Reports of SAEs should be submitted within 15 calendar days; reported orally or by e-mail within seven days and then a written report should be submitted within a further eight days.

Any adverse event, adverse reaction, or unexpected adverse reaction that:

- results in death
- is life-threatening
- requires hospitalization or prolongation of existing hospitalization
- results in persistent or significant disability or incapacity
- consists of a congenital anomaly or birth defect

must be reported.

Important medical events that may not be immediately life-threatening or result in death or hospitalization but may jeopardize the patient or may require intervention to prevent one of the other outcomes listed above should also be considered serious.

The sponsor shall ensure that all relevant information about a suspected unexpected serious adverse reaction (SUSAR) that occurs during the course of a clinical trial in Saudi Arabia and is fatal or life-threatening is reported as soon as possible to the SFDA and the relevant ethics committee. This needs to be done no later than seven days after the Sponsor was first aware of the reaction. Any additional relevant information should be sent within eight days of the report.

Frequency of Submission of Foreign SAEs: For studies conducted outside Saudi Arabia unexpected or unlisted SAEs for registered products should be submitted within 15 days. If SAEs are known for the given product then this should be a part of the periodic safety update report.

Biological Samples Exportation

- Approvals are required.
- Process: The following documents are required to have the SFDA approval for the export of samples:
  - a letter from the site where the study is conducted
  - registration of the trial at the SFDA
  - provision of a copy of the part of the protocol which states that samples are to be exported outside Saudi Arabia
  - approval from the IRB for exporting the samples and stating the reason for this.

One license for export of samples from the different sites is not enough.

12.3.1.3. Lebanon

The Lebanese Ministry of Health is the official regulatory body for clinical trials in Lebanon. At present, the Lebanese Ministry of Health does not require any submission for the conduct of a new trial. As such, the approval of clinical trials is strictly institutional, and normally takes up to six weeks to obtain approval.

The required documents for IEC approval include, but are not limited to, the following:

- study protocol: final version plus amendments (if applicable)
- investigator's brochure
- subject's informed consent form (Arabic; English is not required for submission)
- CVs of participating investigators
- any other patient-related materials (e.g. questionnaire, diary cards) insurance policy.

The import license for new medicinal products is processed with the Ministry of Health and takes up to two weeks. Required documents include:

- a letter from the principal investigator, describing the need for the study drug, in reference to the conducted clinical trial
- a letter from the pharmacist on site, presented for the same purpose
- certificate of analysis
- commercial invoice.

Timelines: In general, timelines in Lebanon are the shortest for initiating a clinical trial.

12.3.1.4. Jordan

The Jordanian Food and Drug Administration (JFDA) is the official regulatory authority for clinical trials in Jordan, and has implemented the law for clinical studies since 2001. The JFDA clinical studies law clearly identifies two types of clinical trials: (a) those performed for non-therapeutic purposes on health volunteers; and (b) therapeutic clinical studies performed on sick and healthy individuals. In case of the latter, IEC approval is required prior to submission of the clinical trial application.

Submission to Local Site Ethics Committee

- study protocol and any approved amendments (English)
- protocol approval letters from any authorities (if any) (English)
- updated investigator's brochure (English)
- any safety updates (English)
- ICF (Arabic and English), in addition to patient information and patient card, if any (Arabic and English)
- updated CVs for the principal investigators and the subinvestigators (English)
- CRF (English).

Approval differs between hospitals and depends on each hospital local IEC.

Submission to the JFDA (Regulatory)

- Clinical trial application for approval plus the JFDA fees (Arabic form can be translated for the study sponsor)
- protocol approval letters from any authorities (if any) (English)
- updated investigator's brochure (English)
- any safety updates (English)
- ICF (Arabic and English), in addition to patient information and patient card, if any (Arabic and English)

- CRF (English)
- a local insurance certificate (fronting policy for the global insurer)/global insurance policy (English)
- updated CVs for the principal investigator and the subinvestigator (English)
- site local IRB approval letter (Arabic or English)
- a copy of the agreements between the sponsor, investigators/site, and CRO (Arabic or English)
- laboratory accreditations for clinical studies, for the diagnostic laboratory (medical laboratory) (Arabic or English)
- a commitment letter that the biological samples will be used as stated in the protocol ONLY (Arabic or English) (this would be stated in the ICF upon adaptation to the local requirements if it was not originally stated in the protocol)
- GMP certificate for the test product (to show that the test products have been manufactured under cGMP conditions) and its certificate of analysis and dissolution profile tables and charts (English)
- GLP certificate for the central laboratories
- drug label (Arabic and English)
- study drug outer pack: a photocopy or/a photograph would be fine
- a commitment letter to say that the active and inactive materials used in the finished dosage form (test product) are within international standards and requirements and thus not harmful or damaging to the patients (safety declaration) (English)
- presentation (PowerPoint presentation) for the study and any previous studies conducted on the drug to be presented by the investigator or the study sponsor at the Clinical Trials Committee meeting (English).

Regulatory Approval Process

- Protocol development and internal approvals (completed by sponsor) take place.
- Sponsor review and approval, when completed, then adaptation of ICF to local requirements and translation will take a period of two weeks.
- Protocol should be submitted to the IRB at least two weeks prior to IRB meetings and the submissions and approvals process will take four to five weeks.
- Protocol should be submitted to the JFDA at two to three weeks prior to meetings and the submissions and approvals process will take four to five weeks.
- The JFDA committee weekly meetings are held every Tuesday.

- Documents are required for ethics and regulatory submission (part a and b)
- The screening procedures will start eight weeks after the protocol is approved.

Regulatory Process

Clinical studies are divided into: (a) therapeutic clinical studies (any clinical study performed on sick or healthy volunteers) and (b) non-therapeutic clinical studies (any study performed on healthy volunteers in terms of effectiveness, kinetics, bioavailability, and bioequivalence).

(1) Clinical studies shall not be conducted unless the conducting authority has obtained an authorization from the Minister upon a recommendation from the Clinical Studies Committee pursuance of the provisions of this law.

(2) Clinical studies shall be conducted by any authority licensed in accordance with the provision of paragraph (1) of this article:

  i. Public and private hospitals, which possess technical potential to provide the required emergency and intensive care in addition to laboratory that carries out clinical tests.

  ii. University academic institutions specialized scientific research institutions and pharmaceutical manufacturing companies, which have the required technical potential in compliance with clause (1) of this paragraph. In case these entities do not have such potential, any of these entities may perform the clinical part of this study at any authorized hospital.

(3) Analyses on biological samples dedicated for clinical studies shall be made by approved laboratories which have the requirements necessary for conducting such analyses and assure they are accurate and precise.

Submission to the JFDA

Clinical studies shall not be performed on human beings unless with the individual's written approval (signed consent form) and after undergoing the medical tests necessary for his or her safety.

Any authority requesting permission to conduct clinical studies shall comply with the following:

- Prepare a protocol for the study to be conducted, provided that such protocol must include the scientific justifications for conducting the study in addition to any other details specified in this law.
- Sign an insurance contract with an insurance company having business within the Kingdom of Jordan to cover any damage sustained by this study

specially those related to the human beings undergoing such study, provided that cases where such contract can be concluded shall be defined with its terms and requirements under instructions to be issued by the Minister upon the recommendation of the Clinical Studies Committee.

Timelines: After attaining hospital ethics committee approval for the protocol, 10 copies of the protocol should be submitted to the JFDA. The Minister grants authorization to conduct clinical studies upon recommendation from the Clinical Studies Committee that the protocol is well written according to ICH-GCP and feasible within registered sites in Jordan. It takes around four weeks to obtain the approval.

Labeling of an Investigational New Drug

Drug labeling requirements:

- sponsor name and address
- study number
- lot number
- product name (if product is dispensed from clinic)
- dosage regimen (if product is administered by subject)
- quantity
- storage temperature/humidity (good storage conditions)
- expiry date
- retest date
- route of administration
- labels of an IND should include a sentence stating that this drug is for solely clinical studies (Arabic)
- the whole label would need be translated into Arabic in cases where patients will be using them at home, except that drugs to be administered to hospitalized patients will not require to be translated except for the following:
  - ICF
  - subject information sheet
  - diary card (if any)
  - written information to be given to study subjects.
  - any other document that will be used by the study patient.

Insurance Certificate

Insurance for volunteer patients must be provided by a company that has a working office in Jordan, but not necessarily local.

Contracts

Drug Importation

For drug import only an invoice is required to be sent to the JFDA. The JFDA will make sure the investigational drug matches the protocol and then will give approval.

12.3.1.5. United Arab Emirates

Comprised of seven Emirates, each UAE state has its own independent health authorities, which all report to the UAE Ministry of Health. As such, clinical research regulations vary from one state to another. The Health Authority of Abu Dhabi (HAAD) currently has the most organized regulations for clinical research; these clinical research regulations have undergone rigorous review and have been published on the HAAD website since July 2009. As such, the following HAAD regulations apply to Abu Dhabi:

- Clinical trial proposals do not need to be approved by HAAD, the regulatory agency. Proposals must be reviewed by the institution's research ethics committees for approval.
- Each institutional research ethics committee will have its own requirements and timelines. In general, approval may take between one and six months, depending on the meeting frequency of the IEC.
- Each institution is responsible for maintaining a research ethics committee and reporting to the regulatory agency. Investigators are not required to make submissions directly to the regulatory agency.

Timelines: Timelines are four to six months for study initiation, including IEC approval and drug import license.

The HAAD continues to apply reforms to the clinical research regulations, all in accordance with ICH-GCP guidelines.

12.3.1.6. Tunisia

The Direction de la Pharmacie et du Medicament (DPM) at the Tunisian Ministry of health is the regulatory authority responsible for the oversight of clinical research in Tunisia. According to the DPM, it is mandatory to obtain IEC approval prior to regulatory submission and authorization.

Ethics Committee Submission

Ethics committee submission should be made locally by the principal investigator for each site. The list and the number of copies submitted may vary from one committee to another. In addition, the time needed for obtaining the opinion varies from one committee to another. In practice, it is between one and two months. Classical documents are needed and include:

- study protocol: final version plus amendments (if applicable)
- investigator's brochure
- subject's ICF (Arabic and French required for submission)
- CVs of participating investigators
- any other patient-related materials (e.g. questionnaire, diary cards)
- insurance policy.

Regulatory Submission

Once ethics committee approval has been obtained, the sponsor or CRO may submit the application to the DPM for approval. The average timeline is 45 days for approval, after which a drug import license may be processed within two to four weeks. In addition to classical documents, the DPM requires translation of the protocol summary into French.

Drug Import

A referent pharmacist should be designated for all sites. Submission of import license is made following regulatory approval.

The following documents should be submitted to the DPM:

- an advanced pro forma invoice
- a request letter made by the referent pharmacist
- certificate of analysis
- a sample of drug label.

Timelines: The overall timeline for study set-up in Tunisia is four months, allowing Tunisia to rank among countries with the shortest timelines in the MENA.

12.3.1.7. Morocco

The DMP is the Ministry of Health division responsible for approving clinical trials in Morocco. One of three national/central ethics committees (CECs) in Morocco evaluates and approves clinical research proposals prior to submission to the DPM. CEC approval takes between six and eight weeks.

Central Ethics Committee requirements

- study protocol: final version plus amendments (if applicable)
- investigator's brochure

- subject's ICF (Arabic and French required for submission)
- CVs of participating investigators
- any other patient-related materials (e.g. questionnaire, diary cards)
- insurance policy.

Regulatory Submission

Once ethics committee approval has been obtained, the sponsor or CRO may submit the application to the DPM for approval. The average timeline is six to eight weeks for approval, after which a drug import license may be processed within two weeks. In addition to classical documents, the DMP requires translation of the protocol summary into French.

Drug Import

An import license request should be submitted by the sponsor or the sponsor's representative to the DMP once regulatory approval has been obtained. An import license will be needed at each shipment. Required documentation comprises:

- a pro forma invoice
- a brief description of imported kits (should be provided by sponsor).

It is mandatory that the investigational product be received by a pharmacist. There are no country-specific requirements regarding storage, which could be done by the investigator or any other designated person (coinvestigator, pharmacist, or other).

Timelines: The timeline for approval is two weeks. Similar to Tunisia, the overall timeline for study start-up in Morocco is up to four months, making the Maghreb a generally attractive region for clinical research.

12.4. PATIENT PROFILE

Patients in the MENA are generally motivated and compliant with physician instructions. The high density in major cities and availability of technology make access to patients relatively quick and easy. Increasing literacy rates also allow for better consenting of patients and better patient compliance. Payment of incentives is possible within the ethical frame. Many IECs currently require sponsors to disclose patient payments in the IEC submission dossiers, and evaluate both the amount and justification of the payments. Moreover, the MENA population is still untapped in terms of clinical research, which translates to a major pool of treatment-naïve patients for upcoming clinical trials. High birth rates also make the patient population ideal for pediatric trials in the MENA.

Some challenges with patient recruitment lie in the illiteracy of the elderly, and the difficulty of obtaining consent from female patients in rural areas. However, both challenges have been significantly declining and may be faced in very rare situations.

12.5. CHALLENGES AND OPPORTUNITIES

As in the case of every emerging market, the MENA presents a number of challenges. Culture is definitely an important challenge where the research culture is not yet well established among the population. Patient compliance may also be questioned during Muslim holidays and Ramadan, which also affects regulatory timelines and slows approval processes. Political instability is another challenge in the MENA, and represents a justifiable factor in deciding whether to run a clinical trial in the MENA or not.

However, the opportunities certainly outweigh the risks:

- The region has a considerable patient population for several investigational diseases.
- Governmental regulations are enforced but not very stringent. Governments also want to attract more clinical research into the region, and are implementing laws that shall contribute to better quality data, all in compliance with ICH-GCP and 21 CFR.
- Costs are lower in terms of investigator fees, hospital fees, and custom duties.
- The infrastructure is capable of accommodating clinical research needs.
- Investigators are educated, motivated, and capable of running trials and providing quality data.
- The clinical trial staff is available, qualified, and motivated.

REFERENCES

[1] World Health Organization (WHO). Regional Office for the Eastern Mediterranean, www.emro.who.int; 2008.
[2] André F. Hepatitis B epidemiology in Asia, the Middle East and Africa. Vaccine 2000;(Suppl. 1):S20–2.
[3] Tufenkeji H. Hepatitis A shifting epidemiology in the Middle East and Africa. Vaccine 2000;(Suppl. 1):S65–7.
[4] GLOBOCAN 2008.

CHAPTER 13

Clinical Trials in South Africa

Victor Strugo, Lynn Katsoulis, Havanakwavo Chikoto, Tracy Southwood, Marianne Coetzee
Triclinium Clinical Trial Project Management, Sandown, South Africa

Contents

13.1. INTRODUCTION

South Africa's long and distinguished history of medical research includes Max Theiller's 1951 Nobel Prize for his work on yellow fever and the first heart transplant by Christiaan Barnard in 1967. When clinical trials started to expand globally in the mid-1960s, the strong research ethos of

Global Clinical Trials
ISBN 978-0-12-381537-8, Doi:10.1016/B978-0-12-381537-8.10013-5

universities and the disproportionally high number of key opinion leaders in various medical fields attracted trials to the country. Since then, clinical trials have developed into a flourishing and well-regulated industry.

13.2. GEOGRAPHY

Located at the southern tip of Africa, South Africa's area is one-eighth the size of the USA and twice the size of France. The country is divided into nine provinces, in turn segmented into 52 metropolitan and rural municipal districts. Most pharmaceutical companies and contract research organizations (CROs) are located in Gauteng, the smallest province by area but most populous. It has 22.1 percent of the 2010 estimate of 49.9 million people [1], centered around the twin cities of Johannesburg, the country's economic powerhouse, and Pretoria, the administrative capital.

The term "Rainbow Nation" has been coined to describe South Africa's rich ethnic diversity, comprising African (79%), Caucasian (9%), mixed race (9%), and Indian/Asian (3%) groups [1]. Females comprise 51 percent of this still significantly treatment-naïve clinical research population. Children under the age of 15 years and adults older than 60 years represent 31 percent and 7.6 percent of the population, respectively. Migration is an important demographic influence with an estimated influx of 1.3 million Africans, mainly from neighboring countries, and the emigration of 440,000 Caucasians since 1996 [2].

There are 11 official languages [3]. English is ranked sixth in home usage but is nevertheless the language of business, politics, and modern communication media [4]. South Africa has the most extensive transport network of any African country with about 754,000 km of roads and 145 licensed airports with paved runways enabling easy access to all communities [5]. Most clinical trial sites for diseases common in the Western world tend to be concentrated around major cities with national airports (Figure 13.1). Many rural sites have also been developed, mostly geared to conduct trials in diseases associated with poverty such as human immunodeficiency virus (HIV)/acquired immunodeficiency syndrome (AIDS), tuberculosis (TB), and malaria. The country has several airlines. Intercity domestic flight durations are 1–2 hours. Johannesburg is also the African city with the largest number of regular commercial flights to other African countries.

The climate is Mediterranean in the south-west, temperate in the interior, subtropical in the north-east and arid in the north-west [6]. South Africa's

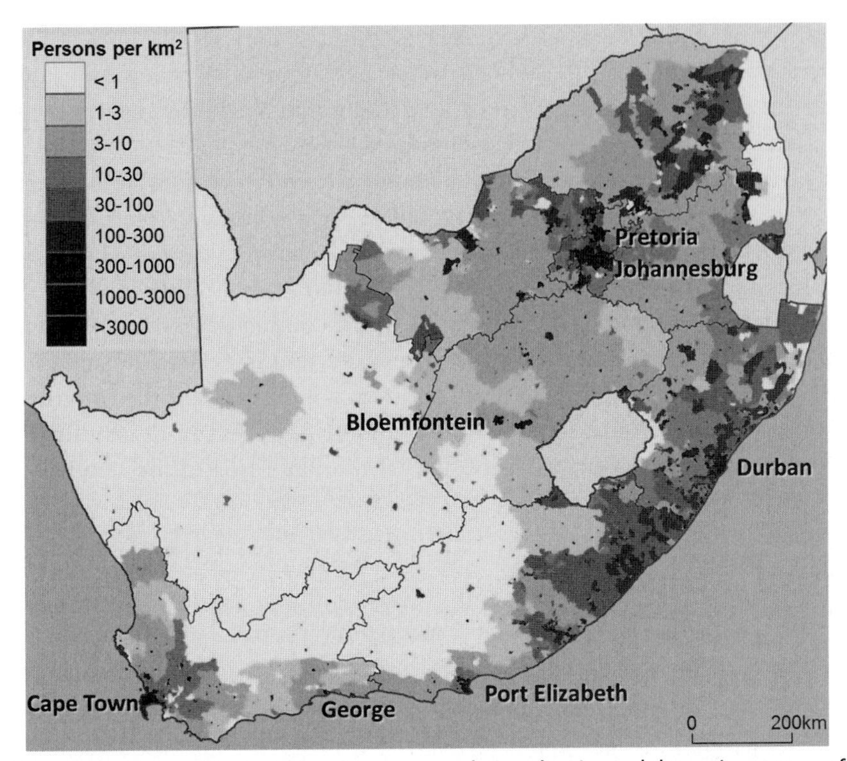

Figure 13.1 Map of South Africa showing population density and the major centers of clinical trial activity *(Source: Adapted from Statistics South Africa)*. (Please refer to color plate section)

inverse seasons relative to the northern hemisphere offer the possibility of year-round enrollment for multinational trials in seasonal indications.

13.3. HEALTHCARE SYSTEM AND INFRASTRUCTURE

Healthcare is provided through a dual public/private system [7], served by some 11,237 general practitioners and 8097 medical specialists and equivalent to 23 and 17 per 100,000 patients, respectively [8]. The management of state-funded public healthcare was reinvigorated by health ministerial changes in 2008 and again after the 2009 general election.

13.3.1. Public Healthcare

South Africa's public healthcare system provides free and low-cost essential healthcare to more than 41.5 million people, mostly from the lower income sector. The country is served by a range of primary to advanced tertiary healthcare facilities, with between 20 and over 300 facilities in each of the 52 districts [9]. Basic day-to-day healthcare is provided by mobile clinics, primary healthcare clinics, and community healthcare centers, with referral to specialized care when indicated.

The public sector healthcare pyramid currently provides 83,626 beds across a tiered system of general and specialist hospital superstructure, supporting an infrastructure of 3174 primary healthcare clinics, 927 mobile clinics and 296 community healthcare clinics (Figure 13.2) [9].

Trials in the public sector are typically conducted in hospitals with access to trained researchers and appropriate resources. Parallel approvals are required from their institutional ethics committee and the provincial authority that ensure that research funding fully exempts institutional resources from subsidizing procedures that are not part of the standard of care.

13.3.2. Private Healthcare

South Africa's well-developed and rapidly growing private healthcare system provides a standard of medical care comparable to advanced Western countries. Patients access healthcare mainly through medical insurance provided by 124 registered medical insurance companies that currently

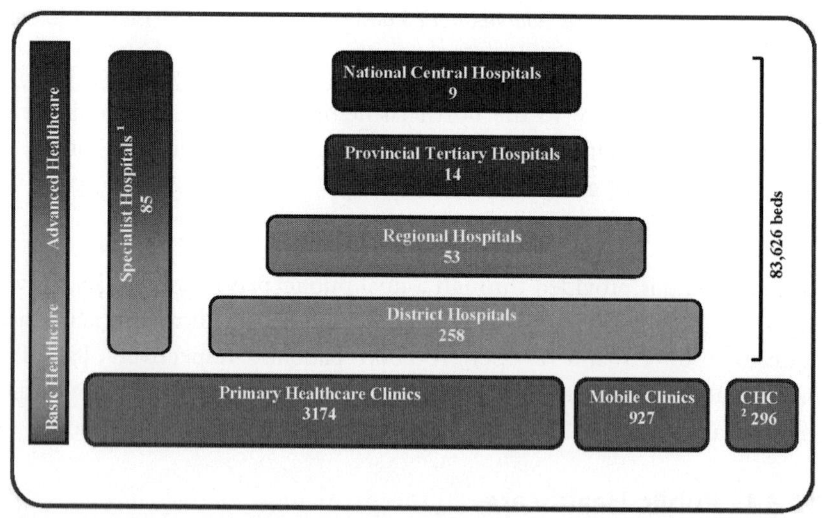

Figure 13.2 Structure and size of South Africa's public healthcare infrastructure [1]. Specialist hospitals include TB, psychiatric/mental health units, and other specialties [2]; CHC: community healthcare clinic *(Source: Adapted from District Health Barometer, 2008/09 [10])*. (Please refer to color plate section)

cover only 14 percent of the population [10]. Providers generally do not cover trial-specific treatment and diagnostic costs.

It is estimated that 60 percent of the country's health expenditure rests in the private sector [11]. The country has four major private hospital and clinic groups, whose 149 hospitals contain 28,226 beds including many high-care, cardiology, stroke, renal, spinal, psychiatric, rehabilitation, and other specialized units [12–15]. Research-experienced investigators, often with consulting rooms adjacent to hospitals that are equipped with modern diagnostic facilities, make this sector highly attractive for both outpatient and inpatient trials.

13.3.3. Clinical Trial Distribution Across Sectors

While no national statistics exist on comparative overall enrollment in public hospitals and private sites, a review of Food and Drug Association (FDA) inspections of South African sites shows that of 37 listed sites inspected, 16 (43%) were at public hospitals and 21 (57%) at private sites [16]. Whereas public sites may have the capacity to enroll larger patient numbers (especially in communicable diseases), private sites are more numerous and generally better organized. The national regulatory agency encourages sponsors to utilize public sector sites, as part of a long-term capacity-building objective. There is also a degree of cross-over cooperation: some private research sites and semi-private/academic partnerships tap into regional networks of primary care and community clinics for referral of patients who would otherwise not have access to participation in trials.

13.4. DISEASE PROFILE

South Africa's total burden of disease per capita has been estimated to be four times higher than that of developed countries and almost double that of some developing countries [17]. The most comprehensive prevalence data available are from Statistics South Africa, which collates the causes of death captured on death certificates. Figure 13.3 shows the percentage distribution of causes of deaths by groups classified according to International Classification of Diseases (ICD 10) codes. The top-ranking group of causes of death in 2007 was a group of infectious and parasitic diseases which accounted for a quarter of all deaths (25.4%), predominantly attributable to overt or underlying HIV and/or TB, followed by diseases of the respiratory and cardiovascular system, which together accounted for another quarter of

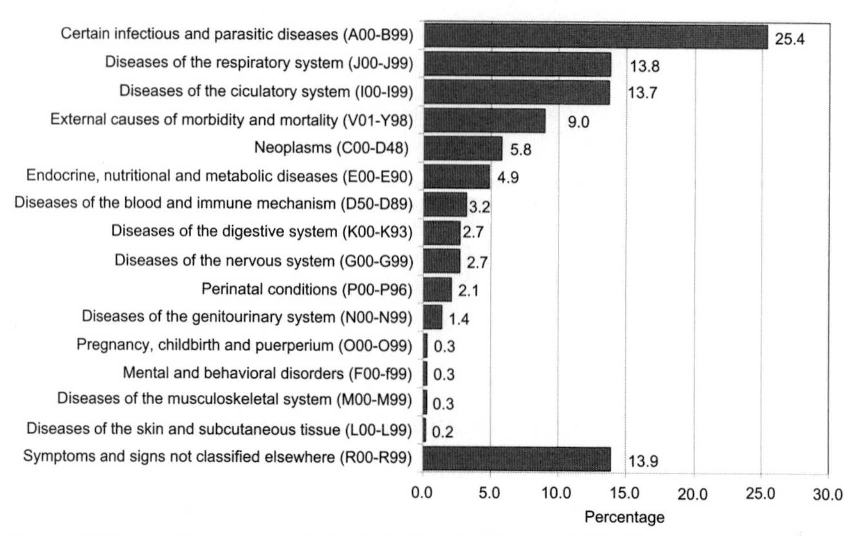

Figure 13.3 Leading causes of death in South Africa in 2007 *(Source: Statistics South Africa, 2009)*

the deaths (27.5%). Next highest are unnatural causes (mostly accidents and violence).

The heavy burden of HIV/AIDS is widely acknowledged; 5.24 million people are estimated to be living with HIV, a prevalence of 10.5 percent across all age groups [18], though this approximation is difficult to quantify owing to a tendency to report indicator conditions or co-morbidities on death certificates, rather than underlying (and especially stigmatized) causes [19].

The past 15 years have also seen increased incidences of non-communicable diseases such as cardiovascular disease, maturity-onset diabetes, cancer, and chronic lung disease in both rural and urban areas, with the highest incidences recorded in poor people living in urban settings [20]. A likely explanation is the rapid adoption of an increasingly Westernized lifestyle without commensurate improvements in diet and living conditions. The World Health Organization (WHO) estimates the incidence of these diseases as being double or treble that of developed countries [21].

13.5. HISTORY OF CLINICAL TRIALS IN SOUTH AFRICA

The establishment of the Medicines Control Council (MCC) in 1965 and the country's first international symposium on drug development in 1972 [22] both stimulated clinical research activity. The strong local generic industry also influenced the development of an independent contract research unit at the University of the Orange Free State in Bloemfontein, integrating admission facilities for healthy volunteer trials with drug bio-analytic facilities under the leadership of Professors Otto Müller and Hans Hundt. In 1975, it evolved into the Hoechst Research Unit for Basic Clinical Pharmacology, graduating from bioequivalence to Phase I trials. By 1978, its widespread reputation for early-phase excellence attracted many of the world's foremost experts to Pretoria for an international benchmark conference on Biologic Availability [23]. Now acquired by an international CRO, it has evolved into a full-service network with over 180 beds at three locations and broadened its competence to proof-of-concept trials.

In the earliest therapeutic trials in patients, most investigators were senior academics. From the early 1980s, an increasing number of physicians took the research experience acquired as registrars and consultants in academia into the private sector. This created self-funding private research sites as viable parallel resources to increasingly overloaded state hospitals.

The proliferation of medical departments within pharmaceutical companies during the 1980s and the subsequent advent of CROs was a reflection of steady growth in research volume. National registry figures since 2005 suggest that about 270 new trials start every year and an estimated 550 trials are active in the country at any given time across all phases and multiple therapeutic areas [24]. Many are part of large multinational preregistration trials conducted to support global marketing applications.

It is noteworthy that two clinical development programs were conducted entirely (trials, bioassay, pharmacovigilance, data management, and statistics) in South Africa during the 1990s. Rifapentine (Hoechst Marion Roussel) was approved by the FDA in 1998 [25] and Tryptorelin (Debiopharm) in 2001 [26]. These achievements indicate South African ability to generate data of acceptable quality to the most stringent regulatory authorities. This capacity is far beyond that of most emerging countries, the category in which South Africa is often misclassified.

13.6. REGULATION OF CLINICAL TRIALS

The Medicines and Related Substances Act, 1965 (Act 101 of 1965) made provision for the establishment of the MCC, a statutory body tasked with regulating pharmaceutical products in South Africa, advised by 11 technical committees, one of which is the Clinical Trials Committee (CTC), responsible for review of clinical trial applications. Owing to funding limitations and a finite national pool of expertise, CTC members and evaluators are contracted per review or task [27]. Administrative support for the MCC and its substructures is provided by the Medicine Regulatory Authority (MRA) cluster, headed by the Registrar of Medicines. The MRA encompasses four units, including an inspectorate [conducting good manufacturing practice (GMP) and good clinical practice (GCP) inspections] and the Clinical Trials Unit, which effectively functions as the CTC's secretariat and is the primary interface for all clinical trial applicants.

CTC evaluators review all trial applications involving unregistered medicines (i.e. Phases I–III) and also trials involving off-label indications or dosage regimens of marketed products. The three-section Clinical Trial Form 1 (CTF1) clinical trial application [28] consists of (a) administrative information about the trial, applicant, sponsor, and investigators; (b) condensed summaries of the protocol, product information (pharmacology, toxicology, clinical experience, and chemical manufacturing control); and (c) the applicant's own detailed analysis and justification of the key scientific, ethical, and safety aspects of the proposed research, as well as the motivation for conducting the research in South Africa. This third section, developed between 2001 and 2003 by a joint task force of CTC, MRA staff, and 12 industry members, is a well-inspired instrument intended as a best available surrogate for face-to-face hearings, for which no mechanism exists in South Africa owing to resource limitations. Section III is seen as the critical part of a successful application, but also the most common cause of delays and rejections if poorly compiled.

For ease of communication, MCC has a strong preference for clinical trial applicants to be locally based. Applications are completed in English. Formatting must comply with CTC specifications regarding dossier formatting, number of paper and electronic copies, signed sponsor and investigator declaration forms, and critical trial-specific documents (protocol, consent form, investigator's brochure, supporting literature, proof of trial insurance, certificate of analysis) [28].

The MCC does not certify sites. Instead, all applications must contain signed CVs and numerous other documents and summaries to demonstrate the applicant's due diligence in verifying the qualifications, experience, professional certification, malpractice insurance, available capacity, co-investigator back-up, and proof of GCP training within the last three years for all clinical and key supporting site staff, including nurses, pharmacists, and laboratory personnel. To be approved as principal investigator for a site, a clinician must have subinvestigator experience in two or more previous trials.

For applications involving multiple sites, the MCC requires that at least 80 percent of proposed sites be identified in the initial CTF1 application. Applicants must also designate a locally based study chairperson (misleadingly termed the "national principal investigator") whose responsibilities are similar to those of a medical monitor. As part of ongoing national capacity building, the agency encourages sponsors to utilize the public sector and other sites where investigators and staff from previously disadvantaged populations will benefit from training and gain experience.

A CTF1 approval recommendation must be ratified at an MCC meeting, of which there are currently six each year. For this reason, MCC issues an annual calendar of submission cycles of approximately three months' duration. Each cycle starts with a submission deadline and ends with the corresponding Council meeting. Timing of submissions is critical because application deliveries that miss a deadline are deferred to the next cycle, which starts two months later. During a review, an applicant receives via the MRA a checklist acknowledging completeness or requesting missing documents (week 1) and a query letter after the mid-cycle CTC meeting (week 6–7, requiring applicant response within seven days). If satisfied by the applicant's response, CTC will recommend approval, issued by letter a week after the MCC meeting.

Applications for protocol amendments (CTF2) and site or investigator changes or additions (CTF3) are not subject to these cyclic deadlines and may be lodged at any time. Review time varies from a few days to several weeks, depending on complexity of content and evaluator availability. Application fees are currently under US $1000 for a CTF1 and much lower for CTF2 and CTF3 applications. Modest increases are expected in 2011.

There are three exceptions to the above process:

- Bioequivalence protocols involving healthy volunteers are reviewed not by the CTC but by the Medicine Evaluations and Research (MER) division of the Department of Health, which governs the use of generic medicines. The same application forms and fees apply, also submitted through the MRA, but MER has 11 annual cycles (December recess) and a usual review time of two to four weeks.
- As South Africa is a signatory of the Cartagena Protocol, governing activities with genetically modified organisms (GMOs), clinical trials involving gene therapies, vaccines, or other biologicals containing recombinant DNA or RNA require, in addition to MCC, a parallel approval from the Biosafety Directorate, located in the Department of Agriculture, Forestry and Fisheries (DAFF) [29]. GMO applications are

preceded by public notification of intent. Dossiers focus on product characteristics and demonstrating adequate measures for contained use and prevention of environmental contamination [30]. Review occurs in parallel with MCC (and ethics committee) and follows similar review frequency and duration, though cycles are not always exactly in phase. The GMO import permit must be presented together with the MCC approval letter to obtain customs clearance of the trial product.

- Medical devices are not regulated in South Africa; so trials on medical devices (unless they incorporate an unregistered medicine) do not require an MCC approval, making South Africa an especially attractive location for device trials by combining regulatory expediency (ethics committee approval usually takes four to six weeks) with high clinical standards (all national trials must comply with SA GCP [31]). The only devices that do not enjoy regulatory exemption are those that emit non-ionizing radiation (classified as high- and medium-risk electromedical devices) which fall under the jurisdiction of the Directorate of Radiation Control [32].

Assented to in April 2009, the Medicines and Related Substances Amendment Act, No. 72 of 2008 [33] decreed that within the next two to five years, the MCC will be replaced by a new parastatal regulatory authority, the South African Health Products Regulatory Authority (SAHPRA), functioning on behalf of government but outside its infrastructure. SAHPRA is earmarked to appoint expert reviewers as full-time permanent staff, a development welcomed by industry as likely to accelerate regulatory processes substantially and create an agency consultation mechanism. SAHPRA is also expected to formalize the regulation of medical devices and complementary medicines.

13.7. ETHICS COMMITTEES

The first human research ethics committee was established in October 1966 at Johannesburg's University of the Witwatersrand; many other universities soon followed suit with ethics committees that govern research conducted at the institution and its affiliated hospitals. In 1979, South Africa's Medical Research Council issued research ethics guidelines. Properly constituted ethics committees were established by the South African Medical Association in 1992 and another independent organization in 1995 to oversee trial sites in the private sector. The Bill of Rights, contained within the 1996 Constitution of the Republic of South Africa, explicitly entrenched the principle of informed consent in research.

Twenty South African ethics committees have for many years been registered with the Office for Human Research Protection (OHRP) of the Department of Health and Human Services of the United States [34], indicating that ethical review standards in South Africa compare with global norms.

National harmonization began when the National Research Ethics Council (NHREC) was created as a statutory body responsible for establishing guidelines and setting norms and standards for ethical review of health research in South Africa under the National Health Act No. 61 of 2003 [35]. Its mandate includes the audit of ethics committees, though implementation of this process has been slow. In 2004, the Department of Health issued Research Ethics Guidelines establishing mechanisms for the ethical review of human research protocols [36].

Although subscribing to universal principles such as the Declaration of Helsinki, the ethics committees situated in medical schools that are relatively new to clinical research have limited resources for reviewer training and standard operating procedure (SOP) development. These ethics committees may have difficulty in implementing NHREC requirements [37]. Submission requirements vary between ethics committees, so applicants should scrutinize individual ethics committee guidelines when compiling applications.

Mirroring the trend advocated by the WHO and already adopted by many countries, the Department of Health declared it mandatory to register on the South African Clinical Trials Register (SANCTR) all clinical trials that enrolled participants from July 1, 2005 [24]. A unique SANCTR number is obtained after a trial receives ethics committee and applicable regulatory approvals. As the SANCTR does not yet fulfill the International Committee of Medical Journal Editors (ICJME) mandate for clinical trials registration (because it is not WHO endorsed), sponsors and investigators should consider parallel registration of trials on another WHO-recognized primary registry to ensure that they can be considered for publication in ICJME journals.

Most ethics committees meet at least monthly and post annual calendars on their websites. Although application is defined as an International Conference on Harmonisation (ICH)-GCP investigator obligation, it is common (especially with multicenter trials) for the sponsor or CRO to draft applications for the principal investigator's review and signature. Ethics committee application fees tend to be slightly higher than for MCC. Regulatory and ethics reviews run in parallel, so ethics committee approval is seldom a rate-limiting factor. There is some overlap in MCC and ethics committee dossier requirements, though the latter's application forms are less extensive. For trials with a data and safety monitoring board (DSMB), sponsors are expected to appoint at least one member with first hand understanding of conditions in developing countries.

Only the master (English) version of a consent form requires ethics committee approval. It is then translated into the languages applicable to the target population [31] according to the geographical distribution of sites; most trials do not require more than three or four translations, which are sent to the ethics committee as a notification.

13.8. GOOD CLINICAL PRACTICE

While significant growth in South Africa's clinical trial activity during the 1990s paralleled the global roll-out of ICH-GCP, the Department of Health recognized the need for locally customized guidelines that specifically protect vulnerable population groups. This realization resulted in the release of the South African GCP guidelines in 2000. The second edition (2006) is currently in effect [31]. The document is closely based on ICH E6 and interprets how GCP should be applied in the local context, in particular:

- research involving vulnerable communities (minors, women, illiterate patients, persons highly dependent on medical care, etc.)
- HIV/AIDS clinical and epidemiological research
- multicenter studies (to ensure that local realities are considered in the study design)
- timelines for interim progress and safety reports
- participant incentives
- MCC inspections
- competency and responsibilities of the principal investigator.

The MCC and ethics committees require investigators and key site staff to have certification of basic or refresher GCP training in the three years immediately preceding participation in a trial. In view of the unique attributes of trial conduct in South Africa, some ethics committees only accept certification from courses tailored to South African requirements. Numerous courses are run locally by CROs, universities, and independent consultants. The GCP subcommittee of the NHREC is in the process of evaluating 23 training providers. In the future, it is anticipated that ethics committees may recognize only courses containing pertinent South African content. This may necessitate the supplementation of internationally recognized GCP courses with local content.

13.9. CLINICAL TRIAL STATUS

The existence of a first world infrastructure and increasingly Westernized lifestyle within urbanized regions means that apart from widespread HIV and its sequelae, South Africa has a parallel pattern of disease distribution that is not dissimilar to many developed countries. It is therefore not surprising that large numbers of clinical trials are conducted across a wide range of therapeutic areas, as seen in the distribution of SANCTR registered trials by therapeutic indication shown in Figure 13.4. In total, 1630 trials were registered in the 63 months between July 1, 2005 and October 8, 2010 [24]. The leading four indications, central nervous system diseases, oncology, cardiovascular diseases, and diabetes, are chronic diseases prevalent in the Western world and account for almost half (48%) of all studies registered. Infectious diseases including HIV and TB comprise 18 percent. Of the total, 76 percent were regional contributions to multinational trials, 9 percent were multicenter trials in South Africa only, and 15 percent were trials conducted at a single registered site, i.e. mostly early-phase and bioequivalence trials. The breakdown by phase/type is shown in Figure 13.5.

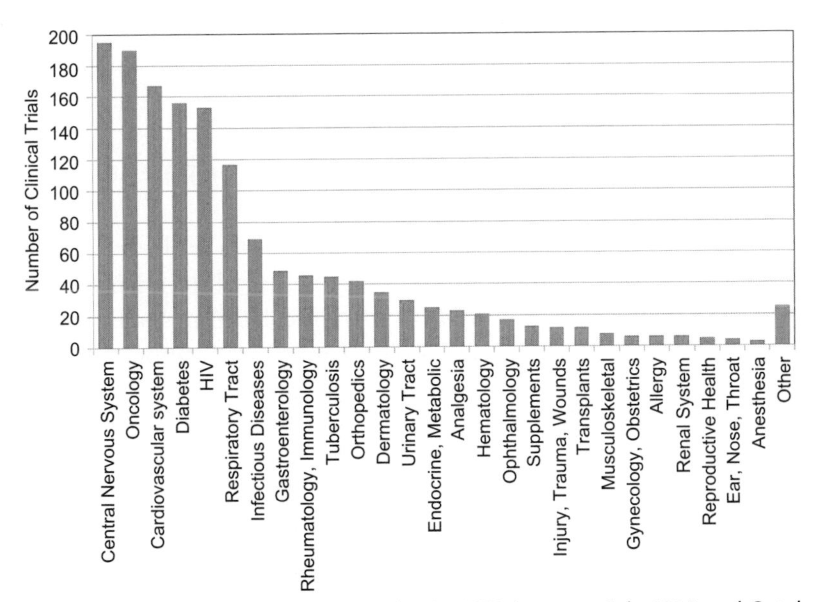

Figure 13.4 Clinical trials registered on the SANCTR between July 2005 and October 2010: breakdown by therapeutic indication *(Source: Data extracted from SANCTR [24])*

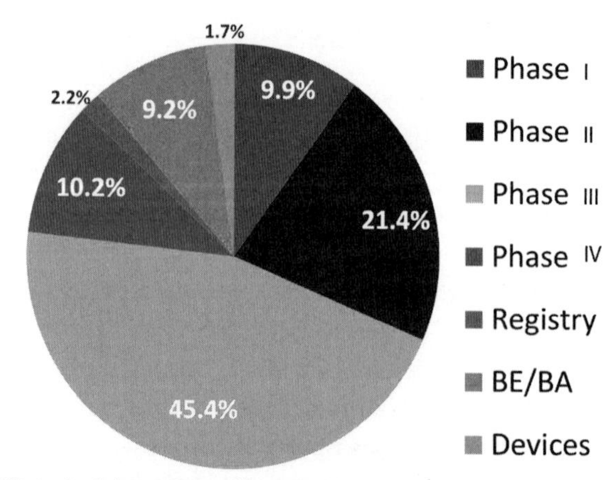

Figure 13.5 Clinical trials registered on the SANCTR between July 2005 and October 2010: breakdown by phase/type *(Source: Data extracted from SANCTR [24]).* (Please refer to color plate section)

13.10. PARTICIPANT PROFILE

The preceding discussion of disease patterns noted that there is a wide diversity of patient profiles in South Africa, across broad ethnic and socioeconomic spectra. The impact of rapid urbanization on lifestyle in South Africa is similar to most other countries. More interesting to sponsors who wish to place trials among foreign cultures are the attitudes to research and specific local challenges affecting recruitment and retention, all of which may impact profoundly on enrollment rates.

13.10.1. Factors Motivating Patients to Participate in Clinical Trials

A Cape Town site recently studied the motives for trial participation of 250 participants of modest means and educational level and reliant on public healthcare [38]. The results were markedly different from similar studies in developed countries where participants harbor expectations of medical benefits. The most common motivating factors here were (a) access to otherwise unaffordable medical care, with regular follow-up; (b) the opportunity to learn more about their condition; and (c) altruistically contributing towards scientific understanding of their disease. Remuneration (for transport), which is often cited as a potential perverse incentive among vulnerable populations in the developing world, did not emerge as a strong motivator (94 percent said they would participate even without payment).

In the early 2000s, the MCC recommended an amount of R150 ($20) per clinic visit as a standard minimum remuneration for transport and other out-of-pocket expenses. They later conceded to the individualized approach preferred by ethics committees, based on factors such as visit duration, distance from site, and median earnings of the target community.

For research to be carried out at community level (e.g. vaccine trials), the SA GCP guidelines recommend consultation with civil organizations within affected communities. This is usually achieved by establishing community advisory groups (CAGs), usually composed of respected members in the target community. By communicating the aims, methods, ethics, benefits, and risks of the research to CAGs during the planning and conduct of trials, researchers gain the confidence of communities, and the willing participation of their members. Voluntary counseling and testing (VCT) programs have also been set up to ensure (before trial screening and consent procedures) that potential volunteers understand that trials with unregistered vaccines do not guarantee protection or replace responsible lifestyle choices [39]. Congruent with this approach in HIV/AIDS vaccine trials, MCC reviewers have advocated a minimum education level as an inclusion criterion.

13.10.2. Local Hurdles for Patient Recruitment and Retention

Comparative enrollment figures in multinational trials often list South African sites as high contributors, in spite of tangible challenges. The main hurdle for low-income patients is often transportation over large distances to attend frequent clinic visits. Sites that recruit patients from a wide geographical area with poor public transport have found two solutions to this problem: either moving participants to the site or placing sites among target populations. Several large trials have boosted enrollment significantly by dedicating their own vehicles and drivers to transport patients to and from the clinic. In other cases, it has been found more effective to set up mobile trial units that travel to either the workplaces or residential areas of target populations.

A common challenge to predicting enrollment rates is the continued reliance of most public and even some small private sites (due to lack of resources and skills) on paper-based databases. Although South African sites generally have high enrollment and retention numbers by global norms, enrollment could probably be increased significantly through maintenance of comprehensive electronic databases of patients with chronic conditions. The most successful recruiters in the private sector are usually specialist consortia and investigators who develop wide referral networks. South

African sites generally enroll well by making a concerted effort actively to inform colleagues of a specific study and the criteria required for enrollment.

13.10.3. South African Successes

Data captured in internal databases of several companies that coordinate multinational clinical trials show that a high proportion of South African sites enroll the target number of patients [40]. One large multinational company's data across all global studies (mostly in first world diseases) conducted in 2008, South African figures compared with rest of world (ROW) sites were: average 10 participants per site (ROW 3–6); 51 percent of sites enrolled more than 10 patients (ROW 18–34%) and 4 percent of sites did not enroll (ROW 15–19%) [41].

Enrollment rates for communicable diseases are often higher. One solution used in outpatient HIV and TB vaccine trials is to position satellite sites using converted shipping containers (fully equipped, air-conditioned, and securely lockable) adjacent to existing community clinics that have no space to house investigators. By placing satellite sites at mine shaft entrances, one research organization included 70,000 miners in a TB study. Of these, 27,861 were enrolled into the interventional arm [42].

A successful strategy used by another large multinational company for conducting 13 large trials to measure steady-state pharmacokinetics over several weeks in specific patient populations (e.g. Parkinson's disease, schizophrenia) enrolled up to 120 patients by advertising nationally and flying groups of patients to the research unit where home-style accommodation was temporarily set up, creating fully controlled conditions that were more acceptable to patients than typical Phase I wards, thus minimizing dropouts and fostering camaraderie among sufferers with shared chronic conditions [43].

13.11. INSPECTIONS

Since most clinical trials conducted in South Africa are sponsored by international companies, trial sites, local offices of pharmaceutical companies, and CROs are all subject to audit by international regulatory authorities as well as the MCC's inspectorate. The latter has a long history of inspecting manufacturing facilities, but only began inspecting clinical trial sites in 2003, when 11 inspections were conducted, gradually increasing to 26 in 2009. Through these six years, 46 percent (58/117) of inspections

Table 13.1 Categories of FDA inspection findings at trial sites in South Africa's public and private healthcare sectors

	NAI	VAI	OAI	Total
Public	9 (56%)	6 (38%)	1 (6%)	16
Private	7 (33%)	14 (67%)		21
	16	20	1	37

FDA: Food and Drug Administration; NAI: No Action Indicated; VAI: Voluntary Action Indicated; OAI: Official Action Indicated.

were conducted at Phase I sites that had conducted bioequivalence trials for marketing applications. The balance comprised mostly routine inspections and a small number of for-cause audits. The most common findings were deficiencies in source records, consenting procedure, product labeling, and site SOPs [44].

In addition, the Biosafety Directorate inspects all investigational sites included in applications to conduct trials involving genetically modified vaccines and other biologicals, specifically to verify the adequacy of measures for contained use and prevention of environmental contamination.

The FDA conducted 37 site audits in South Africa from 1994 to 2009 [16]. The most common deficiencies were "failure to follow the investigational plan" (cited at 15 sites) and "inadequate and inaccurate records" (eight sites). One single inspection conducted in 1997 resulted in an "Official Action Indicated" report, with the deficiency being cited as "inadequate and inaccurate records" (Table 13.1).

13.12. PHARMACOVIGILANCE

Safety monitoring and reporting obligations [45] are similar to those in other countries. Sponsors are expected to provide prompt notification of safety alerts (wherever they occurred) involving investigational products undergoing clinical trials in South Africa to the MCC and to participating investigators, who in turn are responsible for the clinical management and reporting of expected and unexpected adverse events at their sites to the sponsor and to their governing ethics committee. Investigators must also inform their ethics committee and trial participants of significant new safety alerts disseminated by sponsors.

Clinical trial protocols should include a risk management procedure, including unblinding procedures, for dealing with serious unexpected events or reactions. The composition and terms of reference of any independent DSMB should be established prior to trial initiation and filed in the CTF1 application [31]. For expedient communication, it is advisable that sponsors identify a local representative (CRO, medical monitor or the national principal investigator) as the primary local pharmacovigilance contact.

Sponsors must notify the MCC and ethics committees as follows:

- Expedited reporting is required for serious unexpected suspected adverse reactions (SUSARs) occurring in South Africa, initially reported on the MCC serious advert event (SAE) form or an equivalent form containing all required data elements. Related laboratory or other elucidatory information should be attached. Follow-up reports are expected until event resolution.
- Biannual study progress reports must include an update in line listing format of pertinent safety information occurring during all global clinical trials; these are the South African equivalent of the European Medicines Agency's (EMA's) Development Safety Update Report (DSUR).

Report templates are available on the MCC's website [45]. Timelines are summarized in Figure 13.6, but may be revised according to

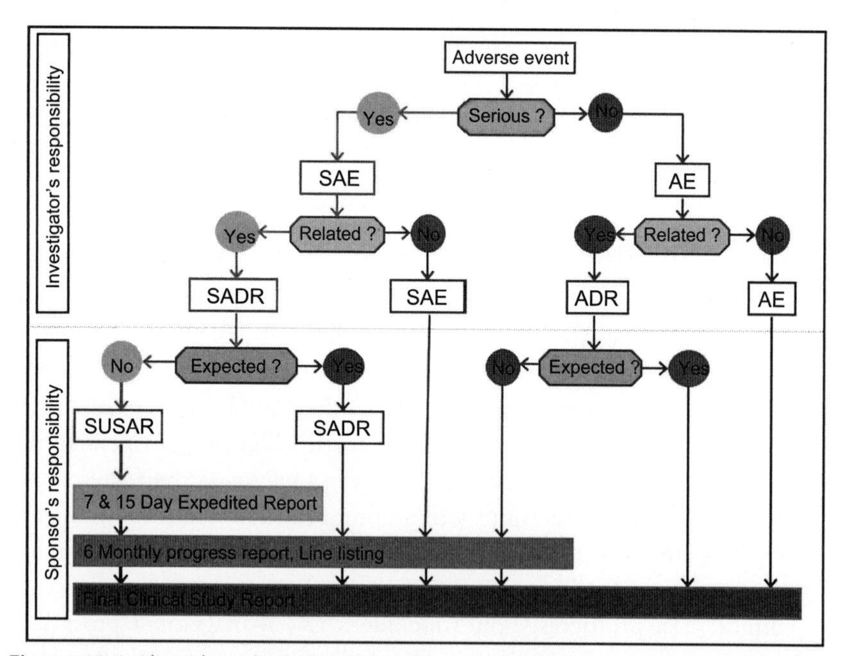

Figure 13.6 Algorithm showing MCC safety reporting requirements and timelines. BE/BA: *(Source: Adapted from MCC regulations [45]).* (Please refer to color plate section)

recommendations of a National Assessment of Medicine Safety Systems and Pharmacovigilance survey, conducted at government and industry levels during 2010 [46].

13.13. STAFFING OF CLINICAL TRIALS

13.13.1. Investigational Sites and Site Management Organizations

At present no South African authority maintains a register from which to obtain an accurate estimate of the number of experienced trial sites in the country (SANCTR records go back only to 2005). The largest online contact database for the medical profession [8] lists registered medical practitioners but there is no field to indicate research experience. A published estimate found 553 South African sites on clinicaltrials.gov until 2007. Not all trials are registered on this database [47], so this is a conservative estimate. Their corresponding computation of trial site density as 11.9 sites per million people for the whole country is considerably higher than for large countries such as Russia (7.7), India (0.7), and China (0.4), but is effectively an underestimate since almost all sites and half the national population reside in densely populated urban areas.

To obtain MCC approval to conduct a trial, sites must have a principal investigator with experience in at least two previous clinical/interventional trials (though in the interests of capacity building a waiver may be granted if there is access to an experienced mentor and intensive monitoring support) and at least one subinvestigator as back-up. If neither investigator has a valid dispensing license, a registered pharmacist must be available. Site coordinators are usually nurses working on trials in a full- or part-time capacity. Key site staff must have attended a GCP course within the last three years.

The site management organization (SMO) is uncommon in South Africa, but exists in various forms such as clusters of independent local researchers, dedicated trial-coordinating functions within general and specialist practice networks, and the affiliate of at least one international SMO.

In the public sector, faculty research offices have been formed to expedite the administrative and marketing aspects of academically affiliated clinical trial sites (e.g. teaching hospital departments). These offices typically provide support in training of site staff, completion of ethics applications and compilation of research budgets. They levy an institutional fee of approximately 10 percent [48,49]. In addition, large independent non-profit institutes set up by prominent academics to specialize in research of national importance (e.g. HIV and TB treatment and prevention) have created site complexes in the midst of mostly underprivileged urban [50], mining [51], and farming [52] communities that have a massive collective enrollment potential for large late-phase trials in target indications.

13.13.2. Clinical Research Organizations and Clinical Research Associates

In a survey conducted in September 2000 [53], the number of "trained persons" employed by CROs and the pharmaceutical industry to manage trials was estimated at 2367, comprising mainly clinical research associates (CRAs) but also personnel in medical affairs, pharmacovigilance, and regulatory support for clinical trial applications, which in South Africa are mostly managed by clinical trial staff as distinct from marketing authorization dossier specialists. The extent of growth in the last 10 years is difficult to estimate, but the doubling of membership of the South African Clinical Research Association (SACRA) may be considered a rough indicator [54]. The educational background of most CRAs ranges from a basic nursing diploma or science degree to PhDs and medical graduates.

A recent pharmacovigilance survey [46] identified 25 CROs within South Africa that are either full service or primarily focused on monitoring clinical trials (as distinct from research units, dedicated suppliers of bioanalytical, data management, and other trial-related services). Of these, 19 are international companies whose affiliates range in size from more than 200 (offering a full range of services) to a single home-based CRA (allowing them to flag their African presence on the world map). For local site monitoring, these microaffiliates usually in-source independent CRAs (former pharma and CRO employees), most of whom are based in major cities.

Since the late 1990s, a further six privately owned and mostly owner-managed CROs have been set up in South Africa. The largest local CRO has a staff of about 45, making it the country's fourth largest overall.

Smaller CROs that do not employ staff, independently contract medical experts to provide local expertise in protocol development, interim safety reviews, and ongoing review of SAEs, and as local contributors to DSMBs.

13.14. LOGISTICS

Although South Africa has very few big "one-stop" CROs, a dynamic, well-established network of specialized vendors exists, providing the services necessary to support a full clinical development. It is relatively easy for a lead CRO to assemble rapidly a full or partial-service solution that is tailored to the exact requirements of any project, operates seamlessly and is competitively and transparently priced. Services available from a range of experienced niche providers include biostatistics, 21 CFR 11-compliant data management, medical writers, CRF developers (paper and electronic), GMP-GCP compliant drug distribution depots, recruitment facilitators and certified translators (for consent forms and patient diaries; investigational product need only be labeled in English).

Competent logistic support also exists for CRF printing (as an alternative to incurring the high duty levied on imported "stationery"), local and international courier companies (accustomed to cold chain management), and customs clearing agents.

Although often provided under a sponsor's global policy, country-specific clinical trial insurance is readily obtained through local brokers who are fully conversant with local law. Regulatory guidelines require that, in addition to investigator malpractice insurance, trial sponsors must hold a no-fault insurance policy that guarantees compensation to trial participants for any injury incurred as a result of protocol-related procedures or treatments. The NHREC has adopted the UK's Clinical Trial Compensation Guidelines as the national standard [55].

For all forms of advertisements prior ethics committee approval of text is required. There is also a number of easily contactable patient organizations through which significant numbers of chronic disease sufferers may be accessed. In 2009, an index of clinical trials was launched as a quarterly supplement to the country's Monthly Index of Medical Specialities, aiming to connect suitable patients to ongoing trials, advertised therein by sponsors [56]. Trial monographs contain key information that would allow subscribing doctors to refer potentially eligible participants, though the publication reaches few patients directly.

South Africa has a number of public-funded and private clinical laboratories that can readily conduct safety analysis of clinical trial samples. These are typically ISO 17025/GLP accredited by the South African National Accreditation System (SANAS). They also adhere to international standards which include ICH-GCP and the Clinical Laboratory Improvement Amendments of 1988 (CLIA'88). Some are accredited by the College of American Pathologists (CAP). Samples are not always shipped out of South Africa as local testing can be performed on instrumentation equivalent to that of international laboratories, and data are consequently integrated into a single global database with identical reference ranges and reports.

Academic research units that specialize in major infectious diseases have developed specialized laboratories in their specific disciplines, offering state-of-the-art analytical and technical support to clinical, epidemiology, pharmaceutical, and translational science.

South Africa's largest bioanalytics and biomarker laboratories have the capability to perform thousands of different drug and biomarker assays with high output volumes and so provide services at global as well as regional levels.

13.15. CLINICAL TRIAL INDUSTRY ASSOCIATIONS

The most visible association in the country's clinical research industry is SACRA. Established in the early 1980s as a forum for CRAs, SACRA's membership (currently over 600) now also encompasses senior managers from most pharma companies and CROs, trial physicians and site staff, laboratories, and other service organizations. SACRA is a non-profit organization, managed by an executive committee of annually elected volunteers [54]. SACRA's main purpose is to disseminate information pertinent to the local industry. This is accomplished through its website, quarterly meetings in Johannesburg and Cape Town, and an annual conference.

The Informal Clinical Trial Network (iCTN) is a smaller group of managers from pharmaceutical companies, large trial sites, SMOs, and CROs who meet every two to three months to discuss new challenges and work collectively toward resolving issues that affect the industry nationally. The Association of Clinical Research Professionals (ACRP) has a chapter in South Africa, but its membership is very small.

13.16. CHALLENGES AND OPPORTUNITIES

In the last decade, South Africa's regulatory timelines were generally considered a drawback in comparison with other countries with faster approval periods. Canceled MCC meetings, unreachable MRA secretariat staff, an erratic database, and general inefficiency contributed to these concerns. Significant recent strides have been made to improve the situation. A three-month turnaround on clinical trial applications is still not ideal, especially with only six annual Council meetings to ratify an increasing number of applications. However, systematic improvement has been observed in recent years and should continue with the much-awaited creation of the new authority (see section 13.6 above) with more permanent staff and less reliance on part-time evaluators.

Challenges to establishing trial sites in rural areas include the difficulty of finding trained staff willing to work in less developed areas, complex laboratory logistics, and variable Internet connection quality. Under the latter circumstances paper CRFs remain more reliable than web-based technologies.

The high quality of South African medical training creates expectations of similar research quality, but also carries the risk of seeing increasing numbers of trial physicians lured abroad by new and more lucrative opportunities [57]. This may partly be explained by a perception that investigator fees are not keeping pace with the rising costs of running and staffing a trial site to a level that complies with continually intensified regulations [58]. The effective negotiation of realistic site budgets has thus become a vital skill for investigator survival.

Nothwithstanding these concerns, and its distance from the main axis of clinical development in the northern hemisphere, South Africa remains highly regarded for timeously contributing quality data across all phases of research, offering many advantages:

- a strong bimodal (public and private) community of experienced investigators at well-equipped sites
- a still large percentage of treatment-naïve patients in both first world and third world indications
- the predominance of English as the language of clinical research
- inverse seasons, allowing all-year enrollment in multinational trials
- convenient time zone (GMT +2 hours) with manageable overlap of business hours with America and especially Europe
- favorable exchange rates against the major currencies
- the only country in which GCP is not only a guideline but a legal requirement and where informed consent is stipulated in the Bill of Rights in the national Constitution
- a growing recognition at government and parastatal levels of the benefits of augmenting clinical research in the population and the commensurate need to support training [59,60]
- a robust system of strict protocol review within manageable and improving timelines
- rapid initiation of device trials due to their exemption from regulatory review

- manifold opportunities for nutritional research (in an impoverished population) and trials with non-prescription drugs
- competitive environment for conducting generic bioequivalence trials
- globally respected network of Phase I research units
- well positioned by location and experience as a gateway for setting up and coordinating clinical trials in the African region
- globally competitive trial start-up times because all trial preparatory processes (regulatory, ethics contractual negotiations, etc.) run in parallel
- a proven enrollment track record, with significant contributions of high-quality research data across a wide range of therapeutic areas over the past three decades.

REFERENCES

[1] Statistics South Africa. Mid-year population estimates. Statistical Release P0302, http://www.statssa.gov.za/publications/P0302/P03022010.pdf; 2010 (accessed August 19, 2010).
[2] Statistics South Africa. Census 2001; key results, http://www.statssa.gov.za/census01/html/Key%20results_files/Key%20results.pdf; 2001 (accessed August 19, 2010).
[3] South African Constitution. Founding Provisions, Chapter 1, Section 6, http://www.info.gov.za/documents/constitution/1996/96cons1.htm#6; 1996 (accessed August 20, 2010).
[4] Brand South Africa. The Languages of South Africa. Big Media Publishers, http://www.southafrica.info/about/people/language.htm (accessed August 19, 2010).
[5] Brand South Africa. South Africa's Transport Network. Big Media Publishers, http://www.southafrica.info/business/economy/infrastructure/transport.htm (accessed August 19, 2010).
[6] SA-Venues.com. South Africa Weather and Climate. Net-Focus Interactive, http://www.sa-venues.com/no/weather.htm (accessed August 19, 2010).
[7] Pelzer K. Patient experiences and health system responsiveness in South Africa. BMC Health Services Research:117, http://www.biomedcentral.com/1472-6963/9/117, 2009;9 (accessed August 18, 2010).
[8] MEDpages, http://www.medpages.co.za (accessed August 25, 2010).
[9] Day C, Monticelli F, Barron P, Haynes R, Smith J, Sello E. District Health Barometer 2008/2009. Health Systems Trust, http://www.hst.org.za/publications/864; 2010.
[10] Council for Medical Schemes. About us, http://www.medicalschemes.com/Content.aspx?1 (accessed 11 Oct 2010).
[11] Mooney GH, McIntyre DE. South Africa: a 21st century apartheid in health and health care. Medical Journal of Australia 2008;189:637–40.
[12] Life Healthcare. Investor relations, http://www.lifehealthcare.co.za/IR/Default.aspx (accessed August 25, 2010).
[13] Mediclinic Private Hospital Group. About our clinics, http://www.mediclinic.co.za/hospitals/Pages/default.aspx (accessed August 25, 2010).
[14] Netcare. Company profile, http://www.netcare.co.za/2390/company-profile (accessed August 25, 2010).
[15] National Hospital Network. List of members, http://www.nhn.co.za/ReadContent.aspx?id=36 (accessed August 25, 2010).

[16] US FDA/Center for Drug Evaluation and Research. Clinical Investigator Inspection List: for investigational new drug studies. Last updated, http://www.accessdata.fda.gov/scripts/cder/cliil/index.cfm July 8, 2010 (accessed September 8, 2010).
[17] ECONEX. National Health Insurance Note 2: South Africa's burden of disease, http://www.econex.co.za/images/stories/ECONEX_NHInote_2.pdf; 2009 (accessed September 10, 2010).
[18] Statistics South Africa. Mortality and causes of death in South Africa, 2007: findings from death notification. Statistical Release P0309.3, http://www.statssa.gov.za/publications/P03093/P030932007.pdf; 2009 (accessed September 10, 2010).
[19] Norman R, Bradshaw D, Schneider M, Pieterse D, Groenewald P. Revised burden of disease estimates for the comparative risk factor assessment, South Africa 2000, Methodological Note. Cape Town: South African Medical Research Council, http://www.mrc.ac.za/bod/RevisedBurdenofDiseaseEstimates1.pdf; 2006 (accessed September 9, 2010).
[20] Mayosi BM, Flisher AJ, Lalloo UG, Sitas F, Tollman SM, Bradshaw D. The burden of non-communicable diseases in South Africa. Lancet 2009;374:934–47.
[21] Househam KC. Africa's burden of disease: the University of Cape Town Sub-Saharan Africa Centre for Chronic Disease. South African Medical Journal 2010;100:94–5.
[22] Snyman HW, editor. Medicines in our Time: From Concept to Therapeutic Reality. Pretoria: Pretoria University Press; 1972.
[23] Offermeier J, editor. Seminar on Biologic Availability of Medicines. Pretoria: CSIR; 1978.
[24] Department of Health, South Africa. SA Clinical Trial Information, http://www.sanctr.gov.za (accessed October 8, 2010).
[25] Roehr B. FDA approves rifapentine for the treatment of pulmonary tuberculosis. Journal of the International Association of Physicians in AIDS Care 1998;4(8):19–25.
[26] Debiopharm Group. Press Release. Lausanne: Switzerland, http://www.debiopharm.com; July 5, 2001 (accessed August 25, 2010).
[27] Matsebula T, Goudge J, Gilson L. Regulating the pharmaceutical sector: Coping with low capacity while maintaining regulatory independence. Health Economics and Financing Programme: Working Paper 01/05. London: London School of Hygiene and Tropical Medicine, http://www.hefp.lshtm.ac.uk/publications/downloads/working_papers/01_05.pdf; 2005 (accessed August 25, 2010).
[28] Department of Health, South Africa. Medicines Regulations in South Africa, http://www.mccza.com (accessed August 25, 2010).
[29] Department of Agriculture Forestry and Fisheries, http://www.daff.gov.za (accessed August 25, 2010).
[30] Government Communications. Genetically Modified Organisms Act. [No. 15 of 1997]. Government Gazette:18029, http://www.info.gov.za/acts/1997/act15.htm, 1997;Vol. 383 (accessed August 25, 2010).
[31] Department of Health. Guidelines for Good Practice in the Conduct of Clinical Trials with Human Participants in South Africa. Pretoria: Department of Health; 2006.
[32] Department of Health. Directorate Radiation Control, http://www.doh.gov.za/department/radiation/01.html (accessed August 25, 2010).
[33] National Gazette, http://www.sacra.za.net/site/files/5909/medicine2.pdf; April 18, 2008 (accessed September 9, 2010).
[34] United States Department of Health and Human Services. Office for Human Research Protections (OHRP) Database for Registered IORGs & IRBs, http://www.hhs.gov/ohrp (accessed August 20, 2010).
[35] Department of Health. The National Health Research Ethics Council, http://www.doh.gov.za/nhrec/index.php (accessed August 19, 2010).
[36] Department of Health (2004). Ethics in health research: principles, structures and processes, http://www.doh.gov.za/docs/factsheets/guidelines/ethics (accessed August 20, 2010).

[37] Strugo JL. The impact of regulatory research structures on smaller ethics committees in South Africa. MTech Pharm Sci dissertation. Tshwane University of Technology; 2006. 76–7.
[38] Burgess LJ, Sulzer NU, Hoosain F, Leverton N, Blignaut S, Emanuel S. Patients' motivation for participating in cardiovascular clinical trials: a local perspective. Cardiovascular Journal of Africa 2009;20:220–3.
[39] Perinatal HIV Research Unit. Voluntary counseling and testing projects, http://www.phru.co.za/phruprojects/prevention/vct-projects.html (accessed September 8, 2010).
[40] Katsoulis LC. Regulatory timelines and enrolment potential in South Africa starting to improve. Journal of Clinical Studies 2010;2010:10–1.
[41] Venter J. How do we compare? Presentation at SACRA Quarterly Members' Meeting, June 2009.
[42] CREATE. Thibela TB, http://www.tbhiv-create.org/about/studies/thibela (accessed September 8, 2010).
[43] Richardt D. Innovative recruitment and retention strategies in Ptrun -1arexel Early Phase studies. Presentation at SACRA Members' Meeting 2009.
[44] Bonthuys L. Regulatory framework for GCP inspections in South Africa. Presentation at 4th Annual Clinical Trial Conference, SACRA 2010. September 2010.
[45] Medicines Control Council, South Africa. Reporting of adverse drug reactions in South Africa, http://www.mccza.com; 2003 (accessed September 1, 2010).
[46] Adam C. National assessment of medicine safety systems and pharmacovigilance. Survey conducted by Management Sciences for Health (as yet unpublished) 2010.
[47] Thiers FA, Sinskey AJ, Berndt ER. Trends in the globalization of clinical trials. Nature Reviews Drug Discovery 2008;7(1):13–4.
[48] Wits Health Consortium, http://www.witshealth.co.za/Pages/default.aspx (accessed September 10, 2010).
[49] University of Cape Town Research Office, http://www.researchoffice.uct.ac.za/ (accessed September 10, 2010).
[50] Perinatal HIV Research Unit. About us, http://www.phru.co.za/about-phru.html (accessed September 10, 2010).
[51] Aurum Institute. Aurum Institute Branches, http://www.auruminstitute.org/branches/index.php (accessed September 10, 2010).
[52] South African Tuberculosis Vaccine Initiative. About SATVI, http://www.satvi.uct.ac.za/about-us/about-satvi.html (accessed September 10, 2010).
[53] Joffe M. Survey of pharmaceutical industry staff involved in clinical research. Wits Health Consortium, unpublished data 2000.
[54] South African Clinical Research Association, http://www.sacraza.com (accessed 06 September 2010).
[55] Association of the British Pharmaceutical Industry. London: Clinical Trial Compensation Guidelines; 1991.
[56] Manyike P, editor. Quarterly index of clinical research trials and sites; 2009. MIMS 49(7).
[57] Just Landed. Public healthcare: South Africa's healthcare system, http://www.justlanded.com/english/South-Africa/South-Africa-Guide/Health/Public-Healthcare (accessed August 25, 2010).
[58] Burgess LJ, Sulzer NU. Editorial: The growing disparity between clinical trial complexity and investigator compensation. Cardiovascular Journal of Africa 2010;21:272–3.
[59] Siegfried N, Volmink J, Dhansay A. Does South Africa need a national clinical trials support unit? South African Medical Journal 2010;100:521–4.
[60] Academy of Science of South Africa. Consensus report on revitalising clinical research in South Africa: a study on clinical research and related training in South Africa. Pretoria: Academy of Science of South Africa; 2009.

CHAPTER 14

Clinical Trials in Latin America

Henrietta Ukwu*, Mariano Parma, Antonio Guimaraes[†], Carlos Fernando de Oliveira[‡], Anna Paula Más[§] and Elizabeth Villeponteaux[¶]**

*Global Regulatory Affairs, PPD, Blue Bell, PA, USA
**Strategic Development Latin America, PPD, Buenos Aires, Argentina
[†]Human Resources Latin America, PPD, Wilmington, NC, USA
[‡]Latin America Pharmacovigilance, PPD, Morrisville, NC, USA
[§]Regulatory Affairs Country Management, Latin America, PPD, São Paulo, Brazil
[¶]Global Medical Writing, PPD, Morrisville, NC, USA

Contents

Global Clinical Trials
ISBN 978-0-12-381537-8, Doi:10.1016/B978-0-12-381537-8.10014-7

14.1. INTRODUCTION

14.1.1. Geography and Demography

The Latin American region is one of the top emerging markets, together with South East Asia and Eastern Europe. The region consists of Mexico, South America, Central America, and the Caribbean nations (which are sometimes considered part of Central America). With an area covering 20 million square kilometers, Latin America has a population estimated at over 580 million [1,2]. Spanish is the predominant language, but Portuguese is spoken in Brazil, the most populous country of the region. French and English are also spoken in some Caribbean countries. In addition, many Native American languages are spoken in parts of several countries, especially in the Andean region of Peru, Bolivia, Ecuador, and Chile. In Peru, Quechua is, along with Spanish, an official language.

Brazil is the largest country both in area (8.5 million square kilometers) and in population (around 200 million) [3]. Mexico is the next most populous, with approximately 112 million people, then Colombia with 44 million, Argentina with 41 million, and Peru with 30 million. There are large urban sections in Latin America; Mexico City and São Paulo are the third and fifth most populated cities in the world, each with approximately 20 million residents.

From a time zone perspective, one advantage of working in the region is that Latin American countries are all within 2 hours of all US time zones. This is ideal for communication with US pharmaceutical companies and may also be relatively convenient for Western European companies.

The population in Latin America shows great diversity. The initial habitants of the region were the Amerindians, and the main cultures were the Aztecs, Mayas, Incas, Tupis, and Caribs. At the end of the fifteenth century the powers then dominant in Europe, Portugal and Spain, began to colonize the region, bringing with them individuals from Africa. Over the next four centuries there were also incursions by the French, English, and Dutch. In addition, during the last century, the region has seen a large migration of Italians, Jews, Japanese, Germans, Lebanese, Syrians, Arabs, Russians, and Polish, thus adding to an already growing melting pot. The predominant religion is Roman Catholic [3]. However, in recent years there has been an increase in the Protestant population.

14.1.2. Economy

Although Latin American is an important emerging market, social disparity remains a significant problem for the region; in almost half of Latin American countries, half or more of the population is below the poverty line. In the region as a whole, one-third of the population live in poverty and 13 percent live in extreme poverty [4]. Nonetheless, the economy has been growing quickly in most of the region.

Brazil is the largest economy in the region. It is the 10th ranked economy in the world and is often considered to be one of the top four future economies in the world, along with China, Russia, and India [3,5].

14.1.3. Healthcare Infrastructure in Latin America

Resources and healthcare infrastructure in the major nations of Latin America are well suited for enabling global clinical trials. In fact, compilation of data from 2000 to 2006 shows that Argentina had 30 physicians per 1000 people, more than the USA, which has 26 [6]. Mexico had 20 physicians per 1000, while the number ranged from 11 to 14 in Colombia, Brazil, Peru, and Chile. Within the medical community, awareness and adoption of good clinical practices (GCPs) and International Conference on Harmonisation (ICH) guidance are widespread.

In 2005, spending on healthcare as a percentage of gross domestic product (GDP) was just above 10 percent in Argentina, between 7 and 8 percent in Colombia and Brazil, between 5 and 7 percent in Chile and Mexico, and just above 4 percent in Peru [6].

Most principal investigators and subinvestigators in the region have completed at least part of their medical training in the USA or Europe, often including exposure to clinical trial participation. As a result, to achieve readiness in conducting clinical trials these investigators generally need only training in GCP and consultation to ensure that facilities are appropriate. A database of approximately 7000 potential principal investigators in Latin America for numerous indications including (but not limited to) oncology, infectious diseases, cardiology, and neurology was analyzed to determine the locations of investigators. The analysis determined that in the major countries in the region (Argentina, $n = 1464$; Brazil, $n = 2920$; Chile, $n = 483$; Mexico, $n = 1576$; Peru, $n = 535$), they are heavily concentrated in a few cities, typically the largest cities of the various states (Table 14.1).

Figures 14.1–14.5 show the locations of available investigators in five major Latin American countries.

Table 14.1 Urbanicity of investigators

Country	Potential principal investigators		% Urban
	Main city in state: *n*	Total in state	
Argentina	Santa Fe: 72	111	64.9%
	Buenos Aires: 937	1081	86.7%
Brazil	Rio do Janeiro: 324	354	91.5%
	São Paulo: 936	1345	69.6%
	Porto Alegre: 317	346	91.6%
Mexico	Mexico: 887	920	96.4%

14.1.3.1. Patient Profile

Patient retention is excellent in Latin American investigative sites because of the ongoing relationships between physicians and patients. Strong local ties in the population and the availability of large hospitals mean that patients often receive treatment at the same hospital or institutions throughout their lives. The enthusiasm that investigators in the region demonstrate also contributes to the strong patient retention rates.

The patient pools are not necessarily limited to those receiving usual treatment from their physicians. Clinical investigators and site staff usually work within a network of referrals, allowing outside patient populations to be enrolled through referrals from physicians who, though not acting as investigators, recognize that trial participation may offer the chance of better treatment.

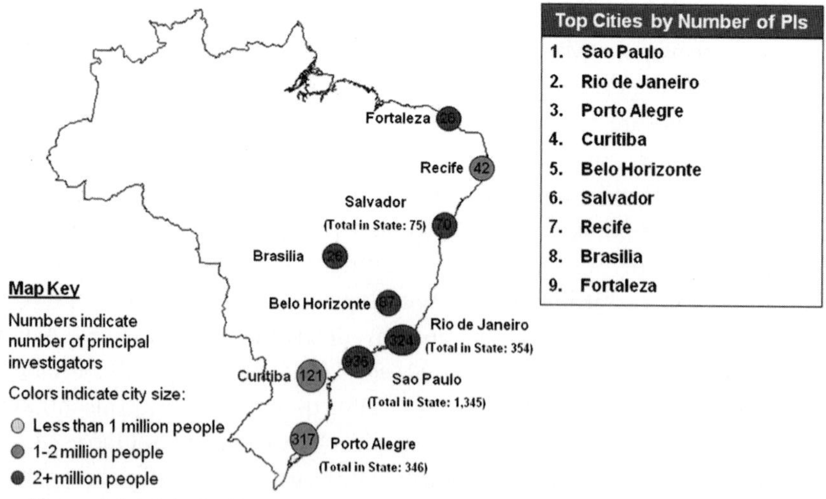

Figure 14.1 Principal investigators in Brazil. (Please refer to color plate section)

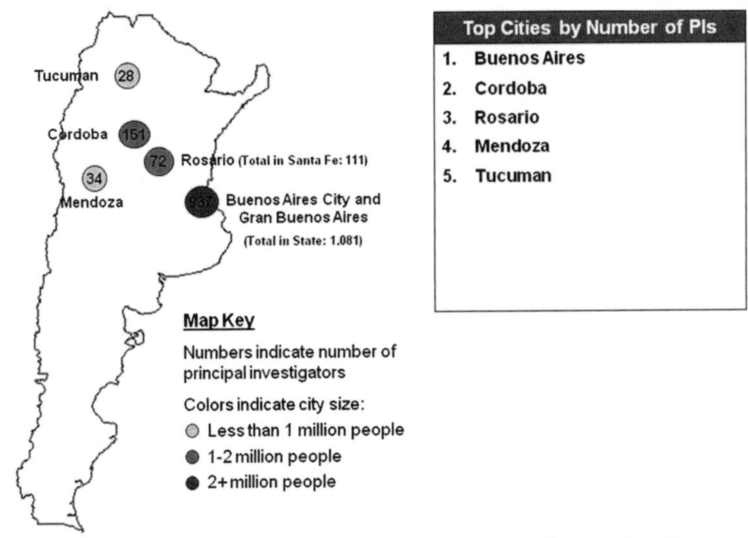

Figure 14.2 Principal investigators in Argentina. (Please refer to color plate section)

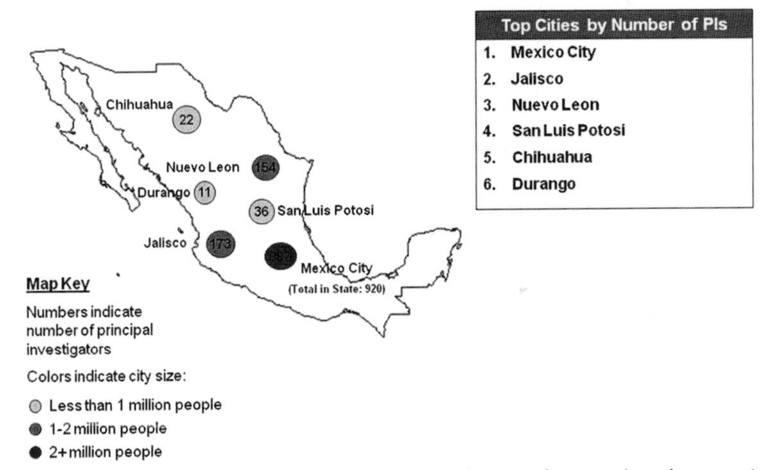

Figure 14.3 Principal investigators in Mexico. (Please refer to color plate section)

Because patients in some more remote areas of the region may have poor access to healthcare, these populations may be vulnerable targets for clinical trial participation. To minimize this risk, ethics committees and national health authorities have taken an active role in ensuring patient protection. Contract research organizations (CROs) and sponsors have also focused on protection of patients' rights to ensure compliance with international regulations. Particular attention has been paid to compensation, to avoid

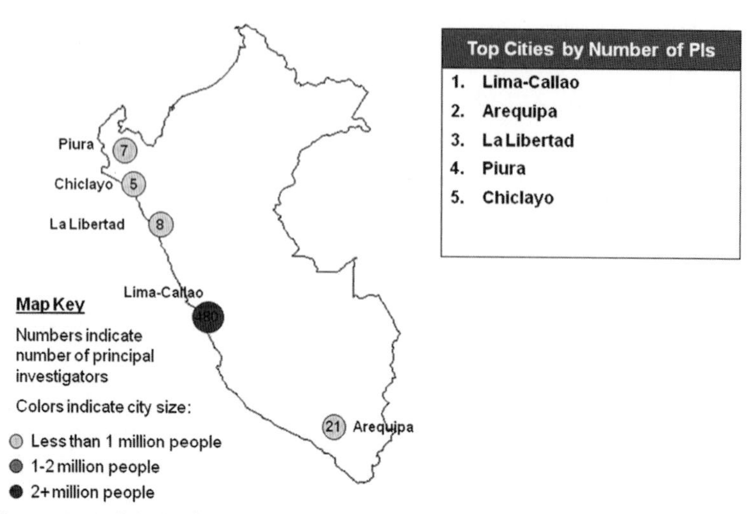

Figure 14.4 Principal investigators in Peru. (Please refer to color plate section)

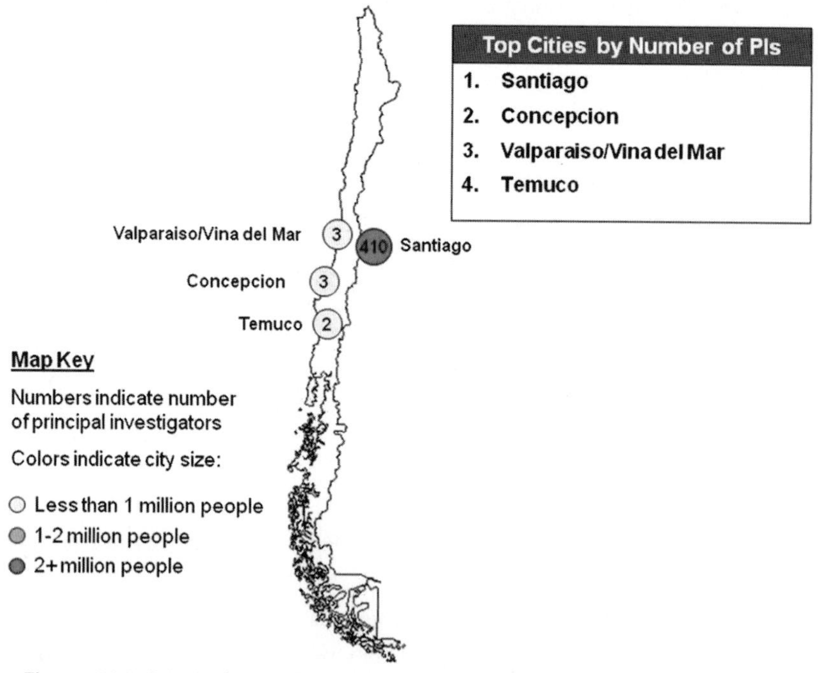

Figure 14.5 Principal investigators in Chile. (Please refer to color plate section)

excessive incentives for trial participation. Amounts of money allotted for travel expenses and meals must be reasonable, compensation plans must be approved by ethics committees, and a clear explanation of the program must appear in the informed consent form (ICF).

Latin America offers high levels of ethnic diversity, with large Hispanic populations as well as European, Asian, and African populations. Mexico, Peru, and particularly Brazil received large numbers of immigrants from Japan from the late 1890s to World War II. Brazil has the largest concentration of ethnic Japanese living outside Japan [7]. Adoption of the ICH Guideline on Ethnic Factors in the Acceptability of Foreign Clinical Data allows Japan's drug approval authorities to accept data from studies that include ethnic Japanese living outside Japan.

Across the Latin American countries, regulations governing ICFs can be particularly strict, yet not necessarily in a uniform manner. The ICF must be tailored for each individual country for regulatory compliance. In general, Latin American regulatory authorities require the use of simplified, patient-level descriptions of the study procedures. Failure to comply can lead to study rejection.

The unique environment in Latin America necessitates a high level of familiarity with the culture. Norms in the heavily Catholic population can be in conflict with protocol requirements; this conflict can lead to rejection of the submission by either the local ethics committees, the regulatory authorities, or both, if the design requires collection of information about a topic that is considered inappropriate. Specific regulations for personal data protection must also be taken into consideration. It is important for sponsors to be or become familiar with the cultural norms to facilitate approval as well as smooth conduct of the trial.

14.1.3.2. Disease Profile

Deaths in Latin American countries are most commonly caused by non-communicable chronic diseases, such as cardiovascular diseases, cancer, diabetes, and chronic pulmonary obstructive disease [8]. Diabetes in particular is increasing in prevalence. In 2009, the prevalence in Central and South America was 6.8 percent, representing approximately 18 million people, and in 2030, it is expected to be 7.8 percent, or 29.6 million people. In Mexico, meanwhile, the prevalence is 6.8 million, or 10.8 percent; this is expected to reach 12.9 percent in 2030 [9].

The Pan American Health Organization (PAHO) has reported that non-communicable diseases, traditionally thought to affect primarily older adults in developed countries, are now often seen in younger, poorer residents in Latin America [10]. The increasing rates of cardiovascular disease in those of working age in the region has an especially great impact on the populace because of the effect on worker productivity. Cerebrovascular diseases, in particular, have a great impact, as the associated mortality rate is four times higher in Latin America and the Caribbean than in North America. The deaths are disproportionately premature, occurring in people less than 65 years old in almost one-quarter of the cases (approximately 22%; 90,000/ 400,000). The premature deaths have the greatest impact among women, and almost 30 percent occur among those in the lowest economic quintile of the population, while only 13 percent fall in the highest quintile.

In Latin America as in the USA and in more developed regions as a whole, the most common cancer site is the prostate among men and the breast among women (Table 14.2) [11]. The similarity in distribution of the oncological diseases further illustrates the appropriateness of conducting clinical research activities in Latin America.

14.2. REGULATORY CLIMATE

14.2.1. Legislation Versus Harmonization

No Latin American country is a signatory of ICH-GCP guidelines; however, regulatory authorities in these nations increasingly employ GCP terminology and have adopted features of GCP within regulations and guidelines.

The capabilities and resources of regulatory agencies in Latin America are often not adequate to meet the demand from the pharmaceutical industry, and regulatory requirements can limit the ability of the pharmaceutical industry to meet competitive timelines.

Table 14.2 Most common cancers (10)

Rank	Latin America	More developed regions	USA
Men			
1 (%)	Prostate (27.1%)	Prostate (21.8%)	Prostate (25.0%)
2 (%)	Lung (10.6%)	Lung (16.2%)	Lung (15.4%)
3 (%)	Stomach (8.9%)	Colorectum (13.1%)	Colorectum (10.6%)
4 (%)	Colorectum (6.9%)	Stomach (6.0%)	Bladder (6.9%)
5 (%)	Liver (3.3%)	Bladder (5.8%)	Melanoma of skin (4.8%)
Women			
1 (%)	Breast (24.9%)	Breast (26.8%)	Breast (26.4%)
2 (%)	Cervix uteri (14.8%)	Colorectum (13.1%)	Lung (14.5%)
3 (%)	Colorectum (7.2%)	Lung (9.4%)	Colorectum (10.8%)
4 (%)	Stomach (5.6%)	Corpus uteri (5.5%)	Corpus uteri (5.8%)
5 (%)	Lung (5.2%)	Ovary (3.9%)	Thyroid (4.4%)

Each country in the region has different local regulatory requirements. Safety reporting requirements, for example, vary among the countries, which all have designated country-specific timelines and reporting processes (see Section 14.2.5). Variation is particularly wide when it comes to reporting of serious adverse events (SAEs) that occur locally. For clinical trials applications, as well, the Latin American countries have independent and different legislations. On the whole, procedures, regulations, and processes are not harmonized within the region.

PAHO, an international public health agency with over 100 years of experience dedicated to improving health and living standards in the countries of the Americas, has worked toward promoting harmonization initiatives in all regulatory aspects, including clinical trials.

PAHO has developed a document called Documento das Americas (Document of the Americas). The objective of the Document of the Americas is to propose guidelines for good clinical practice that can serve as a foundation for regulatory agencies, as well as for investigators, ethics committees, universities, and businesses. It was developed with representation of the regulatory agencies from Argentina, Brazil, Chile, Costa Rica, the Caribbean Community (CARICOM), Cuba, Mexico, the USA, and Venezuela. From other countries, though, groups providing important input into ethics review and other aspects of clinical trial regulation have not been involved; for example, CONEP-National EC in Brazil did not participate, and there were no representatives from Peru and Colombia. The limited scope of country representation may affect the value and applicability of this initiative.

An earlier clinical practice guideline was developed in 1996 by South America's leading trading block, MERCOSUR. Argentina, Brazil, Uruguay, Paraguay, and Venezuela are members of MERCOSUR, and associate members are Bolivia, Chile, Colombia, Ecuador, and Peru. However, this guideline has not been fully adopted as the primary guideline for clinical trial conduct by the involved Latin American countries; only Paraguay and Uruguay have adopted it as the primary guidance for clinical research.

14.2.2. Regulatory Requirements, Legislation, and Related Bodies

Each country in Latin America has its own regulatory agency that, as a part of the Ministry of Health or in association with it, regulates clinical trials conducted in the country. Ethics review is typically an important component, but the details vary among the countries.

Table 14.3 presents an overview of these aspects, including the primary regulations for clinical trials, the type of institutional review board (IRB) and ethics committee review required (i.e. whether local, central, regional, or independent), and specific individual requirements of each country.

Each Latin American country has its own requirements for a clinical trial application (CTA) dossier. The first regulatory step is submission to the ethics committee, which is usually performed by the site staff–principal investigator. However, the procedure is different for the ethics committees in Argentina [independent ethics committee (IEC)] and Peru (central-independent ethics committee), where submissions can be performed directly by the sponsor or CRO.

In the majority of Latin American countries, only two languages, Portuguese and Spanish, are used; however, in the Caribbean and several smaller countries, other languages such as English and French may be spoken.

Submission of the CTA dossier to the regulatory agency is usually performed by the sponsor or hired CRO, although there are some exceptions; for example, the site staff–principal investigator submits the dossier in Central American countries. Typically, completion of country-specific forms may be required along with the dossier. These local forms vary in the level of detail, with some quite detailed and some requiring only basic information. Information that might be required includes chemistry, manufacturing, and control (CMC) information (e.g. study drug qualitative and quantitative information, shelf-life, packaging information); an estimate of study drug and material quantity to be imported for the trial; a report of any biological samples to be exported; a list of countries and country sites participating in the clinical trial; an estimate of the number of enrolled patients per country and per site; and study timelines.

A letter providing formal delegation of responsibilities (also known as power of attorney of transfer of responsibilities) must be included in the CTA in some countries, i.e. Argentina, Brazil, and Chile. All of these countries require specific legalization seals (e.g. Apostille of Hague Convention or Consulate seals). Some other countries may require that similar documentation be kept under clinical trial files for reference.

Obtaining trial approval in Latin American countries can be challenging, with hurdles related to the trial design as well as providing required documents. When the clinical trials program includes a placebo control, the sponsor must provide a rationale for placebo administration that incorporates information from treatment guidelines and patterns of usual care, and

Table 14.3 Regulatory overview

Country	Regulatory agency	Local EC	Central, regional or independent EC	Individual requirements	Main regulations
Argentina	ANMAT (Administracion Nacional de Medicamentos, Alimentos y Tecnologia Medica)	Yes	Independent EC that acts as central for sites with no local EC; GCP compliant	MoJ: ICF must be revised by the MoJ Provincia: each provincia (state) may have its own regulations for clinical trials; most strict: Provincia de Buenos Aires Phase I/II: 1 month longer (INAME – additional body review) Sites not part of main submission: approved through fast track, after EC approval	ANMAT: Regulation Nos 5330 & 6550 MoJ: Law 25.326, Provisions 06/08 & 10/08 MoH: Res 1490/07 & 102/09
Brazil	ANVISA (Agência Nacional de Vigilância Sanitária)	Yes (known as CEP)	Central (national) EC = CONEP (Comissão Nacional de Ética em Pesquisa)	CONEP: national EC that registers and oversees local EC (CEPs), and must also be involved in the ethics approval of any series of trials, including international trials, submitted through a coordinating EC ANVISA and CONEP approval: approves the full list of sites included in the submission (no need for CEP approval). Biological products: more documents required (manufacturing, quality analysis, and controls) Placebo: among the more strict LA countries regarding placebo-exclusive trials (specific CONEP and CFM resolution for non-approval of placebo-exclusive trials, if there is approved treatment available)	ANVISA: RDC 39/08 & 26/99 (EAP) CONEP: Res 196/96, 251/97, 292/99, 346/05, 347/05, 370/07, 404/08 CFM (Federal Council of Medicine): Res 1885/08

(Continued)

Table 14.3 Regulatory overview—cont'd

Country	Regulatory agency	Local EC	Central, regional or independent EC	Individual requirements	Main regulations
Chile	ISP (Instituto de Salud Pública)	Yes: sites may or may not (known as CEI)	Regional EC (CEEC), geographically assigned through the regional divisions of health services	CEI vs CEEC: for sites with no CEI, CEEC review and approval is enough. CEEC: may be responsible for more than one site review CMC: specific requirements for GMPc (or manufacturing authorization) and labels Sites not part of main submission: notified after EC approval Biological products: more documents required (manufacturing and controls)	ISP: Law 20.120, Technical Norm 57 & Authority Guidances 4/09 & 5/10 Data protection: Law 19.628
Colombia	INVIMA (Instituto Nacional de Vigilancia de Medicamentos y Alimentos)	Yes	Not applicable	IPS: each site must be an official health service institution registered as IPS GCP: INVIMA is to certify sites for GCP compliance. Sites must have submitted a GCP gradual plan and request GCP certification visit Sites not part of main submission: notified after EC approval	INVIMA: Res 008430/93, 22378/08, 2010020508/10
Costa Rica	CONIS (Consejo Nacional de Investigación de Salud)	Currently clinical trial regulations have been annulled by the government; clinical trial approvals and enrollment have been halted until a new law is released			

Dominican Republic	SESPAS through CONABIOS (Secretaría de Estado de Salud Pública y Asistencia Social & Consejo Nacional de Bioética en Salud)	Yes	If local EC is not available, an independent EC must be used	EC and CONABIOS: entire and single review and approval process (no fast track) on a site-by-site basis	CONABIOS: Ley General de Salud No. 42/01, Disposicion 004/2004
Guatemala	MSPAS (Ministerio de Salud Pública y Asistencia Social)	Yes	If local EC is not available, an independent EC must be used	Phases I/II: might not be accepted by EC/RA Accredited EC: local ECs must be accredited at RA EC and MSPAS: entire and single review and approval process (no fast track) on a site-by-site basis	MSPAS: Acuerdo Ministerial SP-M-466/07
Mexico	SSA through COFEPRIS (Secretaría de Salud through Comisión Federal para la Protección contra Riesgos Sanitarios)	Yes	If local EC is not available, an independent EC must be used (however, highly recommended to be from same city as the site)	CMC: no specific CMC information is required for submission Insurance: local or medical assistance letter Phase IV: safety reporting requirements are stricter EC and COFEPRIS: entire and single review and approval process (no fast track) on a site-by-site basis	COFEPRIS: Ley General de Salud, Reglamento de la Ley General de Salud en Materia de Investigación para la Salud & NOM-220-SSA1-2002

(*Continued*)

Table 14.3 Regulatory overview—cont'd

Country	Regulatory agency	Local EC	Central, regional or independent EC	Individual requirements	Main regulations
Panama	MINSA through GORGAS (Ministerio de Salud through Comité Nacional de Bioética de la Investigación)	Yes, if available	National EC (that is actually the Comité Nacional de Bioética de la Investigación)	Local EC vs national EC: if local EC is available, approval is required before national EC submission. Otherwise, submission is direct to national EC Local EC and national EC: entire and single review and approval process (no fast track) on a site-by-site basis	GORGAS: Res 390 (Guía Operacional de Bioética en Investigación) & 201
Paraguay	GIGEVISA (Dirección General de Vigilancia Sanitaria; part of MSPBS – Ministerio de Salud Publica)	Yes	Not applicable	CB (Comité de Bioética) vs GIGEVISA: CTA submitted to the RA is submitted for ethics review to a chosen EC by the RA, and this EC is usually the CB from the University of Medicine of Paraguay (Facultad the Medicina) Mercosur: no specific Paraguayan regulations, but adoption of Mercosur GCP guidelines Sites not part of initial submission: expected to be approved through fast track, after EC approval	GIGEVISA: GCP Guidelines from the Mercosur (Res 129/96) & Res 2885/99 accepting Mercosur Guidelines for GCP

Peru	INS & DIGEMID (Instituto Nacional de Salud & Dirección General de Medicamentos, Insumos y Drogas)	Yes	If local EC is not available, an independent (central) EC must be used	INS vs DIGEMID: INS is responsible for CTA approval, while DIGEMID is responsible for IL approval (and IB assessment) CMC vs GMP standards: DIGEMID is very strict in regarding to CMC documentation and usually applies GMP marketing legislation standards on labeling, CoA, GMPc, and study drug-related CMC documentation Sites not part of main submission: approved through fast track, after EC approval Biological products: longer timelines for CTA review	INS: law 28.482, Supreme Decrees 017-2006-SA, 006-2007-SA, 011-2007-SA, 013-2009-SA & Guidance MAPRO-INS-001
Uruguay	CNB (Comision Nacional de Bioetica en Investigacion en Salud)	Yes	If local EC is not available, an independent EC must be used	Biological samples export and drug/material import: approving clinical trials in Uruguay is currently problematic. A law from the Bank of Transplant, which manages biological material use, is holding up clinical trial approval. There are ongoing discussions to resolve the issue	CNB: GCP Guidelines from the Mercosur (Res 129/96)

(Continued)

Table 14.3 Regulatory overview—cont'd

Country	Regulatory agency	Local EC	Central, regional or independent EC	Individual requirements	Main regulations
Venezuela	INHRR (Instituto Nacional de Higiene Rafael Rangel)	Yes	Not applicable	CMC: detailed requirements, including label, GMP, and drug composition Local EC and INHRR: entire and single review and approval process (no fast track) on a site-by-site basis	INHRR: Reglamento de investigación en farmacología clínica

CMC: chemistry, manufacturing and controls; CoA: certificate of analysis; CTA: clinical trial application; EC: ethics committee; GCP: good clinical practice; GMP: good manufacturing practices; GMPc: good manufacturing practices compliance; ICF: informed consent form; LA: Latin America; MoJ: Ministry of Justice; IL: import license; IPS: Institución Prestadora de Servicios Salud; MoH: Ministry of Health; RA: regulatory authority.

must commit to close monitoring and strict criteria for trial withdrawal and provision of rescue treatment.

Restrictions on placebo-controlled trials are most stringent in Brazil, where the National Ethics Committee and Federal Council of Medicine prohibit the conduct of clinical trials in which placebo is administered as sole treatment when an approved treatment is available for use as a comparator. Exceptions may be made only when usual care standards would justify a period with no treatment. Argentina is the Latin American country with the second most stringent restrictions, followed by Chile, Panama, Peru, Venezuela, and Uruguay. Other Central American countries, Mexico, and Paraguay are the region's least restrictive in this regard.

In Latin America, ICFs for clinical trials are typically given rigorous inspection, which can delay trial approval. This critical document requires extensive review on a country-by-country basis. It is common for the ICF to trigger the greatest number of questions from ethics committees and regulatory authorities.

Extensive CMC documentation may also be required, including bovine spongiform encephalopathy (BSE) risk compliance certifications [such as RO-CEP certificates from the European Directorate for the Quality of Medicines (EDQM)], certificate of analysis (CoA), and good manufacturing practice compliance certifications (GMP-c). In addition, study drug labels are of special concern in Peru; regulations list specific items that must be on inner and outer packaging, with no exceptions allowed by the regulatory body in charge of this review (DIGEMID).

In Brazil and Panama, there are additional restrictions on clinical trials related to poststudy access to the drug. If the study doctor considers the study drug to be the best treatment option for a participant, the patient must be assured of continued access to the drug.

Additional challenges include ensuring that site facilities are adequate and that staff members have appropriate levels of experience, and ensuring provision of the study drug, comparators, and background medication required for the conduct of the clinical trial. Sponsors must also often supply insurance as well as written assurance that treatment for injuries related to the study will be provided.

14.2.3. Clinical Trial Application Timelines

The time required to initiate a clinical trial varies among Latin American countries. The time may also vary by type of project, as some countries have extended start-up times for some types of trials. Furthermore, in some countries, only regulatory agency notification may be required in some cases. For example, there are reduced requirements for Phase IV trials in some countries (Brazil, Colombia, and Chile); details of these are described below.

It may take approximately three to four months to initiate a trial in Paraguay and Ecuador, and about five to six months in Chile, Colombia, Dominican Republic, Guatemala, Mexico, Panama, Peru, and Uruguay. The longest start-up periods are seen in Argentina, Brazil, and Venezuela, where eight to nine months may be required.

Individual countries have different timelines for particular kinds of clinical trial programs. For example, for Phase I and II trials, the timeline for initiation is one month longer in Argentina. In contrast, unless an import license is required, regulatory authorities need only to be notified of Phase IV trials in Chile, Brazil, and Colombia; this is true in Argentina as well for a subset of Phase IV trials, and also for observational studies in the latter three countries. Regardless of the need for import licenses, regulatory authorities in Chile and Peru require only notification of observational studies; meanwhile, in Mexico the full regulatory process is required for observational studies.

Timelines are extended by one month for biologicals in Peru, while in Brazil and Colombia, for products with genetically modified organisms (GMOs), trial initiation requires an additional regulatory process, through a biosafety regulatory agency (in Brazil this is called CTNBio – Comissão Nacional de Biossegurança).

Procedures for protocol amendments also vary widely. Ethics committee approval is required in all countries. In addition, regulatory authority notification is required in Brazil and Guatemala, but is required in Chile and Colombia only if the import license is not affected. Otherwise, in the latter two countries, the usual regulatory approval process must take place. Argentina, Paraguay, and Peru have a fast-track regulatory approval process for protocol amendments, while the full regulatory approval process is required in Dominican Republic, Ecuador, Panama, Mexico, Uruguay, and Venezuela.

14.2.4. Import License

An import license is always required in Latin American countries. In some the process is included as part of the CTA dossier, and the permit is granted along with the CTA approval. In others, the import license is obtained through a separate step outside the regular CTA dossier. Another variation is in whether a single license can cover all shipments, as in some countries, or whether each shipment requires a separate permit, which is the case in other countries.

Commonly, several documents are required before the import license is granted. CMC documents such as the product labeling, the GMP certificate, the manufacturing authorization, and the CoA are usually required. In addition, Peru requires that the study drug batch coding system be filed. Brazil, meanwhile, has specific requirements that sponsors provide documentation showing that products are free of BSE. Regulations also require filing of a pro forma invoice, which describes the shipment details including material, shipper, consignee, and weight. Most importantly, the application must state the amount of study drug that will be imported, and a rationale based on the expected enrollment must be provided.

In Argentina, Colombia, Paraguay, and Chile, the process for obtaining an import license is part of the CTA process, and the license issued may be used in multiple shipments; however, a second step is generally required before the drug can be imported. In Argentina, Colombia, and Paraguay, the additional step is performed at a department that handles foreign trade commerce; in Chile, the additional requirement occurs after the shipment has been received, in that the use and destination must be approved. Import licensing must be obtained for each shipment in Brazil, Dominican Republic, Guatemala, Panama, Uruguay, and Venezuela. Mexico, Ecuador, and Peru grant a global import license, which the regulatory agency considers sufficient for multiple shipments with no additional steps; however, as in all Latin American countries there may be additional specific procedures at customs or other specific commerce bodies. In addition, Mexico allows use of an import license waiver for some drugs and material (under foreign trade rule 3.1.33).

14.2.5. Pharmacovigilance and Serious Adverse Event Reporting

Because safety reporting tasks are governed by local rather than harmonized rules in most Latin American countries, sponsors of clinical research in the area must be prepared to act locally, understanding and adapting to rapidly changing requirements and methods.

In the 1990s, a first phase of safety systems development began in Latin America, as a great number of countries joined the network of participating countries of the World Health Organization (WHO) Programme for International Drug Monitoring coordinated by the Uppsala Monitoring

Centre. At this time, the classical methodology of voluntary notification by healthcare providers formed the basis for the postapproval pharmacovigilance systems, with physician-investigators rather than pharmaceutical companies being held responsible for notifying regulatory agencies of safety events.

A second phase of transformation is now underway, as specific legislation has been published incorporating preapproval safety data collection and analysis together with the systems and methods used postapproval. In this new phase of systems development, responsibility is progressively migrating from investigators to the pharmaceutical and biotechnology companies conducting the studies. This transformation is well illustrated by the number of countries (Table 14.4) in the region that have adopted new regulations for safety reporting in clinical trials within the past four years.

As a general rule, even in the countries without specific local safety reporting regulations, regulatory agencies have long expected that clinical research sponsored by companies with headquarters in ICH-participating countries will be conducted in compliance with ICH guidelines.

Dominican Republic and Panama are two examples of countries with no specific regulation of safety reporting in clinical trials. In addition, Costa Rica constitutes an unusual case: clinical trial regulation (decree 31078-S) formerly in place was retired in May 2010, having been found unconstitutional. As a result, authorities in Costa Rica have halted enrollment of ongoing trials and approval of new clinical trials, as they work to develop a law to replace the former decree.

The sections following outline the most prominent matters related to safety reporting. Local regulations include additional country-specific details.

Regardless of local regulations, sponsors should determine the means and frequency with which suspected unexpected serious adverse reaction (SUSAR) reports (local and foreign) will be delivered to the investigators to guarantee compliance with ICH-GCP and local requirements.

Table 14.4 Safety reporting guidelines and dates of publication

Country	Regulation	Year of publication
Argentina	Disposition 6677	2010
Brazil	RDC 39/08	2008
Chile	Guidance 5	2010
Colombia	2010020508	2010
Peru	Decreto 17-2006-SA	2006

SUSAR reporting to regulatory authorities must also comply with local requirements in terms of format, timelines, language, and delivery means.

Local practices and regulations in Latin America usually prevent direct contact between sponsors and ethics committees. For that reason, each investigator is usually responsible for forwarding SUSAR reports received from sponsors to the applicable ethics committee.

Timelines for SUSAR submissions by sponsors to investigators usually follow ICH-GCP guidance, meaning that reports of life-threatening and fatal events should be delivered within seven days, and all other reports within 15 days of first acknowledgment by the sponsor. In most of the Latin American countries, it is common practice to deliver CIOMS/MedWatch formatted reports to investigators in English within these time frames, after which translation of the report into the local language (Portuguese or Spanish), if locally required, is then prepared and submitted a second time to the investigators. After these steps are completed the report is submitted to the ethics committee.

Timelines for SUSAR submissions to ethics committees may vary and are prone to regulation by the ethics committee's standard operational procedures. Special conditions may apply to local events, as described below.

SUSAR submissions to regulatory authorities are in general a sponsor's responsibility, with time frames that may vary and may be more stringent when the event occurred locally. In Argentina, SUSARs must be submitted to the regulatory authorities in an expedited manner within 10 working days and are later submitted, through a periodic report of all SUSARs, every six months. In Brazil, the regulatory agency requires that a study annual report include all safety information updates, including any SUSARs reported to sites and ethics committees. In Chile, all SUSARs and local SAE-related event reports are to be delivered to the regulatory authorities within 15 working days.

SUSAR reports are often delivered to regulatory agencies by hand, as systems for electronic submissions are not in place. However, there have been some recent changes in delivery, as regulatory authorities have begun developing web-based systems for electronic submissions of local SAEs. For instance, Brazil requests that local SAEs be reported through a web-based system called NOTIVISA, and recently Peru has adopted a similar web-based system for reporting of all local SAEs.

Table 14.5 presents details on safety reporting requirements as of November 2010 for reference. However, consultation with local regulatory

Table 14.5 Safety reporting in Latin America

Country	Reporting type	Reporting requirement	Unblinded report required	Format (accepted or required)	Delivery (hand delivery/web-based/e-mail)	Language	Initial timeline for local expedited to regulatory agency	Follow-up timeline for local expedited to regulatory agency
Argentina	Expedited	SUSAR	Y	CIOMS 1, MedWatch, or sponsor's SAE form	E-mail	English or Spanish	10 working days	Not available
	Aggregate	SUSAR	Y	Summary following Provision 6677, item no. 7.4	Hard copy	English or Spanish	Every 6 months	Not applicable
Brazil	Expedited	Local all related	N	Local format (NOTIVISA)	Web-based	Portuguese	7 working days (fatal); 15 working days (others)	15 working days
	Aggregate	SUSAR	N	Line listings and full individual reports	Hard copy	Portuguese	Yearly	Not applicable
Chile	Expedited	Local all related; SUSARs foreign	N	CIOMS 1, MedWatch, or sponsor's SAE form	Hard copy or e-mail	English or Spanish	15 working days	15 working days
	Aggregate	Related national and international AE	N	Line listings (AEs summary); or updated IB	Hard copy or e-mail	English or Spanish	According to study duration. Studies up to 1 year's duration: at half and end of the study. Studies > 1 and up to 2 years' duration: every 6 months and at the end of the study. Studies > 2 years' duration: yearly and at the end of the study	Not applicable

Colombia	Expedited	Local all	N	Local format (INVIMA)	Hard copy	Spanish	7 working days	15 working days
	Under evaluation at INVIMA	SUSAR	N	To be defined	Hard copy	To be defined	To be defined	To be defined
Costa Rica		Regulations for clinical trials have been annulled, and currently there are no valid regulations						
Dominican Republic	Expedited	Local all; SUSARs foreign	N	Local: sponsor's SAE form; foreign: CIOMS 1/MedWatch	Hard copy	English with Spanish translation	Not specified: ASAP	Not available
Ecuador	Expedited	Local all	N	Any format	Hard copy	Spanish	8 calendar days	Not available
	Not specified	SUSAR	Not specified	Not specified	Not specified	Not specified	Not specified	Not specified
Guatemala	Expedited	Local all	N	Sponsor's SAE form	Hard copy	English with Spanish translation	24 hours	Not available
	Aggregate	SUSAR	N	Cover letter attaching full copies of investigator's safety letter or full individual reports (CIOMS 1 or MedWatch)	Hard copy	English with Spanish translation	Yearly	Not applicable
Mexico	Expedited	Local all	N	Local format (COFEPRIS)	Hard copy or e-mail	Spanish	7 calendar days	8 calendar days
	Aggregate	SUSAR	N	Line listings (specific information to be included)	Hard copy	Spanish	At the end of study report (for Phase I–III trials); and every 6 months (for Phase IV trials)	Not applicable

(*Continued*)

Table 14.5 Safety reporting in Latin America—cont'd

Country	Reporting type	Reporting requirement	Unblinded report required	Format (accepted or required)	Delivery (hand delivery/web-based/e-mail)	Language	Initial timeline for local expedited to regulatory agency	Follow-up timeline for local expedited to regulatory agency
Panama	Expedited	Local all; SUSARs foreign	N	Local: sponsor's SAE form; foreign: CIOMS 1/ MedWatch	Hard copy	English with Spanish translation (or Spanish summary for foreign)	Local: 24 hours; foreign: not specified, but ASAP	Not available
Paraguay	Expedited	Local all; SUSARs foreign	N	Any format	Hard copy	Local: Spanish; SUSAR: Spanish and English	Immediately	Not available
Peru	Expedited	Local all	N	Local format (INS)	Web-based	Spanish	7 working days	8 working days
	Aggregate	SUSAR	N	Cover letter with line listings, attached to full individual reports (CIOMS 1 or MedWatch)	Hard copy and pdf	English or Spanish	Quarterly or every 6 months (depending on INS approval)	Not applicable

Uruguay	Expedited	Local SUSARs	N	PI's forms and CIOMS	Hard copy and pdf	Spanish	2 calendar days	Not available
	Aggregate	SUSARs	N	Cover letter, CD with SUSAR form official and CIOMS plus note from the sponsor about causality	Hard copy and pdf	English and Spanish	Every 6 months	Not applicable
Venezuela	Expedited	Local all	N	Local format (CENAVIF) or sponsor's SAE form	Hard copy	Spanish	8 calendar days	Not available
	Not specified	SUSAR	N	Not specified	Not specified	Not specified	Not specified	Not specified

INVIMA: Instituto Nacional de Vigilancia de Medicamentos y Alimentos; SUSAR: suspected unexpected serious adverse reaction; AE: adverse event; CIOMS: Council for International Organizations of Medical Sciences; SAE: serious adverse event; IB: investigator's brochure; COFEPRIS: Comisión Federal para la Protección contra Riesgos Sanitarios; INS: Instituto Nacional de Salud; PI: principal investigator; CENAVIF: Centro Nacional de Vigilancia Farmacológica de Venezuela.

authorities and other local experts in the area is highly recommended because of the frequency with which these requirements are updated.

14.3. CURRENT STATUS AND STATISTICS ON CLINICAL TRIALS

The largest countries in the region are all seeing increases in the number of clinical trials. Thiers et al. analyzed clinical trial participation to determine trends in globalization, determining that from 2002 to 2006, the average annual growth rate was 33 percent in Peru, which tied with Russia as the global market with the highest rate of growth [12]. The growth rate was among the top 10 in Colombia, at 28 percent, and not far behind (no. 12) for Argentina, at 27 percent. It was 22 percent in Mexico, 16 percent in Brazil, and 11 percent in Chile.

Thiers et al. also discussed trial density as an indicator reflecting the percentage of population participating in clinical trials, reporting that in the developed countries, trial density was 29 in the UK, 50 in France, 92 in Canada, and 120 in the USA [12]. Latin American countries have far lower trial densities: 3 in Colombia, 4 in Peru and Brazil, 6 in Mexico, 11 in Chile, and 19 in Argentina. These figures clearly indicate the potential for expansion of trial participation.

As of November 2010, ClinicalTrials.gov showed that 35,156 studies were open worldwide; together, the 904 ongoing in South America, 326 in Mexico, and 322 in Central America compose only 4.4 percent of the total number. Most of these studies were located in the six major nations in the region: Brazil (594), Mexico (326), Argentina (274), Chile (165), Peru (154), and Colombia (141). The activity in emerging markets is also worth noting. In South America, Uruguay, Paraguay, and Bolivia each had fewer than 10 open trials, while Ecuador had 15 and Venezuela, 31. In Central America and the Caribbean, Bahamas, Cuba, El Salvador, Haiti, Honduras, Jamaica, and Nicaragua each had fewer than 10 studies open; Dominican Republic had 12, Costa Rica 24, Panama 31, and Guatemala 39. Puerto Rico led the Central America region with 220 open trials, probably because of the speed of the regulatory process due to its recognition of the US Food and Drug Administration (FDA) process. As these figures indicate, clinical research activity is not just confined to the major countries of the region but is seen in emerging countries within Latin America.

More clinical research data from not only the major countries in Latin America but also from these emerging Latin American countries are expected in the near and long-term future, primarily in indications such as infectious diseases, endocrinology, cardiology, and oncology, where sites and patient populations are readily available, as well as in rare diseases.

In the major nations, available sites and patients may become more scarce if growth continues, but in emerging Latin American countries there is more room for development. CROs, sponsors, governments, and non-profit organizations can facilitate the expansion of participation by investing in finding and developing new sites and resources in the region, now that experience has demonstrated ongoing research success in the major nations.

14.4. OVERVIEW OF INVESTIGATORS AND SITES IN LATIN AMERICA

14.4.1. Inspections

The US FDA is involved in review of clinical trials conducted elsewhere by virtue of the need to ensure the integrity of data that the sponsor submits to the agency to support marketing approval in the USA. The agency expects sponsors to conduct trials in compliance with the laws and regulations that govern the trial sites. Nonetheless, the FDA can perform on-site inspections of investigator sites in all nations participating in clinical trial programs that will yield results to be submitted to the FDA. Through inspection, the FDA can determine compliance with GCP and thus ensure the quality of submitted data. Where local or national laws and regulations impose additional requirements exceeding FDA regulations, the FDA expects compliance with those local or national laws.

The US FDA is working to increase its ability to oversee foreign clinical trials. As part of the "Beyond Our Borders" initiative, the FDA is creating a regional presence in Latin America, currently maintaining three offices in Latin America: in Costa Rica (regional office), Chile, and Mexico. The Beyond Our Borders initiative also emphasizes sharing of inspection reports and other information related to clinical research activities with foreign regulatory authorities.

The FDA has accepted data to support US marketing submissions from many different Latin American countries: Argentina, Bahamas, Brazil, Chile, Colombia, Costa Rica, Dominican Republic, Ecuador, Guatemala, Mexico, Panama, Peru, and Venezuela. The FDA international clinical investigator inspections have increasingly been performed in non-US countries, starting with slightly more than 20 in 2001 and increasing by approximately 100 percent between 2003 and 2004. From 2006 to 2008, the number of international inspections performed by the FDA averaged around 90 per year.

In Latin America, as of March 2009, the greatest number of FDA inspections had been performed in Argentina, with 29 total inspections, in Brazil, with 20, and in Mexico, with 18. The FDA has also inspected sites in Chile (eight), Costa Rica (eight), and Peru (six), as well as Guatemala (three), Venezuela, Panama (two each), Dominican Republic, Colombia, Ecuador, and Bahamas (one each). From 2005 to 2009, Latin American trial sites saw much greater success in FDA inspections than in the 24 years before (Table 14.6) [13].

The FDA GCP inspections bring opportunities and challenges. Investigative sites with GCP knowledge and experience should easily pass an FDA inspection, while those that are new to GCP or not willing to comply with ICH-GCP standards will face challenges. In general, the most common findings during inspections occur in comparable proportions between foreign (outside the USA) and domestic (USA) sites. These include protocol deviations (outside USA: 37%; USA: 41%), record-keeping/documentation (29% vs 27%), drug accountability (12% vs 11%), informed consent (5% vs 6%), and issues related to adverse events (3% vs 2%). To avoid deviations, sites should focus on training to ensure that documentation and all proceedings follow GCP.

As FDA inspection findings show, participants in Latin American clinical trials are all taking measures to minimize the risk of deviations from ICH-GCP. Additional assurance comes from provision of routine inspections by ministries of health in countries such as Argentina, Brazil, Chile, Colombia, and Peru. The governments of Latin American countries are committed to preserving patients' rights and to securing high-quality data. Ethics committees in Argentina, Peru, and Costa Rica

Table 14.6 Results of Latin American FDA site inspections

FDA observations	1980–2004	2005–2009
No Action Indicated (NAI)	20%	60%
Voluntary Action Indicated (VAI)	67%	38%
Official Action Indicated (OAI)	13%	2%

FDA: Food and Drug Administration.

also have programs of site inspection in accordance with internal processes.

14.4.2. Site Certifications and Facility Requirements

In Peru, Brazil, Colombia, and Argentina, facility requirements for sites are specific. Principal investigators must comply with certain requirements (e.g. licenses and training) before receiving authorization to act as a principal investigator.

In Argentina, several private (e.g. Universidad Austral) and public (e.g. Universidad de Buenos Aires) universities, as well as non-profit organizations such as SAMEFA (Argentinean Medical Association of the Pharmaceutical Industry, www.samefa.org.ar) and FECICLA (Foundation for Clinical Research's Ethics and Quality, www.fecicla.org.ar), offer clinical trial training programs for site staff, clinical research associates, and ethics committees; these programs also serve as a forum for professional discussions. Several other investigator associations are also very active in the promotion of the conduct of clinical trials, in states including Provincia de Buenos Aires and Ciudad de Buenos Aires.

In Brazil, the SBMF (Brazilian Medical Society of Pharmaceutical Medicine, www.sbmf.org.br) as well as some other non-profit organizations like the SBPPC (Brazilian Society of Clinical Research Professionals, www.sbppc.org.br) and the newly created SOBEPEC (Brazilian Society of Clinical Research Nurses, www.sobepec.com.br) play a significant role in training activities. Sites must establish that they have the facilities required for the conduct of a specific trial, as specified in ANVISA's norm number 4 (local regulation). The local ethics committees must be registered and approved in CONEP (Comissão Nacional de Ética em Pesquisa).

In Peru, sites must be registered at the Ministry of Health and licensed according to a specific facilities profile. Principal investigators must be graduated physicians who are registered and licensed by Colegio Medico; they must also have formal training in GCP and be trained in the medical research area.

In Mexico, although there is no formal certification requirement for investigators, multiple GCP training programs are available, as well as academic, institutional, and private forums addressing the topic. The Ministry of Health considers participation of Mexican subjects in international clinical trials to be an advantage for any new marketing application.

In Guatemala, there was no association of principal investigators as of November 2010. All investigators are legally required to complete a GCP course authorized by the Ministry of Health, which also oversees all clinical trials. At least five independent, ICH-compliant ethics committees are recognized and registered by the Ministry of Health.

In Panama, the Instituto Conmemorativo Gorgas and the National Ethics Committee require that all investigator staff members participating in a clinical trial have GCP training and the education level needed to conduct the research. A public institution (INDICASAT) that is part of the Ministry of Technology of Panama offers investigators and their staff regular GCP courses and other training related to clinical research.

In Costa Rica, the Ministry of Health began a review of clinical research regulations at the beginning of 2010 that was ongoing as of November. A new regulation is expected in early 2011. In the meantime, no new clinical trials will be approved by the Ministry of Health, although ongoing trials are allowed to continue with enrolled patients.

Across Latin America, ethics committees are taking a central role in the development of clinical trials, in response to requests from the governments. The ethics committees not only are responsible for the review of the protocol and other documentation, such as the investigator brochures and ICFs, but also have been requested to continue monitoring the course of the studies to secure patients' rights and welfare with an official monitoring plan.

Pharmaceutical companies and CROs are expending great efforts and resources in training for investigator sites and especially to develop adequate human resources, such as study coordinators, pharmacists, and clinical research associates. While in some locations a solid, established infrastructure exists to support the conduct of clinical trials, other locations are undergoing accelerated development to meet demand. These less experienced countries are learning from those with more trial experience to develop regulations, train staff, and secure high-quality data.

The terms of the contract with the site and with the investigators can vary among locations. Typical contract terms may place Latin American countries at an advantage in global clinical trial programs. Average investigator fees in the Americas span a large range, with the US investigators receiving the highest fees in most cases. In this evolving, highly dynamic environment, Latin American investigator fees are widely variable from one country to another, and may depend on several factors that vary among the countries as well as the indication. In some locations, the number of competitive trials forces sites to expend greater efforts and resources to comply with the enrollment rate estimated at the project planning stage. However, in general, investigator fees in Latin America are lower than investigator fees in the USA and Europe. In situations where investigator fees in Latin America are higher, there will be greater efficiency as compensation, brought by higher patient enrollment, fewer sites, and shorter timelines.

Site financial and contractual requirements add some complexity to this process. CROs and sponsors are working to develop best practices to ensure compliance with site and sponsor needs. Fixed costs at sites are increasing in a changing economy, so flexibility and openness are imperative for those involved in the conduct of clinical trials.

Several other advantages add to the efficiency brought by Latin American trial locations. Not only are investigator fees and salaries often lower, but individual sites typically achieve higher recruitment rates than US sites, reducing overall trial management costs. Recruitment is usually accomplished more quickly as well, adding to the cost savings.

14.4.3. Clinical Trial Staff

Despite the efforts already made to develop experienced and trained human resources, more resources and efforts are necessary in all of the countries. The constant growth of clinical trial activity in the region has led major CROs and sponsors to focus on identifying and developing new sites and site staff.

While development of site facilities requires only investment by public and/or private institutions, the development of the needed human resources will take more time. All involved in the pharmaceutical industry, including regulatory authorities, need experienced personnel. For several reasons, sponsors and CROs sometimes struggle to balance these needs with the current availability of trained professionals willing to join the clinical research business. Staff turnover due to the high demand for trained personnel has a further impact, intensifying the challenge of identifying adequate personnel. CROs and pharmaceutical companies along with non-profit organizations are helping to address this gap by focusing on education and training to develop the human resource pool.

14.5. CONTRACT RESEARCH ORGANIZATIONS

14.5.1. Perspective of Contract Research Organizations

As clinical research became a global activity in the early 1990s, the first CROs in Latin America were established to meet a high demand for patients to comply with tighter project timelines expectations. In the major countries, companies with a reach of only one or two countries were established first. Shortly thereafter, the CROs that already had a global presence, with headquarters in the USA or Europe, were established in the region.

In line with a trend toward globalization of clinical research in the decade, several global CROs first established legal entities and offices in Argentina, Brazil, or Mexico and from there expanded operations into the rest of the region. More entrepreneurial companies continued appearing in the market, with the result that as of 2010 the presence of CROs is well established in Latin America. Currently, CROs in the region range from large, international full-service organizations to small, niche specialty groups.

The CRO industry faces several challenges in the current era. Traditionally, CROs are expected to have local knowledge of regulations and sufficient resources in the countries to compensate for the lack of sponsor presence; they also must be able to provide experience in the studied indication or disease and multilingual staff, including clinical research associates. Multicenter, multinational trials with more complexity and larger patient populations have increased the demand, and CROs have expanded their operations to meet the need for a presence in more countries.

The demand for local knowledge of regulations, language skills, and appreciation of local culture has continued to increase. Some companies have addressed these demands by opening new offices, while others have chosen to employ regional resources, flying long distances between the various countries. The regional approach may bring additional cultural challenges, as individuals from one country are responsible for monitoring and managing sites in another.

As in all regions, the CRO team must have chemistry with the sponsor's team, experience in clinical trial execution, a proactive attitude, investments in technology, flexibility, and the capabilities the clinical trial program demands.

Some local CROs maintain a presence in one or two countries in the region as niche companies satisfying specific needs, such as oncology experience or resourcing expertise.

14.5.2. Perspective of Sponsors

The importance of Latin America in the pharmaceutical market is well known in the industry. With some market fluctuations, Argentina, Brazil, and Mexico have been leading the market over the past several decades. However, other countries such as Chile, Peru, and Colombia have maintained a place of importance for large pharmaceutical companies.

All large global pharmaceutical companies have some type of presence, whether an affiliate or a local company acting as a sales representative (i.e. a Latin American pharmaceutical company), in each of the countries.

Because the affiliates of large global pharmaceutical companies have been performing clinical research studies for marketing or research and development purposes for more than a decade, many have fully staffed, highly professional teams of clinical research professionals. Others have decided to concentrate on core activities, such as site selection and regulatory submissions, choosing to outsource activities such as monitoring, project management, pharmacovigilance, and/or study drug management and trial logistics. The services outsourced therefore range from stand-alone services, such as monitoring or pharmacovigilance, to full service, including protocol design, regulatory services, data management, analysis, and reporting.

Some biotechnology companies are established in the region, although most are based in the USA or Europe and hold no legal or physical presence in Latin America. Biotechnology companies in particular, therefore, require partnerships with CROs to achieve the knowledge and expertise necessary to perform clinical research in the region.

Some Latin America companies with an origin in one country only have become regional or even global, or are currently undergoing expansion. With recent changes in local regulations, these companies, which in the past could not afford global CRO costs, are becoming potential CRO clients. Global CROs have identified and responded to this need, and are developing strategic or business development departments to fulfill it. Some companies of Latin American origin are expanding to other regions such as Eastern Europe, the Middle East, and/or Asia, thus incurring a need for the truly global expertise of a global CRO.

14.5.3. Outsourcing Strategies

As large global pharmaceutical companies have continued determining the best outsourcing model to fit their needs and company culture, the dilemma has been catalyzed by the economic crisis of 2008 and 2009.

Outsourcing models and strategies are a dynamic scenario of sponsor–CRO relationships. In Latin America, the transactional model remains the choice of many US and European biotechnology companies, while others, along with large global pharmaceutical companies with a strong presence in the region, choose preferred supplier and partnership models. Strategic partnerships are consolidating in the region, with deals secured at the global level between CROs and large pharmaceutical companies.

In summary, CROs are adapting their services to meet sponsor demands in the region, and are changing to meet the needs of this dynamic business environment. In addition to the traditional CRO capabilities, CROs operating in the region need a global reach, the flexibility to adapt to different outsourcing models and strategies, and efficient processes to minimize clinical trial costs.

14.5.4. Clinical Trial Organizations

Pharmaceutical associations are in place in each of the Latin American countries. Although these were originally formed with the goal of general promotion and support for the pharmaceutical industry, the focus has turned to clinical research because of the rapid growth in this field.

Table 14.7 shows the associations of pharmaceutical companies in each of the countries and the newly created associations of CROs (similar to the Association of Clinical Research Organizations in the USA, www.acrohealth.org). CRO associations play an important role in business-related

Table 14.7 Pharmaceutical organizations and CRO associations in Latin America

	Pharmaceutical organization	CRO association
Argentina	CAEME* (Cámara Argentina de Especialidades Medicinales): secretariageneral@caeme.org.ar; www.caeme.org.ar	CAOIC (Camara Argentina de Organizaciones de Investigación Clinica): www.caoic.org.ar
Brazil	Interfarma (Associacao da Industria Farmaceutica de Pesquisa): www.interfarma.org.br	ABRACRO (Associação Brasileira de Organizações Representativas de Pesquisa Clínica): www.abracro.gov.br
Bolivia	ASOFAR* (Asociación de Representantes, Importadores y Distribuidores de Fármacos): asofar@asofar.org	None

Table 14.7 Pharmaceutical organizations and CRO associations in Latin America—cont'd

	Pharmaceutical organization	CRO association
Centroamérica	FEDEFARMA★ (Federación Centroamericana de Laboratorios Farmacéuticos): www.fedefarma.org	None
Chile	CIF★ (Cámara de la Industria Farmacéutica de Chile AG): www.cifchile.cl	Under formation
Colombia	AFIDRO★ (Asociación de Laboratorios Farmacéuticos de Investigación): contactenos@afidro.com; www.afidro.com	None
Ecuador	IFI★ (Corporación de la Industria Farmacéutica de Investigación): www.promesa.com.ec	None
Mexico	AMIIF★ (Asociación Mexicana de Industrias de Investigación Farmacéutica AC): www.amiif.org.mx	CRO Alliance AC
Peru	ALAFARPE★ (Asociación Nacional de Laboratorios Farmacéuticos): alafarpe@alafarpe.com	APOICC (Asociación Peruana de Organizaciones de Investigación Clinica por Contrato): web page under construction
Uruguay	CEFA★ (Camara de Especialidades Farmaceuticas y Afines): cefa@adinet.com.uy	None
Venezuela	CAVEME★ (Cámara Venezolana del Medicamento): caveme@caveme.org; www.caveme.org	None

CRO: contract research organization.
★ Member of FIFARMA (Latino American Federation of Pharmaceutical Companies).

activities, engaging in meetings with the community and with regulatory and health authorities, as well as providing training.

Both the pharmaceutical and the CRO associations are helping to build long relationships with all stakeholders in the industry while spreading information to the community in general. Some of the organizations (denoted with an asterisk in the table) are members of the Latino American Federation of Pharmaceutical Companies (FIFARMA); all pharmaceutical organizations mentioned are members of the International Federation of Pharmaceutical Manufacturers and Associations (IFPMA). In the future, it is likely that CRO associations will become integrated into a pan-American or Latin American federation of CROs, in order to harmonize processes and practices and to work more collaboratively.

In the countries with a longer history of participation in clinical research, CRO associations were formed between 2008 and 2010. Others are under formation, and the remainder of the countries are expected to begin pursuing CRO associations in alignment with the growth in clinical research activities in their countries.

14.6. CHALLENGES AND OPPORTUNITIES

14.6.1. Challenges

The sharp increase in Latin American clinical trial activity has created competition for participants, both patients and investigators. The high population concentration in the cities is partly outweighed by the limited number of investigative sites within those cities. This complicates investigator recruitment and also results in high workloads at those sites, which necessitates highly skilled support staff. Identification and training of new sites and clinical research associates in the major countries as well as in the countries with lower activity should be a priority.

Regulatory challenges to conducting clinical trials in Latin America arise owing to the variability in regulatory requirements among the countries; a high level of expertise is necessary to master the specific requirements of each. While start-up times are competitive in several Latin American countries, including Mexico, Peru, Colombia, Panama, Costa Rica, and Guatemala, regulatory timelines can be more extended in Argentina and Brazil. Regulatory authorities, associations of CROs, and the pharmaceutical industry are working together to simplify these processes by adapting regulations to current needs.

Management of investigational products and other clinical trial supplies can be challenging in some remote areas of the region. Several well-known global and regional companies currently offer courier and depot services in the region. Although there is room for improvement in this service, it is possible in the meantime to find companies that have sufficient processes and infrastructure to comply with challenging protocol designs, managing, for example, frozen patient samples and/or samples that require fast and secure logistic procedures.

An import license is necessary in most of the countries before investigational products and study supplies may be brought in. The process of obtaining the license, as well as the procedures necessary to clear the goods at country customs, can be cumbersome, but a high level of familiarity with regulations can assist in successful navigation of these regulations.

Information technology infrastructure remains a challenge in some remote areas of the countries in Latin America. However, the use of the Internet has become more frequent, so electronic data capture and interactive voice response systems have become more available. Efficient site selection processes and project plans are key to the success of the trials.

Although only two official languages are used in the majority of countries in the region, Spanish and Portuguese, the translation of documents such as protocols and investigator drug brochures may be challenging, as may adaptation of the ICFs to local variations of the languages. Just as there are differences between American English and British English, there may be small differences in use of language among the Spanish-speaking countries. Currently, several freelance certified translators and small or medium-sized companies provide language conversion. An opportunity remains for larger, regional companies to provide translation services, applying standard operating procedures and local resources to improve the review cycles.

14.6.2. Opportunities

Clinical studies are becoming more and more complex as the numbers of enrolled patients increases to meet requirements of the FDA and the European Medicines Agency (EMA). Clinical trial designs are also changing, as studies to test risk of cardiovascular events or to evaluate long-term safety are more frequent. Latin America provides study-naïve patient populations and also has larger urban populations than other emerging regions. The disease patterns seen in Latin America reflect those experienced in the Western developed world as well as in the developing world. These population characteristics enable accelerated enrollment, high per-site recruitment levels, high patient retention rates, and simplified patient follow-up. These factors can make a dramatic difference in trial efficiency, in an environment of tighter timelines and funds.

Seasonal patterns in Latin America are the reverse of those in Europe and North America, so including Latin American sites can increase the efficiency of trials in seasonal disease studies, such as large vaccine trials.

Meanwhile, the overlap of time zones and frequency of flights between Latin America and the USA is a particular advantage for US-based sponsors.

The healthcare infrastructure in Latin America, including laboratories, is robust, and the region is a growing pharmaceutical market. The region has many well-trained investigators who are familiar with GCP and ICH guidelines. Experience has proven that the quality of data collected in Latin America is similar to that of North American data, demonstrating the clinical research potential of the region.

REFERENCES

[1] United Nations Population Division. World Population Prospects: The 2008 Revision. Population Database, http://esa.un.org/unpp/ (accessed November 19, 2010).
[2] US Central Intelligence Agency. CIA World Factbook, https://www.cia.gov/library/publications/the-world-factbook/ (accessed November 19, 2010).
[3] United Nations Department of Economic and Social Affairs. 2008 Demographic Yearbook. United Nations, New York, http://unstats.un.org/unsd/demographic/products/dyb/dybsets/2008%20DYB.pdf; 2010 (accessed November 19, 2010).
[4] United Nations. Millennium Development Goals: Achieving the Millennium Development Goals with Equality in Latin America and the Caribbean: Progress and Challenges. United Nations, Santiago; 2010.
[5] Wilson D, Purushothaman D. Dreaming with BRICs: the path to 2050. Goldman Sachs Global Economic Paper No. 99, http://www2.goldmansachs.com/ideas/brics/book/99-dreaming.pdf; 2003.
[6] World Health Organization. World Health Statistics 2008. Geneva: World Health Organization, http://www.who.int/whosis/whostat/EN_WHS08_Full.pdf; 2008.
[7] Hearing Before the House Subcommittee on Immigration. Citizenship Border Security and International Law, 111th Congress, 1st Session. Testimony of Daniel M. Masterson, Department of History, US Naval Academy; 2009.
[8] Pan American Health Organization. Health in the Americas. Washington, DC: Pan American Health Organization; 2007.
[9] International Diabetes Federation. Diabetes Atlas, www.diabetesatlas.org (updated 2010) (accessed November 21, 2010).
[10] Pan American Health Organization Health Information and Analysis Project. Health Situation in the Americas: Basic Indicators 2009. Washington, DC: Pan American Health Organization, http://new.paho.org/hq/dmdocuments/2009/BI_ENG_2009.pdf; 2009 (accessed November 19, 2010).
[11] International Agency for Research on Cancer, World Health Organization. Globocan 2008, http://globocan.iarc.fr/ (accessed November 19, 2010).
[12] Thiers FA, Sinskey AJ, Berndt ER. Trends in the globalization of clinical trials. Nature Reviews Drug Discovery 2008;7:13–5.
[13] Lepay D. FDA requirements for conducting clinical trials in foreign countries: opportunities and challenges. Presented at 6th DIA Latin American Congress of Clinical Research; 2009. September 25, 2009.

CHAPTER 15

Clinical Trials in Central and Eastern Europe

Irina Baeumer*, Yuriy Zuykov, Pavel Lebeslé†, Krisztina Szabo‡, Svetlana Riekstina§, Monika Stepniewska|, Alin Balalau¶, Victoria Datsenko†† and Greg Voinov*****

*Medical and Safety Services, ICON Clinical Research, Germany
**ICON Clinical Research, Russia
†ICON Clinical Research, Czech Republic
‡ICON Clinical Research, Hungary
§ICON Clinical Research, Latvia and Estonia
|ICON Clinical Research, Poland
¶ICON Clinical Research, Romania
††ICON Clinical Research, Ukraine
***Clinical Operations, France

Contents

Global Clinical Trials
ISBN 978-0-12-381537-8, Doi:10.1016/B978-0-12-381537-8.10015-9

15.1. CZECH REPUBLIC

15.1.1. Geography

The Czech Republic is a landlocked country situated in central Europe. It has a population of 10.33 million, the vast majority of whom are ethnic Czechs. The number of inhabitants decreased between 1994 and 2002, but has risen markedly since 2004. Economically, the country has performed well since the Velvet Revolution in 1989 and now is one of the most developed industrialized economies among the new European Union (EU) member states. Life expectancy at birth is increasing and, at 73.82 years for men and 80.30 years for women in 2007, is well above the average for the new EU member states. With 3.79 deaths per 1000 live births, the infant mortality rate in 2007 was among the lowest in the world.

15.1.2. Healthcare Infrastructure

The Czech health system is characterized by relatively low total healthcare expenditure as a share of gross domestic product (GDP) compared to Western Europe. Total health spending accounted for 6.8 percent of GDP in the Czech Republic in 2007, lower than the average of 8.9 percent in Organisation for Economic Co-operation and Development (OECD) countries. The Czech Republic also ranks below the OECD average in terms of total health spending per capita, with spending of US $1626 in 2007 (adjusted for purchasing power parity), compared with an OECD average of $2964.

The population enjoys virtually universal health coverage with a broad range of benefits, and some important health indicators are better than the EU average (such as mortality due to respiratory disease) or are among the best in the word (e.g. infant mortality). In contrast, the standardized death rates for diseases of the circulatory system and malignant neoplasms are above the EU average. Healthcare utilization rates, such as outpatient contacts and average length of stay in acute-care hospitals, are also above the EU average. In summary, there is substantial potential for efficiency gains and improved health outcomes.

15.1.2.1. Organization and Regulation

The Czech Republic has a system of social health insurance (SHI) based on compulsory membership of one of the country's 10 health insurance funds. The funds are quasi-public, self-governing bodies that act as healthcare payers and purchasers. The Ministry of Health's main responsibilities include setting the healthcare policy agenda, supervising the health system and preparing health legislation. The Ministry also administers certain healthcare institutions and bodies, such as the public health network and the State Institute for Drug Control (Statni ustav pro kontrolu leciv; SUKL). The regional authorities and the health insurance funds play important roles in ensuring access to healthcare, the former by registering healthcare providers, the latter by contracting them. Eligible residents may freely choose their health insurance fund and healthcare providers. The health insurance funds must accept all applicants who have a legal basis for entitlement; risk selection is not permitted. SHI contributions are obligatory and based on wage or income; they are paid by employers, employees, and self-employed individuals, among others.

15.1.2.2. Resources (Data From 2007)

There were 3.6 practicing physicians and 8 qualified nurses per 1000 population (OECD average 3.1 and 9.6, respectively). The number of acute-care hospital beds was 5.2 per 1000 population (OECD average 3.8 beds). As in most OECD countries, the number of hospital beds per capita has decreased over time, coinciding with the reduction in the average length of hospital stay and an increase in the number of day surgical procedures. The number of computed tomography (CT) and magnetic resonance imaging (MRI) scanners was 12.9 and 4.4 per million population, respectively (OECD average of 20.2 and 11, respectively). Not all healthcare facilities have been able to keep pace with advances in medicine, and some psychiatric, long-term care, and nursing facilities are outdated and in need of modernization.

15.1.3. Disease Profile

The main causes of death (2006 data, number of deaths per 1000 population) are ischemic heart disease (25), cerebrovascular disease (15), trachea, bronchus and lung cancers (6), colon and rectal cancers (5), and lower respiratory tract infection (2).

15.1.4. Patient Profile

There are approximately 20,000–30,000 patients currently enrolled in clinical trials. The Czech Republic is an attractive location for clinical studies because it benefits from a homogeneous patient population owing to its centralized healthcare system. Patients are motivated to participate in clinical trials to obtain more detailed medical follow-up and significantly better overall medical care with reduced treatment costs. In addition, the high level of trust in doctors encourages a high level of treatment compliance. The Czech population is relatively static, which facilitates study follow-up. There is a high incidence of hypertension, hypercholesterolemia, coronary heart disease, diabetes, schizophrenia, and neurotic disorders.

15.1.5. Regulatory Environment

SUKL (competent authority) and ethics committee procedures fully adhere to International Conference on Harmonisation good clinical practice (ICH-GCP) and EU Directives.

15.1.5.1. Competent Authority Submission

For multicenter clinical trials, a single clinical trial application (CTA) listing the proposed sites and core and country-specific documents is made to the competent authority. The content of submission dossiers fully adheres to EU Directive requirements. On receipt of the application, the competent authority reviews the package for completeness and assigns a SUKL reference number. This preliminary review is completed within 10 days. Following the preliminary review, a written confirmation of receipt of a valid package or request for further documentation is sent to the applicant.

The review starts immediately after the package validation process is completed or as soon as the outstanding documents are received. For all studies, the overall review timelines conform to the EU Directive. Comments or objections are usually provided in writing within 45–55 days of the review start date. As the local legislation obliges the authorities to communicate officially via an electronic system ("data box", a legislative requirement introduced in 2009), the competent authority only issues its opinions to the applicants electronically. The SUKL accepts communications in English or Czech. Initial submissions, as well as the submission of substantial amendments, are subject to fees.

The list of clinical trials (except Phase I) approved since 2008 by the SUKL is available on the competent authority website. In 2008, 226 studies were approved, and in 2009, 274 were approved. The SUKL accepts the Voluntary Harmonization Procedure for the assessment of multinational clinical trial applications. The SUKL performs inspections of sites and ethics committees on a regular basis. For authorization of a clinical trial with a product containing genetically modified organisms, a license or a decision from the Ministry of the Environment is required. If radiopharmaceuticals are involved, the SUKL will either apply for an opinion from the State Office for Nuclear Safety or request the sponsor to obtain this opinion.

15.1.5.2. Ethics Committee Submission

Applications for both multicenter ethics committee (MEC) and local ethics committee (LEC) approval are made in parallel with the competent authority submission. The MEC provides general ethical assessment for a study, while the LEC provides an opinion on the site and the local study team but is not entitled to request any changes to the study design or relevant documents. To date, there are nine MECs and numerous LECs established by hospitals and private medical facilities. MECs are approved by the Ministry of Health and sponsors can choose to submit a study to any MEC. Ethics committees issue their opinions on applications at regular meetings and the legal timelines for the final opinion adhere to ICH-GCP and EU Directives. The final opinion is provided in writing to the applicant, competent authority, and other relevant ethics committees. LEC approval is valid only for the particular site and only if the study is approved by an MEC and the competent authority. Communication with ethics committees must be conducted in the local language but, except for the cover letter, patient documents, and the protocol synopsis, submitted documents can be in English.

The requirements for patient information leaflets (PILs) and informed consent forms (ICFs) comply with the EU Directive and should also comply with the requirements of Section 4.8.10 of the EU Note for Guidance on Good Clinical Practice (CPMP/ICH/135/95). A separate ICF is required for genetic testing, although a sponsor template is acceptable. Ethics committees and the competent authority require PILs and ICFs be less than six to eight pages long, so the information is easy for the patient to understand.

15.1.6. Good Clinical Practice

GCP certificates granted by a sponsor to investigators after an investigator's meeting are sufficient. GCP training can be provided by either a contract research organization (CRO) or a pharmaceutical company.

15.1.7. Clinical Trial Status

The number of clinical studies performed has been increasing during the past decade. Phase II and III trials currently represent 75–85 percent of the studies conducted. The Czech Republic is still an attractive place to conduct clinical studies, although the increase in the number of clinical trial applications has stabilized.

15.1.8. Site

Two types of contract are predominantly used, either a bipartite contract between the CRO or sponsor and the institution (mostly hospitals) or a tripartite contract between the CRO or the sponsor, the hospital, and the principal investigator. Private practices can also be contracted. For hospital sites, the median contract turnaround time is six weeks, while that for private practices is typically shorter, usually four weeks. Hospitals usually request at least 20 percent of the study grant; the usual percentage is 30–50 percent. The principal investigator is provided with the study grant at an initial stage. Depending on the study procedures, it may be necessary to set up separate agreements with different departments if these procedures are not performed centrally.

15.1.9. Inspection

Local regulatory authority, European Medicines Agency (EMA), and Food and Drug Administration (FDA) inspections take place. Data on these inspections are not available.

15.1.10. Pharmacovigilance and Serious Adverse Event Reporting

Serious adverse events (SAEs) are not required to be reported to the regulatory authority, ethics committee or investigator. Suspected unexpected serious adverse reaction (SUSARs) must be reported according to the local competent authority (guidance KLH 21, version 4).

15.1.10.1. Regulatory Authority: SUKL

All local SUSARs and those from other countries in a study running in Czech Republic must be reported within seven to 15 days (seven days for death or life-threatening cases plus eight days for follow-up/final report). All SUSARs from other countries do not need to be submitted if they have been submitted electronically to the EudraVigilance database. All new findings that increase patient risk and safety concerns must be reported immediately. An annual safety report that includes SUSARs identified from spontaneous reports or the literature is issued with annual investigator brochure revisions.

15.1.10.2. Multicenter and Local Ethics Committees

Local SUSARs must be reported within seven to 15 days to the MEC and the LEC where the SUSAR occurred. All new findings that increase patient risk or safety concerns must be reported immediately. A line listing containing all SUSARS from the study is sent to MEC and LEC every six months and every three months in case of pediatric studies. An annual safety report that includes SUSARs identified from spontaneous reports or literature is issued.

15.1.10.3. Investigator

Local SUSARs must be reported within 15 days (unblinded but without subject and investigating product identification) and SUSARs from other countries within 15 days or in a six-monthly line listing (as decided by the sponsor), plus a cover letter signed by a physician addressing the benefit/risk ratio (at least six-monthly overview plus conclusion) supplemented by immediate "Dear Investigator" letters in cases of urgent safety issues. All new findings that increase patient risk and safety concerns must be reported immediately.

15.1.11. Contract Research Organizations

The CRO industry has enjoyed a 30 percent growth over the past decade and there are currently about 25 CROs. However, the stabilization in the number of clinical trial applications in the past two years and the fragmentation of the CRO market into global, mid-size, and local companies are signs that the market is in the late growth phase.

ICON began operating in 1997, using Czech clinical research associates (CRAs) based in the ICON office in Germany. ICON established an office in the Czech Republic in 2007. Since 1997, ICON has conducted more than 60 studies in over 350 sites. Today, ICON has more than 80 employees and is currently conducting over 70 studies. The major therapeutic areas include cardiovascular, oncology, central nervous system (CNS), endocrinology, gastroenterology, hematology, urology, dermatology, infectious diseases, ophthalmology, orthopedics, and pain control.

The clinical services offered by ICON include Phase II–IV clinical services, feasibility studies, site selection and evaluation, regulatory applications, ethics committee submission coordination, investigator meetings, trial initiation and monitoring, SAE reporting and evaluation, project management, and supplies distribution. Other services can be subcontracted where necessary or can be operated from other ICON offices in Europe. These include laboratory services, storage and archiving of documents, medical, statistics, and data management.

15.1.12. Outsourcing Model

Owing to the strong growth in the number of clinical studies, the industry has to handle the challenges in recruiting new, well-qualified employees. To fill the resourcing gap, large global CROs are hiring entry-level CRAs. Currently, the use of study nurses or site management organizations (SMO) is not common. However, it is expected that this will change in the future owing to the increasing investigator responsibilities and the increasing complexity of study procedures.

15.1.13. Logistics

Sites are familiar with interactive voice response system (IVRS) and electronic data capture (EDC) technologies. Clinical trial insurance can be contracted locally if necessary. Czech regulations require that study medication supplies are received by the pharmacy or pharmacist. Investigatory product storage conditions should be provided and checked by the pharmacy or delegated pharmacist.

15.1.14. Clinical Trial Organization

The Czech Association of CROs (ACRO), a professional association of CROs, was founded in 2005 and is a member of the European Federation of CROs. The goal of the ACRO is to support the development of clinical research and the development of pharmaceuticals in the Czech Republic. The SUKL supports the development of clinical research in the Czech Republic by providing guidance and seminars to the clinical research industry to support high-quality and ethical standards. Several local companies offer local seminars on the various aspects of clinical research in the Czech Republic.

15.1.15. Challenges and Opportunities

The Czech Republic provides several advantages. There are many highly educated, well-trained, motivated investigators familiar with GCP standards and FDA/EMA regulations. The healthcare infrastructure provides access to modern technology, diagnostic equipment and procedures, and treatment techniques. Centralized healthcare centers provide access to a large patient population. Patient compliance is facilitated by the high levels of trust in the medical profession. However, competition is increasing from other Eastern European countries with less developed healthcare systems, such as Russia and Ukraine, which have larger treatment-naïve populations.

15.2. HUNGARY

15.2.1. Geography

Situated in Central Europe, Hungary has a land area of 93,030 km^2 and is bordered by Austria to the west, Serbia, Croatia, and Slovenia to the south and south-west, Romania to the south-east, Ukraine to the north-east and

Slovakia to the north. The population is 10 million: almost 2 million people live in the capital Budapest and about half the population lives in communities of less than 20,000 people. The population has been in continual decline since the late 1970s and aging is significant, with 17 percent of the population over the age of 65 years. Life expectancy at birth is one of the lowest in Europe: 69.97 years for males and 78.25 years for females. The adult literacy rate is 99.4 percent.

15.2.2. Healthcare Infrastructure

The public healthcare system is financed through the Health Insurance Fund (HIF), which is predominantly based on a social insurance system and is primarily responsible for recurrent healthcare costs. Health insurance contributions are collected from employees and employers. Patients make co-payments for certain services, including drugs, dental care, and rehabilitation. The HIF is able to contract freely with providers and reimburses them in various ways. Hospital outpatient clinics are paid on a fee-for-service basis, while acute- and chronic-care centers are paid on a diagnosis-related basis and by length of hospital stay. General Practitioners (GPs) can be employed by the local government but the majority opt for "functional privatization", a payment scheme in which GPs contract with the HIF and are paid a capitation fee based on their patient list.

In the current system, most healthcare is provided by local government. Municipalities own primary care and outpatient clinics and hospitals providing secondary care. County governments run county hospitals that provide secondary and tertiary care. Some private, church-owned hospitals also exist but most still operate under HIF financing. Most pharmacies are private. The national government owns university and specialist hospitals but most healthcare is provided by local government. Currently, there are 70,971 hospital beds (70.8 beds per 10,000 population) and 36,088 doctors (36 doctors per 10,000 population), 6560 of who are GPs. Total healthcare expenditure is 8.3 percent of GDP and healthcare costs are approximately 60 percent of the EU average.

The HIF provides benefits in kind, i.e. health services provided by suppliers financed by the HIF, and benefits in cash. Health services provided free of charge (benefits in kind) include preventive medical examinations, primary healthcare services, dental care, outpatient and inpatient care, maternity care, medical rehabilitation, patient transportation, and accident and emergency care. Cost allowances for healthcare include drugs, medical aids, travel, and international medical costs. Co-payment is charged for orthodontic treatment for children and adolescents (less than 18 years old), teeth maintenance and replacement for adults (18 years old or over), extra meals and accommodation for inpatients, and sanatorium treatment. Cash benefits include sick pay, maternity benefit, childcare fees, disability benefits, accident benefits, and accident pensions.

15.2.3. Disease Profile

Based on some key metrics, the general heath of the population is one of the worst in Europe, with 52.0 healthy life-years for males and 53.9 for females. The leading causes of death are cardiovascular diseases (54%), cancer (27%), gastrointestinal diseases (8%), and respiratory diseases (4%). Hungary has the highest incidence of colorectal and lung cancers in Europe in both sexes. It also has the third highest alcohol consumption in Europe, 30 percent of the population smoke daily, 30 percent of women and 42 percent of men are overweight, and 20 percent of the population is obese. The most common diseases by incidence, based on 2008 hospital discharge statistics, are cardiovascular diseases, cerebrovascular diseases, respiratory diseases, gastrointestinal diseases, and musculoskeletal diseases.

15.2.4. Good Clinical Practice

GCP has been mandatory in Hungary since 1995, so investigators are usually very well informed about the professional requirements. The most experienced investigators have been involved in clinical trials for 10–15 years. Since Hungary joined the EU, ICH-GCP standards are the norm.

15.2.5. Clinical Trial Status

During the past five years, 250–300 CTAs have been submitted. The majority of clinical trials are performed in university clinics, national institutes, and general hospitals or their associated outpatient clinics. The number of primary care studies has increased significantly during the past three to four years and the number of private practices involved in studies is also increasing. There are several specialized private research organizations that perform only clinical trials. The largest hospitals and institutions usually have a clinical pharmacology research unit or clinical trial coordination unit. Phase I study sites have to be accredited and, to date, 10 sites have received certification from the National Institute of Pharmacy (NIP). Clinical research sites, especially those based in larger institutions, are generally well equipped with modern diagnostic and therapeutic equipment.

15.3. LATVIA

15.3.1. Geography

Latvia is a European state, which is situated on the coast of the Baltic Sea. One of the three Baltic states, Latvia is bordered by Estonia to the north, Lithuania to the south, Russia to the east, Belarus to the south-east, and the Baltic Sea to the west. The population (2009) is 2,261,294. Major ethnic groups (2008) include Latvians (59.2%), Russians (28.0%), Belarusians (3.7%), Ukrainians (2.5%), and Poles (2.4%).

Latvian is the only official state language. Besides Latvian, Russian is spoken fluently by most of the people, since Latvia was part of the USSR.

Latvia is one of Europe's smallest countries, and faced a difficult transition during the 1990s from being part of the USSR to developing a market economy. Of the three Baltic republics, Latvia has the strongest ties to Russia; a sizeable minority of the population is Russian, and the country has a high level of dependency on Russia for energy supplies and other raw materials. However, since the 1998 Russian economic crisis, tough government reforms have led to an improved economic performance and impressive levels of GDP growth. Latvia became a full member of the EU in May 2004 and its medical device legislation is in line with the relevant EU Directives.

15.3.2. Healthcare Infrastructure

The healthcare system has undergone a number of reforms since independence. The initial reforms of the early 1990s saw a wide degree of decentralization. This proved chaotic, and a firmer degree of central control is now in place. The overriding problem for the healthcare system is funding; a system of social insurance was envisaged but has never been introduced, with the result that public healthcare is funded almost entirely from general taxation.

15.3.2.1. Primary Healthcare

The primary healthcare model continues to be implemented. There is an ongoing effort to make a clearer separation between primary and secondary care.

The state encourages primary care practices to register as independent contractors. The capitation system for GPs is being improved and outpatient services are increasing.

Improving the management capacity and the information systems of health institutions is considered a priority. Defining a minimal package of healthcare services to be provided by the state has been a priority for the past few years.

15.3.2.2. Secondary and Tertiary Healthcare

There are two types of public hospital: state (accountable to the Ministry of Health) and municipal (with half of the country's hospital beds). Specialized hospitals are concentrated only in the capital and in the largest cities. Public hospitals are contracted by the regional branches of the State Compulsory Health Insurance Agency. Most hospitals are non-profit-making entities or joint-stock companies.

15.3.2.3. Private Sector

The private sector is increasing rapidly the range and number of services it provides, and in just four years, the number of private doctors has tripled.

15.3.3. Disease Profile

The main causes of death per 100,000 in 2008 were cardiovascular diseases (728.8), oncological diseases (261.7), trauma (108.2), gastrointestinal disease (49.0), and diseases of the respiratory system (32.0).

15.3.4. Patient Profile

Patients are not paid for participation in Phase II–IV trials, although travel expenses to sites regarding protocol-related visits may be reimbursed by the sponsor.

There are no set regulations on what patient recruitment materials are acceptable. Materials of an informative nature and study-related items (e.g. patient bags) are mostly approved by ethics committees.

15.3.5. Regulatory Environment

The regulations for the conduct of clinical trials were adopted in September 2000 and amended in February 2006, January 2008, and March 2010. The current regulation of the Cabinet of Ministers, "Regulations on Conducting Clinical Trials and Non-Interventional Studies and Labelling of Investigational Medicinal Products, and Procedure for Conducting Inspections on Compliance with the Requirements of Good Clinical Practice", was released on March 23, 2010, and came into force on April 1, 2010.

There are five ethics committees, one central and four other independent committees. Members of these committees are proposed by the chairperson of the committee and accepted by the Ministry of Health.

The timelines for the assessment of single and multisite studies are 30 days for ethics committees and 60 days for the State Agency for Medicines (SAM).

If substantial amendments are submitted during the review process, they must be submitted to the ethics committee and to the SAM. The ethics committee shall provide an opinion and the SAM shall take a decision not later than 30 days after receipt of an amendment.

The ethics committees assess the suitability of investigators and sites only for certified healthcare institutions.

Insurance is mandatory for the submission. The sponsor is responsible for the insurance that covers any injury to trial subjects caused by a trial, including insurance that covers liability of the investigator and sponsor (this is stated explicitly and also has to be mentioned on the ICF).

15.3.5.1. Regulatory Authority

The SAM is a regulatory authority under direct jurisdiction of the Ministry of Health. It was founded on October 9, 1996.

The main objective of the Agency is to implement local and international pharmaceutical legislation to ensure that human and animal medicines as well as medical devices used in healthcare and their availability are in conformity with the requirements, and in addition to provide objective information on medicines.

The submission dossier should be prepared in Latvian. If the CRA is not Latvian, communication in English would be accepted. Core documents do not need to be translated, only patient document translations need to be provided. If regulatory bodies in other countries have not started to input data in the EudraCT database, and the protocol is problematic, then in their letter the regulatory authority can ask for proof in writing that the study is approved in other countries. If the protocol is not problematic, they will sometimes ask for this verbally.

Documents are reviewed on a regular basis, with no official submission deadline and no meeting dates. The timeline for review of the initial application is 60 days.

Submission to the ethics committee and the SAM is done in parallel.

15.3.6. Good Clinical Practice

GCP principles are set out in Cabinet Regulation No. 289, Riga, March 23, 2010 (Regulations on Conducting Clinical Trials and Non-Interventional Studies and Labelling of Investigational Medicinal Products, and Procedure for Conducting Inspections on Compliance with the Requirements of Good Clinical Practice), which implements EU Directive 2005/28/EC. GCP training for investigators is run by CROs, pharmaceutical companies and the SAM.

15.3.7. Clinical Trial Status

Local registration in Latvia is similar to EU processes as EU legislation is implemented in local regulations.

15.3.8. Site

Clinical trial applications are reviewed by one central and four independent ethics committees in Latvia, all located in Riga.

Ethics committee meetings are usually held once per month, and it is advisable to hand in submissions one to two weeks before the meeting. Documents can be submitted in English; only patient document translations need to be provided. The timeline for approval is 30 days, and written approval is issued.

At present, there are 131 social care institutions in Latvia with 11,792 beds, of which 11,360 are functional. The number of nursing homes increased from 42 in 1994 to 62 in 1998. There are state- and community-owned public homes for the elderly.

15.3.9. Inspection

No data are available regarding the number of local inspections in Latvia. There have been nine FDA inspections in Latvia since 2003 (http://www.accessdata.fda.gov/scripts/cder/CLIIL/).

15.3.10. Pharmacovigilance and Serious Adverse Event Reporting

SAEs are reported as set in country regulations. SAEs that caused patient death or are life threatening are reported by the sponsor on the EudraVigilance EVCTM module not later than seven days after receiving the information from site. For investigational product-related SAEs that caused death or are life threatening and occurred in sites in Latvia, the sponsor must inform the SAM, provide the EVCTM address to the ethics committee, and supply an additional report within eight days after the initial report. Other SAEs that are not life threatening must be reported by the sponsor on the EudraVigilance EVCTM module and to the SAM and the ethics committee within 15 days.

The sponsor is required to provide SUSARs and line listings to the sites on a regular basis. The SAM and the ethics committee require only an annual safety report.

There are no specific language requirements or local forms. Correspondence with the State Agency of Latvia should preferably be in Latvian, but reports do not need to be translated.

The investigator sends the SAE list to the ethics committee, the sponsor, and the regulatory authority.

15.3.11. Contract Research Organizations

Several global CROs and local CROs are operating in Latvia. The first clinical studies were conducted in 1993. A number of freelancers (CRAs, project managers) supply their services to pharmaceutical companies.

15.3.12. Outsourcing Model

Owing to the strong growth in the number of clinical studies, the industry has to handle the challenges in recruiting new, well-qualified employees. To fill the resourcing gap, large global CROs are hiring entry-level CRAs. The use of study nurses or SMOs is not currently prevalent in Latvia. However, this is expected to change in the future owing to the increasing amount of investigator responsibilities and the increasing complexity of study procedures.

15.3.13. Clinical Trial Staff

CRAs should have a bachelor's degree, or local equivalent, in medicine, science, or a related discipline. In the 1990s and early 2000s, the majority of CRAs had an MD or a pharmacy or medical nursing background. Advanced CRA/project management and on-job-training are usually done within the CRO.

The structure of study staff is almost the same as in other CEE countries, comprising a principal investigator, subinvestigators, and a study coordinator: this last role can be delegated to a medical nurse or doctor, or to a pharmacist. Depending on the hospital's local regulations, the role of the pharmacist can be delegated to either a study nurse or another study investigator, or it can be delegated to a pharmacist from the hospital pharmacy department.

15.3.14. Logistics

Investigators have experience with major and small IVRS (e.g. ClinPhone, ICOPhone) and EDC (e.g. Rave, Medidata, DataTrack) systems providers.

Depot companies can be used for the storage of investigational medicinal products (IMPs); however, since all IMPs are shipped from a distribution company in the EU, it takes two days to deliver the IMP to the sites. In addition, an adequate amount of IMP can be kept at the sites according to the applicable regulations.

15.3.15. Clinical Trial Organizations

There is no organization similar to the Association of Clinical Research Professionals (ACRP) in Latvia. All clinical staff are trained locally by the CRO or the sponsor. Apart from instructor-led training, global CROs use online training systems with acknowledged assessment. A score of at least 80 percent is needed to pass the test.

15.3.16. Challenges and Opportunities

There are significant advantages to performing clinical studies in Latvia. There are many highly educated, well-trained, motivated investigators familiar with GCP standards, local, FDA, and EMA regulations. The healthcare infrastructure provides access to modern technology, diagnostic equipment and procedures, and treatment techniques. Participation in international clinical trials provides doctors and scientists with the opportunity to access know-how in new therapies and treatment standards.

15.4. ESTONIA

15.4.1. Geography

According to data from Statistics Estonia, the resident population of Estonia on January 1, 2008, was 1,340,602 people, 54 percent of whom are women.

15.4.2. Healthcare Infrastructure

15.4.2.1. Healthcare

The main source of healthcare finance is the public health insurance, which accounts for approximately 66 percent of total expenditure on healthcare, with people's own contributions comprising 21 percent. Other public sources include state and municipal budgets, accounting for approximately 8 and 2 percent of total healthcare expenditure, respectively. So, healthcare in Estonia is mainly financed publicly. The total costs of healthcare in Estonia in recent years have been 5.0–5.4 percent of GDP (5.1% in 2005).

15.4.2.2. Investigators and Sites

A new reform, Estonian Hospital Development Plan 2015 (HNDP), was initiated in 2000 (updated 2003) for 15 years, to reorganize the hospital network with the underlying idea of concentrating high-technology specialist care into major centers to increase the efficiency and quality of the services. The plan determined the list of regional, central, general, local, and specialty hospitals based on access criteria to ensure sufficient population pools with necessary service volume (at most 60 minutes' travel by car to reach a hospital). The number of hospitals was thus reduced to 51 by the end of 2002. The list also included investments for building, renovating, and reprofiling the hospitals, which has been and still is a critical area for Estonian policy makers.

At the same time, new regulations for healthcare organizations were established under the 2001 Health Care Services Organization Act, which required all public hospitals in Estonia to be incorporated under private law as foundations (trusts) or joint-stock companies by 2003. This led to a situation where hospitals remain in the public sector but are being run as companies according to private law, granting them full managerial rights over assets, full residual claimant status, and access to financial markets.

15.4.3. Disease Profile

Cardiovascular mortality makes up over the half of all the deaths in Estonia (60% among women, 46% among men). The levels of premature morbidity and mortality, as well as the permanent incapacity to work caused by cardiovascular diseases among middle-aged people, are high in Estonia compared with Western Europe and Scandinavian countries.

Cancer mortality made up over 20.7 percent of all deaths in 2006 in Estonia (19% among women, 22% among men). Most commonly, the deaths are caused by lung cancer, colon cancer, rectum cancer, and stomach cancer.

External causes of morbidity (injuries and poisonings) made up 9.4 percent of all deaths in 2006 (4.5% among women, 14% among men). Despite the increased safety brought about by developments in society, the rate of injury-related deaths in Estonia in 2005 was three times higher than in the EU member states. In 2005, 123 deaths due to external causes in Estonia and 42 deaths in the EU member states per 100,000 population were registered. The most common causes of death in Estonia are suicide, traffic injury, alcohol poisoning, and freezing.

15.4.4. Patient Profile

Different methods are used for patient motivation, including reimbursement of travel expenses, and provision of study materials (e.g. pocket cards, bags, boxes for storage of study drugs).

In Phase I clinical studies, patients receive appropriate payment for their participation in the trials. All incentives have to be approved by the ethics committee and competent authority.

Advertising can be also used but text should be also approved by the ethics committee and competent authority. All materials intended for patients should be translated into Estonian and Russian (if Russian-speaking subjects are to be involved).

15.4.5. Regulatory Environment

15.4.5.1. Regulatory Process

Estonia regained its independence 15 years ago and started to develop its society following Western standards. This change in direction meant that the Oath of the Soviet Physician was abandoned and the debate over ethical issues in medicine, especially regarding autonomy of the patient, broadened. The first ethics committees in Estonia, established in 1990, were set up not because of the need for ethical guidance in the clinical setting but to carry out clinical trials in accordance with Western standards. The first clinical ethics committee was established by Tallinn Children's Hospital in 1997 and was soon followed by the clinical ethics committee of the University Clinics of Tartu. So far, these are the only two clinical ethics committees in Estonia. The population of Estonia is about 1.4 million and there are about 10 major hospitals.

The members of these ethics committees are mainly appointed by the board of the hospital upon proposal of the chairman of the committee. To a large extent, members are physicians working in the same hospital, but may also include nurses, laypeople, lawyers, and so on.

Clinical ethics committees in Estonia fulfill similar tasks to these bodies in Western Europe.

The principal laws that regulate clinical trials in Estonia are The Medicinal Products Act, passed on December 16, 2004, entered into force on March 1, 2005, and Conditions and Procedure for Conducting Clinical Trials of Medicinal Products, Regulation No. 23 of the Minister of Social Affairs of February 17, 2005, entered into force on March 25, 2005, amended by the Regulation of October 7, 2005, entered into force on October 21, 2005.

The average time for approval in Estonia is 60 days for the ethics committee and 30 days for the State Agency of Medicines (SAM), after receipt of all requisite documentation. The timeline for approval of amendments is 35 days after the application is submitted.

15.4.5.2. Regulatory Authority

The Estonian Centre on Medicines (ECM) was set up by the Ministry of Health in May 1991, before Estonian independence was restored. Through the Nordic Council on Medicines, a small educational project supported by the Nordic Council of Ministers was launched to educate ECM staff during 1991–1992. By creating the SAM in 1993, the government finalized the structure of the drug regulatory authorities in Estonia.

Currently, the SAM is an agency under the Ministry of Social Affairs, since during the governmental reform the former Ministry of Health was joined with the Ministry of Social Affairs. The SAM has the following obligations: marketing authorization and quality control of medicinal products including biological products, evaluation and approval of applications for clinical trials, import and export authorization of medicinal products, control of licit use of psychotropic and narcotic substances, control over precursors, drug information, advertising and promotion control, and pharmaceutical inspection. In 2000, the control of medical devices and veterinary medicines, and in 2010, the area of medical devices were added to the Agency's responsibilities.

The SAM conducts a review of the submission package on a regular basis. For Phase II–IV trials approval is granted 30 days after receipt of a complete document package (for Phase I studies within 60 days and for gene therapy or somatic cell therapy, immunological medicinal products, or medicinal products containing genetically modified organisms within 90 days after receipt of the application and required documentation), if no queries are raised by the SAM. If queries are raised, there is no deadline for answer; however, the clock will be stopped.

The amendment approval timeline is 35 days after the application is submitted.

All submission documents may be submitted in English, except for subject-related materials, such as the subject information sheet (SIS)/ICF. Large documents, such as the protocol, investigator's drug brochure (IDB), and investigational medicinal product dossier (IMPD), are submitted only electronically, on CD.

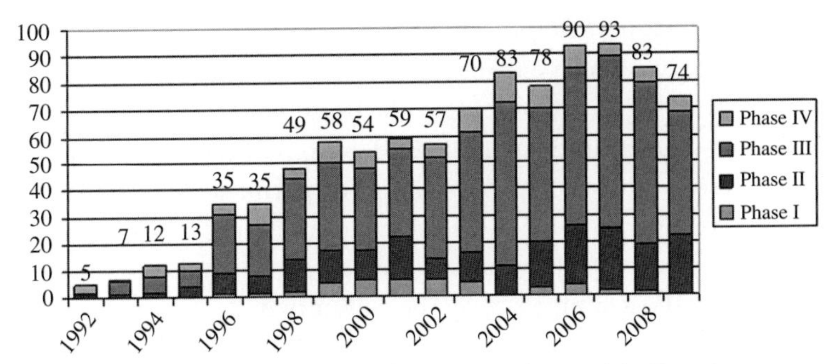

Figure 15.1 Number of clinical trial applications submitted in Estonia per year per phase.

15.4.6. Good Clinical Practice

The regulation of clinical research is in full compliance with the relevant EU legislation.

The trials should be performed and the data generated in compliance with GCP to ensure the ethical and scientific integrity of the trial. GCP code close to the Nordic document by the Nordic Council on Medicines has been mandatory since 1991; later, the ICH-GCP was adopted and is currently directly referred to by the legislation.

GCP training for physicians and investigators is organized by the SAM.

15.4.7. Clinical Trial Status

Figure 15.1 shows the number of clinical trial applications submitted per year for Phase I–IV studies in Estonia.

15.4.8. Site

In the period 2000–2006, the number of hospital beds decreased by 20 percent, from 9828 beds to 7588 beds, and the structure of beds by specialty has changed significantly. In 2006, acute-care beds made up 69.7 percent of the total number of inpatient hospital beds. Of all hospital beds, the share of nursing care beds has increased, and the percentage of psychiatric beds has decreased. Two of the three regional hospitals (secondary and tertiary care) serve an area with approximately 500,000 people. One of them, Tartu University Hospital, covers all specialized services for the southern part of Estonia as well as the North Estonian Regional Hospital in Tallinn. The North Estonian Regional Hospital and University of Tartu Clinic are the largest hospitals in Estonia. Many clinical trials are conducted there.

15.4.8.1. Certification Requirements and Process

There are two independent ethics committees in Estonia: Tallinn Medical Research Ethics Committee (TMREC), located in Tallinn, and the Ethics Review Committee on Human Research of the University of Tartu, located in Tartu. The document package for permission to conduct a clinical trial is submitted to one of the ethics committees, depending on which region the biggest part of the sites is located in.

Clinical trial documents may be submitted in English, except for subject-related materials such as the SIS/ICF and specially required ethics committee application form, which must be completed in Estonian and signed by each site head of hospital and staff.

The average time for approval is 60 days. The timeline for amendment approval is 35 days after the application is submitted.

15.4.9. Inspection

No data are available regarding the number of local inspections in Estonia. There have been six FDA inspections in Estonia since 2003 (http://www.accessdata.fda.gov/scripts/cder/CLIIL/).

15.4.10. Pharmacovigilance and Serious Adverse Event Reporting

International standards of safety reporting apply: all SUSARs from any country should be reported by the sponsor in accordance with ICH E2A; the Council for International Organizations of Medical Science (CIOMS) form is preferable.

Minister of Social Affairs Regulation No. 26 of February 17, 2005, "Procedure for Reporting Serious Adverse Events Occurring in Clinical Trials", regulates SAE reporting in Estonia.

Only SUSARs should be reported to the competent authority and the ethics committee when they happen (within seven days for fatal and life-threatening reactions, within 15 days for others). All other SAEs that occur in Estonia and in other countries are to be reported only once a year.

Protocol violations should be submitted to the ethics committee as soon as possible, if they relate to patient safety.

15.4.11. Challenges and Opportunities

There are significant advantages to performing clinical studies in Estonia. There are many highly educated, well-trained, motivated investigators familiar with GCP standards, and local, FDA, and EMA regulations. The healthcare infrastructure provides access to modern technology, diagnostic equipment and procedures, and treatment techniques. Participation in international clinical trials provides doctors and scientists with the opportunity to access know-how in new therapies and treatment standards.

15.5. POLAND

15.5.1. Geography

Poland is the 34th most populous country in the world and the sixth in Europe. The official language is Polish. Figures for 2008 gave a population of 38,482,919, and an average life expectancy of 73.1 years for men and 80.0 years for women.

15.5.2. Healthcare Infrastructure

According to the Polish Central Statistical Office (GUS), 78,229 doctors were registered as working directly with patients in 2007, 15 percent of whom were internal disease specialists, 10 percent pediatricians, more than 15 percent surgeons, more than 10 percent GPs, and less than 10 percent were neurologists, anesthesiologists/intensive therapists, ophthalmologists, or psychiatrists.

In 2007, there were 748 hospitals, 578 of which were public, with 67.8 inpatient beds per 10,000 population. There were 6,850,000 inpatient admissions, with an average patient stay of 6.2 days. In 2008, there were 14,836 outpatient departments (3297 public facilities and 11,539 private) and 6516 private medical practices (4898 in urban areas and 1618 in rural areas). In total, 290,553 consultations were provided, of which 22,754 were in private medical practices.

Private medical practices, which have not signed a contract with a National Health Fund or with a healthcare facility (i.e. functioning exclusively within the framework of private funds), also operate within the scope of outpatient healthcare. It is estimated that in 2007, 29,683,000 medical consultations were provided in 301,000 private medical practices.

15.5.3. Disease Profile

The two most common diseases are cardiovascular disease and cancer. Other common conditions include allergy, asthma, diabetes, gastrointestinal diseases, depression, osteoporosis, and thyroid diseases. The Report of

Table 15.1 Standardized death rate (SDR) for all ages per 100,000 population in Poland in 2006

Disease	SDR
Cardiovascular diseases (hypertension, acute coronary syndrome)	372.2
Malignant tumor of respiratory tract	210.2
Colon cancer	21.5
Breast cancer	12.2
Respiratory system diseases	40.7
Digestive system diseases	37.7

National Institute of Hygiene (2008) analyzes the standardized death rate (SDR) per 100,000 population (Table 15.1).

The data from the WHO-selected statistics for Poland refer to an SDR of 836.3 for all causes and all ages, per 100,000 population in 2007. The same source refers to a tuberculosis (TB) incidence of 39.4 per 100,000 population in 2008.

15.5.4. Patient Profile

Over 400 new clinical studies are registered in Poland each year with over 50,000 study participants.

Motivational factors for patients to participate in clinical trials are:

- easy and free access to the latest medications and treatment
- easy access to medical consultations and frequent examinations
- access to standard international therapies that may not be available in the healthcare system in Poland
- participation in medical research as a contribution to the progress of science.

A monetary incentive is only possible in Phase I trials; it is prohibited in Phase II, III, and IV studies. Travel costs, time compensation, and meal costs can be reimbursed upon competent authority/ethics committee approval.

Patient recruitment and retention is similar to that in other EU countries. Recruitment rates may vary in different therapeutic areas and where there is an ongoing competitive study.

15.5.5. Regulatory Environment

15.5.5.1. History of Regulatory Affairs

The Central Register of Clinical Trials (CRCT) unit was founded on April 15, 1994, as a division of the Office for Registration of Medicinal

Products, Medical Devices and Biocidal Products, which is a government agency responsible for the evaluation of quality, efficacy, and safety of such devices and products. The Office is headed by a president, appointed for a five-year term, and by the Minister of Health, who speaks on behalf of the Office. The Office reports directly to the Minister of Health. The Office has its own scientific resources, as well as a network of external experts. It is involved in working groups with the Ministry of Health, EMA, and European Commission, as well as in conducting training for the pharmaceutical industry.

15.5.5.2. Role of Government in Clinical Trials

The government issues legal regulations that describe in detail the process of clinical trials. Relevant decrees have been published that cover the processes of clinical trial registration and approval, clinical trial conduct in accordance with GCP obligations, insurance, and the reporting of adverse events. Government institutions that are crucial in the clinical trial process include the Office for Registration of Medicinal Products, Medical Devices and Biocidal Products, and the Appeal Ethics Committee. Legal regulations are based on EU Directive 2001/20/EC.

15.5.5.3. Function and Structure of Organization

The tasks of the CRCT unit include the scientific and formal evaluation of clinical trial applications and associated documentation for medicinal products or future medicinal products (excluding veterinary medicinal products) for inclusion in the CRCT and collecting information on unexpected and serious adverse events that occur during the course of a trial.

15.5.5.4. Challenges and Common Problems

According to Polish law, the CRCT unit does not have a time limit for validating that a submission package fulfills the formal requirements. The response to validation comments is not included in the 60-day review period. Therefore, the preparation of a clinical trial submission dossier that is fully compliant with Polish and EU legislation is critical to avoid an extended evaluation process and receive full approval within the standard time.

15.5.5.5. Insurance

Pursuant to the laws and regulations in force, the sponsor's and study doctor's third party liability has to be covered by a study- and country-specific third party liability insurance. The insurance covers the sponsor's and study investigator's liability for causing physical injury, detriment to health, and/or death of a study subject occurring during the conduct of the study for the term of the insurance period. Both sponsor and investigator are covered by a minimum sum insured that depends on the number of planned randomized patients. All Polish citizens are covered by government insurance provided by ZUS (Social Insurance Institution), which guarantees access to the public healthcare system free of charge. Insurance has to be provided by an insurance company in Poland by the day of submission to the competent authority/ethics committee.

The insurance document must be in Polish or bilingual. The full site address and investigators' names are not required but the following statement must be included: "Researchers who signed the contract in order to participate in the below mentioned study". It is recommended to include both the screened and randomized number of patients on the policy.

15.5.6. Good Clinical Practice

In 1992, GCP was translated into Polish and the Minister of Health recommended that all clinical trials should be conducted in accordance with GCP from 1993. The Polish version of ICH-GCP was released in 1998 and on March 11, 2005, the Minister of Health issued a Decree regarding GCP that is the current binding GCP ordinance in Poland.

GCP training may be provided by a CRO or the pharmaceutical industry but this is not required by Polish law for investigators and/or study staff. Many investigators, especially in the private setting, are GCP certified because many CROs and sponsors require this for participation in a clinical trial.

15.5.7. Clinical Trial Status

Poland is ranked third in the EU for the total number of clinical trials conducted per year. The main therapeutic areas are oncology, endocrinology, ophthalmology, psychiatry, pediatrics, infectious diseases, dermatology, urology, gastroenterology, rheumatology, cardiology, neurology, and surgery.

According to an EMA Report, the number of pivotal clinical trials for marketing authorization applications (MAAs) submitted to the EMA from 2005 to 2008 was 131, ranking Poland in seventh place within the EU/European Economic Area (EEA)/European Free Trade Association (EFTA)

Table 15.2 Number of clinical trial applications from the CRCT unit in Poland in 2004–2009

Year	2004	2005	2006	2007	2008	2009
No. of clinical trial applications	408	396	454	447	486	476

countries. The total number of patients recruited in these pivotal trials was 60,600, with an average number of 15 patients per site. In total, 955 sites participated in clinical trials in the years 2005–2008. According to the most recent projection available, the Polish market for clinical trials will increase by 5 percent in 2010 and will be worth 718 million PLN (180 million euro). The number of clinical trial applications to the CRCT unit from 2004 to 2009 is shown in Table 15.2.

15.5.8. Sites

For multicenter clinical studies in Poland, the sponsor selects the national country investigator (coordinator) out of the investigators involved. The study is submitted only to the bioethics committee of the national coordinator's site (the central bioethics committee), but decisions issued by the central bioethics committee are binding for all study sites. The legal timeline for approval is 60 days. However, in practice bioethics committees usually grant approvals on a monthly basis.

Parts of the submission dossier (cover letter, application form, protocol synopsis, SIS/ICF, other patient documents and materials) must be written in Polish. Core study documents (protocol and investigator brochure) can be in English. No specific certification is required for participation in a clinical trial. For submission of a clinical trial application, contracts with investigators and relevant institutions are required. Contracts should be signed and delivered to the competent authority (CRCT) and relevant central bioethics committee before the study is approved. A study can only start at sites for which contracts are delivered to the competent authority (CRCT) and relevant Central bioethics committee.

Grants are paid to the principal investigator and the institution. The grant allocation depends on the internal regulations of the institution.

15.5.9. Inspection

FDA, EMA, and CRCT inspections take place in Poland. According to an EMA report, 13 GCP inspections were conducted from 1997 to 2008.

15.5.10. Pharmacovigilance and Serious Adverse Event Reporting

The timelines for SUSAR reporting to the CRCT unit and central bioethics committee are seven and 15 days. Reports are sent unblinded. Blinded SUSAR reports are sent to investigators. An annual safety report must be provided to the CRCT unit and the central bioethics committee. For products that have market approval, periodic safety update reports (PSURs) and expedited reports must be provided to the Office for Registration of Medicinal Products, Medical Devices and Biocidal Products. The CIOMS format is accepted for expedited reporting. The language for reporting is English.

15.5.11. Contract Research Organizations

There are more than 100 CROs conducting clinical trials in Poland. The outsourcing models vary between sponsors and CROs. Partnerships have developed in some cases based on long-standing cooperation. Most CRAs in Poland have medical and/or science degrees (biology, chemistry, or biotechnology), and some have PhDs.

15.5.12. Logistics

Polish sites and investigators are experienced in new technologies such as EDC and IVRS, which are commonly used in clinical trials. The shipment of drugs and biological samples by specialized companies is common practice. Clinical trials insurance is arranged by local insurance brokers who are familiar with all the regulations.

15.5.12.1. Association for Good Clinical Practice in Poland

The Association for Good Clinical Practice in Poland (GCPpl) was founded in 1998 on the initiative of representatives of pharmaceutical companies (both foreign and Polish) and CROs conducting international clinical trials in Poland. GCPpl is a non-profit, member-driven organization of representatives from over 30 foreign and Polish pharmaceutical companies, numerous CROs, and members of the medical, scientific, and academic communities.

The association comprises the following working groups:

- ethics committees
- training
- quality and GCP inspections
- legal controls and EU integrations

- public relations and publications
- pharmacovigilance and drug safety reporting.

The members of the Association Executive Board and the working groups operate voluntarily. The mission of the Association for GCPpl is to create and support initiatives promoting ethical principles and regulatory compliance in clinical trials, particularly for compliance with GCP, and to create a discussion forum for all parties involved in clinical research.

15.5.13. Challenges and Opportunities

- **Knowledge**: Participation in international clinical trials provides Polish doctors and scientists the opportunity to access new therapies and treatments.
- **Financial capital**: Participation in international clinical trials provides the opportunity to profit from financial funds. It is an additional financial benefit for the hospitals, clinics, institutions, and investigators who conduct clinical trials.
- **Human capital:** Clinical trials also create new job opportunities in the labor market for highly qualified doctors, who can work as investigators, CRAs, or auditors, and for auxiliary staff. Investigators can participate in training courses including ICH-GCP, to increase their qualifications.
- **Legal factors**: The revisions to the pharmaceutical legislation that came into force in 2004 increased the emphasis on the ethical standards required of clinical trials. Further changes to the law governing clinical trials are expected and the draft act was submitted to the Ministry of Health in June 2010 for review.

15.6. ROMANIA

15.6.1. Geography

Romania, which has been a member of the EU since 2007, is one of the larger countries in the region, with a population of 21.4 million, more than half of whom reside in urban areas. The official language is Romanian. Life expectancy at birth is 76 years for females and 69 years for males, which is below the European average. The adult literacy rate is 98.4 percent.

15.6.2. Healthcare Infrastructure

The public healthcare system, which is funded by a combination of employer and employee contributions to the National Health Insurance

Fund (NHIF), a central quasi-autonomous body, and direct allocations from the state budget, is still predominant. The mandatory insurance-based healthcare financing model involves employer and employee contributions. The health insurance system is administered and regulated by the NHIF.

From 1998, when the country adopted the mandatory social health insurance system, the roles of the main participants in the health system have changed, the relationships between different organizations have become more complex and the number of stakeholders has increased. The system is organized at two main levels: national and district. The Ministry of Public Health is responsible for developing national health policy, regulating the health sector, setting organizational and functional standards and improving public health. The representative bodies of the Ministry of Public Health at the district level are the 42 district public health authorities (DPHAs). Also at district level, 42 District Health Insurance Funds (DHIFs) are responsible for contracting services from public and private healthcare providers according to the rules set by the NHIF.

The private healthcare sector is developing rapidly and during the past 15 years a large number of well-organized and well-equipped private healthcare units have been established. There are approximately 137,984 hospital beds (6.6 beds per 1000 population) in 457 hospitals and about 50,238 physicians (23.7 physicians for 10,000 population). Specialist physicians typically also maintain their private practice so both public and private centers facilitate patient recruitment for clinical trials. Owing to limited resources, access to modern medicines and diagnostic methods is still unavailable for all patients and some rural areas have only limited access to pre-existing resources. The Health Department budget was reduced by almost 20 percent from 2008 to 2009, to 3.2 percent of GDP.

15.6.3. Disease Profile

The 10 most common causes of death are ischemic heart disease, cerebrovascular disease, hypertension, liver cirrhosis, lung cancer, lower respiratory tract infection, chronic obstructive pulmonary disease, colon cancer, gastric cancer, and breast cancer. The most common diseases (by incidence) reported by general physicians are respiratory (48%), gastro-intestinal (9%), bone and muscular (7%), circulatory (5%), skin (5%), infectious and parasitic (5%), and endocrine, nutritional and metabolic (2%).

15.6.4. Patient Profile

Patients are frequently treatment naïve and are usually diagnosed at a much later stage of disease than patients in Western Europe. Patient recruitment is enhanced by the high incidence of many of the diseases commonly investigated in clinical trials. Most patients have limited health coverage and limited access to modern medicines and diagnostics. Patients are therefore highly motivated to participate in clinical trials and compliance is high. Patients respect and trust their healthcare professionals, which also facilitates recruitment and compliance.

15.6.5. Regulatory Environment

15.6.5.1. History of Regulatory Affairs

The Institute for the State Control of Medicinal Products and Pharmaceutical Research (ICSMCF) was created in 1956 and was responsible for the inspection and assessment of medicinal products. In 1973, the Institute also became responsible for coordinating pharmacovigilance, which was carried out by the Medicinal Product Commission, thus becoming a WHO Collaborating Centre for adverse event reporting. In 1984, in recognition of its outstanding research in the field of medicinal plants, the WHO designated the ICSMCF a WHO Collaborating Centre for traditional medicine. Even before its accession to the EU, Romania made significant efforts to align its legislative and regulatory framework with EU directives. Since 1996, a specialist from the ICSMCF/National Medicines Agency (NMA) has represented Romania as an observer to the European Pharmacopoeia Commission of the Council of Europe.

The NMA was created on January 1, 1999, as a public institution under the Ministry of Health and took over the responsibilities of the ICSMCF. The EU Clinical Trials Directive of April 4, 2001, was incorporated into local legislation in July 2004. Romania was thus one of the first countries outside the EU to adopt the Directive as law. This was followed by a number of laws to implement the 2001/20/EC guidelines Directive. The NMA is responsible for approving clinical trials of pharmaceuticals and biologicals, while the Ministry of Health is responsible for approving clinical trials of medical devices.

15.6.5.2. Approval Process

The regulatory process in Romania is transparent and relatively quick: the time from submission to approval is 60 days. The NMA and the

National Ethics Committee (NEC) are responsible for reviewing an application within the timeframe of 60 days. If the NMA has objections, the sponsor can only modify the application once, taking the objections into consideration. If this is not completed successfully within 30 days, the study cannot start. Exceptions to the 60-day rule are clinical trials of genetic, somatic or xenogeneic cell therapy, and all drugs containing genetically modified organisms, for which an extra 30 days (90 days in total) are allowed. In these cases, it is compulsory to obtain the opinion of an expert group. There is no time limit for the approval of clinical trials involving xenogeneic cell therapy. The submission dossier can be submitted in Romanian or English.

CTA documents that must be translated into Romanian are patient information and ICFs, medication labels and instructions, scales and questionnaires intended for patient use, and advertisements and documents used for patient recruitment. Patient insurance is usually covered by the sponsor.

15.6.6. Good Clinical Practice

International ICH-GCP guidelines became law in Romania in 1997. GCP training is conducted by pharmaceutical companies and CROs, and awareness of ICH-GCP principles is continuously increasing among healthcare professionals.

15.6.7. Clinical Trial Status

The number of CTAs received has increased from 76 in 2000 to 424 applications in 2008. Phase III trials represent about 60 percent of the applications and Phase II trials about 30 percent.

15.6.8. Site

Sites should be certified to perform clinical research by the Ministry of Health. It is the responsibility of the institution to submit a certification dossier to the Ministry. Authorization for clinical trials issued by the Ministry of Health is valid for two years and must then be renewed.

15.6.9. Inspection

The Commission for the inspection of GCP, good manufacturing practices (GMP), good laboratory practices (GLP), and good analytic laboratory practices (GALP) is part of the NMA. The Commission examines inspection reports drafted by NMA inspectors, reports relating to compliance of inspected sites with GMP, GLP, GALP, and GCP and/or other problems regarding the activity of the Pharmaceutical Inspection department. The Commission mediates if an inspector's decision is disputed by the inspected unit. In 2008, the Commission conducted 17 sessions to examine 180 inspection reports.

15.6.10. Pharmacovigilance and Serious Adverse Event Reporting

In accordance with Directive 2001/20/EC, sponsors have to submit all SUSARs from ongoing clinical trials in Romania to the NMA. The NMA has implemented an electronic data exchange system for suspected adverse reactions through the web trader component of the EudraVigilance Web Application (EVWEB). The NMA requires sponsors of clinical trials initiated or ongoing as of January 1, 2007, to submit SUSARs electronically in the form of individual case safety reports, ICSRs, in full compliance with the E2B(R2)/M2 guidelines and all medical information coded in MeDRA.

The ethics committee should only receive expedited individual reports of SUSARs that occurred in subjects recruited in Romania. It is strongly recommended that all SUSARs from other member states and, where applicable, from other countries, are periodically reported at least every six months as a line listing accompanied by a brief report by the sponsor highlighting the main points of concern. These periodic reports should only include SUSARs reported within the period covered by the report. A copy of the report should be sent to the NMA. Any changes increasing the risk to subjects and any new issues that may adversely affect the safety of the subjects or the conduct of the trial should also be provided as soon as possible, but no later than 15 days. In addition to expedited reporting, sponsors have to submit once a year throughout the clinical trial or on request a safety report to the NMA and the ethics committee, taking into account all new available safety information received during the reporting period.

15.6.11. Contract Research Organizations

In addition to international CROs, there is a number of small and medium-sized local CROs. The major therapeutic areas include cardiovascular, oncology, CNS, endocrinology, gastroenterology, hematology, urology, dermatology, infectious diseases, ophthalmology, orthopedics, and pain control.

The clinical services offered by CROs include Phase II–IV clinical services, feasibility studies, site selection and evaluation, regulatory applications, ethics committee submission coordination, investigator meetings, trial initiation and monitoring, SAE reporting and evaluation, project management, and supplies distribution. Other services can be subcontracted where necessary or can be operated from other CRO offices in Europe. These include laboratory services, storage and archiving of documents, medical, statistics, and data management.

15.6.12. Clinical Trial Staff

Over 40 percent of CRAs are doctors, with the rest having a life sciences background.

15.6.13. Logistics

The majority of sites have Internet access and investigators are familiar with IVRS and EDC systems, which are centrally contracted from specialist companies in Western countries. If drugs have to be imported from outside the EU, there are local authorized companies that can arrange customs clearance, storage, and distribution. International courier companies specializing in the shipment of biological samples have been established in Romania for over 10 years. Study insurance, arranged locally or globally, should be transmitted to the NMA and ethics committee(s), together with all other requested study documents for submission.

15.6.14. Clinical Trial Organization

The Romanian Association of CROs (ACRO), a professional association of pharmaceutical companies and CROs, was founded in 2009. The process of affiliation to the European CRO Federation (EUCROF) is ongoing. The main objective of ACRO is to support the development of clinical research and the pharmaceutical industry in Romania. The association is recognized by the NMA as a professional partner, working together to provide guidance to the clinical research industry and to promote high quality and ethical standards. Several local and international companies organize national and international workshops on conducting clinical research in Romania.

15.6.15. Challenges and Opportunities

Romania provides a large patient population, including a significant number of treatment-naïve patients. Investigators are well educated and trained and are familiar with ICH-GCP standards. The clinical research cost per patient is approximately 80 percent of that in most other European countries. The centralized healthcare system provides good access to patients, particularly in cardiology, oncology, CNS, and endocrinology. Patients are highly motivated to participate in clinical trials owing to the provision of free treatment, close medical follow-up, and better overall medical care.

15.7. UKRAINE

15.7.1. Geography

Ukraine is the second largest country in Europe and is situated strategically at the crossroads between Europe and Asia. It is bordered by Belarus, the Russian Federation, the Republic of Moldova, Romania, Hungary, the Slovak Republic, and Poland. Kiev is the capital and the largest city. With an estimated population of more than 46 million, Ukraine is the second most populous Eastern European country after Russia and the majority of the population (67%) is urban. However, the population is shrinking at a rate of 150,000 per year due to the lowest birth rate combined with one of the highest death rates in Europe. Life expectancy at birth is 61 years for males and 73 years for females. The number of elderly people is increasing while there is a decrease in the number of young people. The state language is Ukrainian, although Russian is widely spoken, especially in the east and south.

15.7.2. Healthcare Infrastructure

The public healthcare system is predominant. Private healthcare facilities are growing rapidly but are still poorly developed. The Ministry of Health is responsible for setting national health policies and certain specialized healthcare institutions are directly managed and funded by the Ministry. Ukraine is divided administratively into 27 regions, each of which has a health administration that is accountable to the Ministry of Health for national health policies within its territory. The regional health directorates own and manage a range of healthcare facilities, including multispecialty and specialized hospitals and dispensaries. At a district and local level, primary-care facilities and hospitals are owned by the various tiers of local government: municipal, city, village, and rural councils.

In 2008, there were 29,000 medical institutions with 437,000 hospital beds (95.1 per 10,000 population), 88,000 outpatient facilities, 222,000 doctors (48.2 per 10,000 population), and 464,000 paramedical staff (101 per 10,000 population).

There is no nationwide insurance system for health-related expenses. The role of voluntary health insurance is relatively small, largely because of the high costs of commercial insurance premiums, which are unaffordable for the majority of the population. Thus, most people pay for their treatment and high-quality medications are rather expensive. Free-of-charge treatment, which is guaranteed by the country's constitution and provided by most government hospitals, is limited, uses generic drugs, and is often not up to date. Although there are some government- and/or pharmaceutical company-funded programs for certain diseases (e.g. cancer, HIV, TB, and cystic fibrosis), these do not cover the needs of the whole population. Routine screening and prophylaxis programs are inefficient so the rate of untreated or late-stage diseases is very high.

15.7.3. Disease Profile

The main causes of death are diseases of the circulatory system (54%), neoplasms (12%), injury and poisoning (9%), gastrointestinal diseases (4%), and respiratory diseases (3%). As in many other European countries, smoking accounts for a considerable proportion of the burden of disease, particularly among men, with recent estimates suggesting that about 100,000 deaths in Ukraine might have been attributable to smoking, 67 percent of which were among men aged 35–69 years. Another important factor is hazardous alcohol consumption.

15.7.4. Patient Profile

Ukraine has a large patient population with many treatment-naïve patients and a high prevalence of several diseases. The centralized healthcare system makes it easy to search for patients with specific diseases. Patients are eager to participate in clinical trials and there is good trust-based relationship between patients and investigators. A clinical trial often provides a much better standard of care for a patient than the national health services, as it provides not only study medication but complete healthcare services including free laboratory testing, diagnostics, careful monitoring, and inpatient care. For many people, a clinical trial may be the only way to obtain modern diagnostic procedures and potentially effective treatment. This is likely to make potential patients motivated to enter a trial and helps compliance with study protocols. Studies in late-stage cancer, untreated hypertension, and L-dopa-naïve Parkinson's disease have the potential to recruit the necessary number of patients on time.

15.7.5. Regulatory Environment

15.7.5.1. History of Regulatory Affairs

The enterprise "State Center of Expertise of the Ministry of Health of Ukraine" (the competent authority; here "State Center of Expertise") is a specialized expert institution authorized by the Ministry of Health of Ukraine to register and control the circulation of medicinal products. The center's key functions include the expert evaluation of materials related to development, manufacture, preclinical study, and clinical trials of medicinal products; the development of draft regulations on the requirements for entities which develop and test medicinal products and their submission to the Ministry of Health for approval, the inspection and audit of study and manufacturing sites, the preparation of regulatory documents, pharmacovigilance, and postregistration surveillance.

The Central Ethics Commission of the Ministry of Health was set up in 2006 to ensure adherence to international recommendations of multicenter clinical trials conducted in Ukraine. The main principle of the Commission activity is the respect of life, dignity, and rights of an individual (whether sick or healthy), bearing in mind that the interests of the individual should always prevail over those of society and science.

15.7.5.2. Role of Government in Clinical Trials

Both the State Center of Expertise and the Central Ethics Commission must approve a study before it can be conducted. These decisions are taken independently.

15.7.5.3. Function and Structure of the Organization

Expert groups, which review the studies, consist of internal and external (mostly health professionals specializing in a particular disease) experts. The State Center of Expertise holds two meetings per month to review submitted materials, which have previously undergone specialized review by individual experts. Initial submissions are only reviewed at one of these two meetings, which is normally held at the end of the month. The Central Ethics Commission consists of about 10 members headed by a chairperson. The Commission holds two meetings per month to review submitted materials, which have previously undergone specialized review by individual experts. Initial submissions can be reviewed at either of these two meetings.

15.7.5.4. Submission Dossier

The covering letter and standard application form should be in Ukrainian. The study protocol, investigator's brochure and investigational product labels should be in either Ukrainian or Russian. Patient information should be in Ukrainian and Russian. Other documents may be accepted in English.

15.7.5.5. Challenges and Common Problems

All the submissions are manual and done during personal meetings. To perform a submission to the State Center of Expertise, a face-to-face meeting with the authority official should be scheduled beforehand (normally a week before the meeting). Electronic submission is not possible.

15.7.5.6. Insurance

Country-specific insurance is required. All patients and healthy volunteers intending to participate in clinical trials are subject to insurance protection. The validity period of the insurance contract should not be shorter than the duration of the clinical trial concerned. Premature termination of the insurance contract is not possible without notifying the Central Ethics Committee and the State Center of Expertise.

15.7.6. Good Clinical Practice

Local GCP guidelines, approved by the Ministry of Health in February 2009, comply with European and international GCP and ethical guidelines.

GCP training at a country level is conducted by the International Foundation for Clinical Research in collaboration with the State Center of Expertise. CROs can also conduct training, if authorized, normally as part of investigator meetings.

15.7.7. Clinical Trial Status

Although Ukraine is a relative newcomer to the clinical trials industry, it is an attractive location for clinical research. A well-developed healthcare system, together with good academic centers, specialized hospitals, and GCP legislation, have contributed to the growth in the number of clinical trials conducted (Table 15.3).

15.7.8. Site

To initiate a clinical trial, the sponsor should submit the study documents to the State Pharmacological Center and the Central Ethics

Table 15.3 Number of clinical trials approved by the State Center of Expertise of the Ministry of Health of Ukraine and the Central Ethics Commission in Ukraine

Year	1996	1997	1998	1999	2000	2001	2002	2003	2004	2005	2006	2007	2008
No. of clinical trials approved	79	190	160	184	182	179	226	274	252	314	347	328	275

Commission of the Ministry of Health. The application for regulatory and central ethics approval can be done in parallel. The approval time is within 60 days. Import/export permits from the State Pharmacological Center are required for study drugs and other clinical study supplies. The permits, which take up to four weeks to obtain, should be applied for as soon as the study is approved. As study sites cannot receive study drugs and other supplies directly from suppliers outside Ukraine, the early identification of a local customs broker to clear study supplies from customs and distribute to study sites is important in order to minimize the study start-up time.

Study sites must be certified by the Ministry of Health to perform clinical research and it is the responsibility of the respective institution to submit a certification dossier to the Ministry. It is standard practice to obtain two CTAs per site, one with the investigator (investigator CTA) and one with the center (institution CTA). CTAs may take a considerable amount of time to obtain and in some cases can delay site initiation. The investigator grant per patient is usually split between the investigator and institution contracts.

15.7.9. Inspection

The State Pharmacological Center conducts regular clinical trial inspections, which may be scheduled or unscheduled. These are often conducted in collaboration with other state authorities, such as the Control and Revision Office or the Security Service.

15.7.10. Pharmacovigilance and Serious Adverse Event Reporting

Sponsors of clinical trials must report all SUSARs to the State Pharmacological Center and the Central Ethics Committee within seven days. The relevant follow-up information must be given within the following eight days. All other serious unexpected adverse reactions must be reported to the State Pharmacological Center and the Central Ethics Committee within 15 days. Unblinding is not required. The SUSAR form can be submitted in English with a covering letter in the local language.

In addition to the expedited reporting, sponsors of clinical trials must provide an annual safety report (both hard copy and electronic version) to the State Pharmacological Center and the Central Ethics Committee.

15.7.11. Contract Research Organizations

All the major international CROs have offices in Ukraine. There is also a number of small and medium local and regional CROs.

The clinical services offered by most CROs include Phase II–IV clinical services, feasibility studies, site selection and evaluation, regulatory applications, ethics committee submission coordination, investigator meetings, trial initiation and monitoring, SAE reporting and evaluation, project management, and supplies distribution. Other services can be subcontracted where necessary or can be operated from other CRO offices in Europe. These include laboratory services, storage and archiving of documents, medical, statistics, and data management.

15.7.12. Clinical Trial Staff

The majority of CRAs (over 50%) are doctors, while the remainder have a life sciences background.

15.7.13. Logistics

IVRS and EDC services are centrally contracted from specialized companies in the USA and Europe. Several international courier companies, specializing in the shipment of biological samples, have operated in Ukraine for over 10 years.

15.7.14. Challenges and Opportunities

Overall, Ukraine is an attractive location to conduct clinical trials because it benefits from a large patient population and a centralized healthcare system. Patients are highly motivated to participate in clinical trials to obtain better medical care and follow-up at reduced treatment costs, resulting in high average site productivity. In addition, the high level of trust in the medical profession encourages good treatment compliance. The Ukrainian population is static, which facilitates study follow-up. Investigators and site staff are well qualified and trained in GCP, and have gained significant clinical research experience over the past decade. The authorities have made a big effort and succeeded in providing modern and ICH-GCP-compliant legislation, ensuring the quality of approved study sites.

15.8. RUSSIA

15.8.1. Geography

Russia is Europe's largest country, with a population of 142 million. Although there is broad mix of ethnic groups, all people speak Russian, which is the official language.

15.8.2. Healthcare Infrastructure

The Russian healthcare system has a number of unique features. It is highly centralized and specialized and ensures the observation of prospective patients. Healthcare is still free for all people, funded mainly by the government and obligatory medical insurance.

Russia has more physicians, hospitals, and healthcare workers than almost any other country in the world on a per capita basis. However, since the collapse of the USSR, the health of the population has declined considerably as a result of social, economic, and lifestyle changes. In 2006, the government launched a national projects plan that aims to improve four sectors of Russian life, one being healthcare. It approved an additional US $3.2 billion in spending on healthcare to cover salary increases for doctors and nurses, the purchase of new equipment for clinics and the construction of eight high-tech medical centers in outlying regions. There is one doctor per 200 population, 107 beds per 10,000 population. Besides hospitals, there are about 22,800 outpatient clinics in Russia, with a total screening capacity of 3.6 million potential subjects per day. Local experts estimate that 8000, or one out of 86 physicians, are involved in clinical trials. For comparison, in the year 2001, every 14th physician in the USA was involved in clinical research.

15.8.2.1. Standards of Care

In most cases, Russian healthcare follows European and American international guidelines and standards of treatment. In general, patients have access to a modern and effective healthcare system. However, the expensive medications are not always available for free; the availability may vary from region to region and from hospital to hospital. During the past 10 years, the quality of medical care in Russia has significantly improved, and today hospitals participating in clinical trials can provide services that meet international medical standards. For the past few years, outpatient care funding has increased through government initiatives.

15.8.3. Disease Profile

In 2008, 57 percent (1,185,993) of all deaths in Russia were due to cardiovascular disease. The second leading cause of death was cancer (14%, 289,257). Other major causes of death were gastrointestinal diseases (4%), respiratory disease (4%), infectious and parasitic diseases (2%), and TB (1%). The infant mortality rate in 2008 was 8.5 deaths per 1000 population, down from 9.6 in 2007.

15.8.4. Patient Profile

Patients are highly motivated to participate in clinical trials because of the free access to modern treatment, close medical supervision, and the culture of trust in the medical profession.

15.8.5. Regulatory Environment

15.8.5.1. History of Regulatory Affairs

A presidential decree in 2004 founded the new system of federal executive bodies. The Federal Ministries are responsible for the implementation of state policies and legislation. Control and surveillance activities are the responsibility of Federal Services. The state service sector and the management of state property are the responsibility of Federal Agencies. The Federal Service on Surveillance in Healthcare and Social Development (RosZdravNadzor, RZN) is the subordinate executive authority in the Ministry of Health and Social Development of the Russian Federation responsible for control and surveillance.

15.8.5.2. Federal Service on Surveillance in Healthcare and Social Development

The Federal Service responsibilities include the control and surveillance of pharmaceutical activities, compliance with state quality standards for medical care and products, preclinical and clinical trials, the inspection of healthcare organizations, pharmaceutical facilities and pharmaceutical wholesalers, licensing medical and pharmaceutical activities and permissions for clinical trials, and the import and export of medical products and unregistered drugs for use in clinical trials. After September 1, 2010, division of responsibilities between the Ministry of Health and Roszdravnadzor in clinical trials regulation became more complex.

15.8.5.3. IND Approval Process and Submission Dossier

The approval process for clinical trials takes three to five months and the submission process is sequential. The submission dossier documents, Power of Attorney (PoA) form, sponsor protocol, IDB, SIS/ICF, and patient materials must be in Russian.

15.8.5.4. Challenges and Common Problems

On April 12, 2010, the President of the Russian Federation, Dmitry Medvedev, signed the federal law "On Circulation of Medicines". This federal law regulates the development, preclinical testing, clinical trials, expert evaluation, state registration, standardization, quality control, manufacture, compounding, storage, transportation, import, export, advertising, dispensing, distribution, transfer, use, and destruction of medicines, and came into force on September 1, 2010.

The main aspects of clinical trials affected by the new legislation are the new regulatory structure, the new clinical trial application process, stricter requirements for the principal investigator's experience, new requirements for contracts with clinical sites, life insurance for clinical trial participants, central ethics committee structure, and importing of study drugs and comparators.

In the short-term perspective, these massive regulatory changes may have affected the number of studies approved in the fourth quarter of 2010 and first quarter of 2011 and delayed studies start-up. In the medium- and long-term perspective, better insurance of study participants, transparency of investigators' agreements, and a faster study review and approval process declared by the Ministry of Health are positive changes making Russia more attractive for Western sponsors.

15.8.6. Good Clinical Practice

The Russian GCP legislation (GOST P 52379-2005) is identical to ICH/GCP. GCP training for physicians and investigators is conducted by the regulatory body, pharmaceutical companies, and CROs.

15.8.7. Clinical Trial Status

The Federal Service on Surveillance in Healthcare and Social Development approved 577 new clinical trials in 2009. Most trials (60%, 348) were multinational multicenter studies. These clinical trials were sponsored by pharmaceutical companies from 28 countries. Most (191) were initiated by Russian sponsors, followed by American (128), German (51), Swiss (42), UK (35), and Japanese (34) sponsors. Of new studies, 85 percent were in six major therapeutic areas: oncology (104), endocrine and metabolic (66), respiratory (66), cardiovascular (64), musculoskeletal (40), and CNS (38).

15.8.8. Site

Russia is an attractive country for conducting clinical trials because of the high patient recruitment and low investigator costs, which are approximately half those in the USA. Hospitals must be certified to conduct clinical trials and currently there are more than 1000 approved sites, which are listed on the official webpage of the regulatory authority that is updated on a quarterly basis. Sites are well equipped and staff is computer literate. Upon receipt of final study approval from the regulatory authorities, documents, which must be translated into Russian, should be submitted to the local ethics committees, which usually meet once a month. Contracts should be signed with the site and the principal investigator.

15.8.9. Inspection

Thirteen FDA inspections were conducted at study sites in 2009. Their outcomes were No Action Indicated in 10 and Voluntary Action Indicated in three. In addition, clinical sites are regularly inspected by the regulatory authorities.

15.8.10. Pharmacovigilance and Serious Adverse Event Reporting

SUSAR reports should be submitted in the original language (hard or electronic copy) with a cover letter and summary table of the submitted reports in Russian (hard and electronic copy). The timelines for SUSAR reporting are the same as in other countries: all related SAEs that are fatal or life threatening must be reported within seven days and all other related SAEs within 15 days. An annual safety report has to be submitted to the central ethics committee and regulatory authorities. Periodic safety reports (PSRs) should be submitted in the original language in hard and electronic copies with a summary of the safety information and a cover letter in Russian. PSRs can be issued in several forms, for example an Annual Safety Report (ASR), Clinical Trial Safety Update Report (CTSUR), or Development Safety Updated Report (DSUR), and should be submitted at least once a year.

15.8.11. Contract Research Organizations

Currently, there are more than 50 CROs, which conduct more than 50 percent of all international multicenter trials. The annual value of the CRO market is estimated at US $150–300 million, with seven companies (PSI, Quintiles, PPD, PAREXEL, Worldwide Clinical Trials, ClinStar, and ICON) having 60 percent of the market. The CROs operating in Russia can be divided into three basic groups: global, joint-venture, and local CROs. Global CROs are usually able to offer a full range of services, while joint-venture and local companies provide more limited services such as monitoring, training, biostatistics, quality assurance, and logistics.

15.8.12. Clinical Trial Staff

More than 80 percent of CRAs have a medical background. This is beneficial for clinical research in terms of understanding study requirements and developing good working relationships between CRAs and site staff.

15.8.13. Logistics

IVRS services are usually provided by Western suppliers. Many companies provide drug storage and distribution services and courier companies can organize the shipment of samples and customs clearance for clinical trial supplies. In 2010, Fisher Clinical Services, a global clinical trial materials vendor, opened its first drug depot in Russia.

15.8.14. Clinical Trial Organization

The Association of Clinical Trials Organizations (ACTO) is a non-commercial organization of the companies and clinical research professionals engaged in clinical trials in Russia. The objectives of ACTO include the further development of Russia as a leading country for clinical research by engaging in capacity building and activities to shape the professional environment, the generation of awareness of clinical research as a specialty and establishing the industry's reputation among the general public and governmental agencies, the creation of a favorable business environment, and maintaining a constructive dialogue with regulatory authorities to develop stable local legislation for clinical trials. The key objectives of the Association are to promote the optimization of the regulatory process and speed up the approvals process. A regulatory committee monitors the timelines and working patterns of the regulatory system, which helps to identify problems and develop appropriate solutions. The committee also aims to develop a common approach, where appropriate, for ACTO's members to communicate with the regulatory authorities, e.g. setting up standard procedures for document submission.

15.8.15. Challenges and Opportunities

Conducting clinical trials in Russia is very attractive for a number of reasons. Investigators and CRAs are highly motivated and educated; over 60 percent of investigators have a PhD and the majority of CRAs are doctors. Sites usually have a small number of competing trials and a large treatment-naïve patient population. Many can achieve high enrollment rates. The variety of medical institutions available, ranging from small local hospitals to large regional hospitals, is well suited to the needs of clinical trials. The cost of clinical trials is rising but it is still significantly lower than in Western countries.

Challenges include bureaucratic obstacles that can delay study start-up and ongoing changes to the customs procedures.

SECTION 3

Clinical Study Logistics

CHAPTER 16

Design of Clinical Trials for Emerging Countries

Nermeen Varawalla
ECCRO, London, UK

Contents

16.1. INTRODUCTION

The emerging world of clinical development comprises countries beyond the USA, Canada, Western Europe, and Japan that are well suited for global clinical trials. These regions and countries are Central and Eastern

Global Clinical Trials
ISBN 978-0-12-381537-8, Doi:10.1016/B978-0-12-381537-8.10016-0

Europe, South Africa, Latin America, India, China, and South East Asia. Although there are numerous differences in the healthcare environment within this group of countries, they all have important similarities, namely, the potential for rapid subject recruitment and an opportunity for cost savings. However, these emerging countries have relatively nascent clinical trial environments characterized by evolving regulatory approvals processes, less robust systems for subject protection, and smaller, albeit fast growing, domestic pharmaceutical markets. They all present concerns to a varying extent about extrapolation of date accrued from their countries to the more developed regions of the world.

This chapter will examine the opportunities and challenges that emerging countries collectively present for international Phase II–III clinical trials and recommend an approach for the most effective utilization of these countries within global clinical development programs.

16.2. OPPORTUNITIES

The reason for the exponential increase in the utilization of emerging countries for international Phase II–III trials is the opportunity they present for swift, meticulous, and cost-effective clinical trial conduct. Access to patients, investigators, sites, and a cost-effective labor force are the fundamentals that support this.

16.2.1. Access to Patients

The increased complexity of clinical trials and the large pipeline of products in early clinical development have resulted in increased competition for clinical trial participants in North America and Western Europe. Competition is more fierce for patients with cancer, neurological, and cardiovascular diseases. Slow patient recruitment can delay product launch with revenue loss during the precious product patent life. Therefore, access to patients remains critical for the success of clinical development programs. Emerging countries offer this access, hence their inclusion within Phase II–III clinical trials is becoming standard practice.

Emerging countries have delivered the benefits of patient access and there are numerous success stories of patient recruitment rates being up to seven times faster than in North America and Western Europe. This is particularly true for oncology trials. Furthermore, the efficiency of clinical trial conduct is enhanced because the number of patients contributed per site in an emerging country is many times greater than that of their counterparts in North America and Western Europe.

The reasons that emerging world countries are able to recruit clinical trial participants in such a rapid and efficacious manner are as follows.

First, most emerging countries have large numbers of patients with diseases of both the developing and industrialized world. In addition to widely prevalent infectious and tropical diseases, rapid and extensive urbanization in countries such as India, China, South Africa, and Brazil has resulted in disease prevalence similar to that found in North America and Western Europe. For example, India and China have the world's highest prevalence of metabolic syndrome, i.e. individuals with a combination of insulin resistance, hyperlipidemia, and obesity.

Second, the evolving urban centric healthcare systems of emerging countries require patients to travel to centralized hospitals for their healthcare needs. The resulting convergence of large patient numbers at centralized facilities is conducive for swift, efficient, and hence cost-effective patient recruitment.

Third, huge unmet medical need exists in the emerging world. Healthcare provision is a mixture of private and inadequate state-subsidized care. There is low health insurance coverage. Therefore, for a large proportion of the population, participation in clinical trials remains a way to access high-quality healthcare. This is true also in the USA and parts of Europe; however, given larger populations and deficiencies within state healthcare systems, the pool of patients with unmet medical need willing to participate in clinical trials is substantially larger in emerging countries. Furthermore, the authority position of physicians and the respect that patients have for them make patients more likely to participate in a clinical trial if this was recommended by their physician. Strong relationships between patients and their physicians also result in better study retention with lower dropout rates.

16.2.2. Access to Investigators

Emerging world countries have well-trained, motivated physicians well suited to be investigators for global clinical trials. Medical education is often delivered in English and is based on a system similar to that in North America and Western Europe.

Physicians at academic centers are well trained, fluent in English, and computer literate. Many specialists have received postgraduate medical training in North America or Western Europe. The number of good clinical practice (GCP)-trained physicians is increasing.

The motivation for physicians to participate in international clinical trials is more than financial. It is also a means to participate in international clinical research, keep abreast with developments in their field, and access state-of-the-art treatments for their patients. Hence, physicians are motivated to participate in clinical trials and are willing to adhere meticulously to study protocols. As in the USA and Western Europe, not all physicians are well suited to being clinical investigators, so access to the right physicians is necessary.

16.2.3. Access to High-Performing Clinical Trial Sites

The centralized, hospital-based, healthcare delivery systems typical of emerging countries mean that patients travel to relatively few, central, urban hospitals for their healthcare needs. These hospitals that cater to large numbers of patients are usually well equipped with state-of-the-art diagnostic and therapeutic facilities. Furthermore, they are staffed with physicians and academic staff who are well suited to be investigators. This set-up provides for clinical trial sites with the potential to be highly productive. However, site support in terms of staff, processes, and systems is essential for these sites to deliver their full clinical research potential.

16.2.4. Potential for Cost Savings

Clinical trial conduct in emerging countries can be up to 50 percent cheaper than in North America and Western Europe. These cost savings result from lower trial budgets and lower operational costs.

Investigator and site fees are a fraction of those in North America and Western Europe. The lower investigator fees are due to relatively lower physician earnings, a keenness to attract global clinical trials, and comparatively few trials. As the clinical development sector matures, the investigator fees will increase, as has already occurred in Eastern Europe; however, in most emerging countries there will remain a differential before they approach the levels in the USA. Treatment costs are half those in the USA, with lower costs for medication, diagnostic services, and hospitalization. Thus, although there is an expectation for the trial sponsor to cover the costs of hospitalization, comparator, and supplementary treatments and investigations, the amounts involved are never high by US standards.

There is an opportunity to save up to 50 percent of clinical trial execution costs compared to North America and Western Europe. This is because of relatively reduced labor costs for clinical operations personnel such as project managers, monitors, data managers, medical writers, and biostatisticians. Because patient enrollment is concentrated at few sites, more patients are covered per site visit and there is less required travel. This reduced travel requirement, along with relatively cheap domestic travel fares, results in much reduced study-related travel costs. Lower costs of support services such as handling of clinical trial supplies, printing, translation, and local courier services further contribute to cost savings.

16.2.5. Global Social Responsibility

The inclusion of emerging countries in global clinical trials is not a form of exploitation as it is often misrepresented to be, but a way of contributing to the development of their evolving healthcare systems. Hence, the governments, medical institutions, and physicians within emerging countries strive to attract international clinical trials and attempt to ensure that they have the requisite regulatory and ethical frameworks.

The key benefits to the patients and healthcare systems of emerging countries are that trial participants have the opportunity to access cutting-edge, often life saving treatments.

Participation in clinical trials, irrespective of being enrolled in the placebo or treatment arm, is known to improve health outcomes, and this is all the more true for the patient group with otherwise limited access to healthcare. Investigator and site fees are partially utilized for the purchase of equipment and facilities, with benefit to all users. Participation in international clinical research enhances the practice of evidence-based medicine, thorough record-keeping, and better patient communication, and thus contributes to the overall improvement of healthcare standards [1].

16.3. CHALLENGES

The key challenges of conducting international clinical trials in emerging countries are navigating the regulatory approval process in a timely fashion and ensuring that the accrued clinical trial data meet international standards of data quality, scientific acceptability, and ethical integrity. Understanding and addressing these challenges at an early stage of study planning is critical for success.

16.3.1. Regulatory Approval Process

The prolonged and uncertain regulatory approval processes that remain a feature of many emerging countries can delay study start-up and could negate the speed advantage of rapid patient enrollment. However, as the clinical development environment in emerging countries is evolving, the regulatory processes are becoming more streamlined with an improvement in both the speed and predictability of obtaining clinical trial approval. China has the most prolonged regulatory approval process, requiring an average of nine months to navigate a path through six regulatory approval bodies. Latin America continues to be uncertain, with numerous instances of unexpected delays, whereas India has made some progress following implementation of the "fast track" stream, whereby, in situations where US Food and Drug Administration (FDA) investigational new drug (IND) or European Medicines Agency (EMA) clinical trial approval is available, the regulatory approval process is abbreviated and streamlined. Following regulatory approval, there is a process for obtaining the import license for clinical trial supplies as well as one for obtaining ethics committee approval. Processes, documentation, and timelines vary among emerging countries, so local and up-to-date knowledge is essential.

16.3.2. Uncertain Data Quality

There are concerns about the data quality standards in emerging countries which are considered to be relative newcomers to the global clinical development sector as it is perceived that investigators, sites, and clinical operations personnel have limited experience. Furthermore, the number of GCP-trained physicians is limited, though growing. It is also recognized that the constrained healthcare resources require investigators to juggle the demands of patient care and clinical research, limiting their focus on clinical trial conduct [2].

In recognition of these concerns, stakeholders within the clinical trial environment of emerging countries have undertaken numerous initiatives to improve their data quality standards. Clinical research training is being delivered in a number of ways that include study start-up activities, and GCP training of investigators and site personnel.

Local contract research organizations (CROs) and emerging country affiliates of global CROs are fully cognizant of the importance of training and continue to make substantial investment is this area. Accredited training institutes have been set up in countries such as India to meet the demand for trained clinical research personnel, and as a result there are currently 10,000 GCP-trained clinical research professionals and 1500 GCP-trained investigative sites in India.

Furthermore, to ensure a high standard of data quality, CROs and pharmaceutical sponsors well versed with working in emerging countries provide additional site support in the form of site management staff and relatively intensive monitoring schedules. Investigators are often supported by clinical research fellows and nurses, which does not adversely affect study budgets because of lower labor costs. Furthermore, the leading service providers within emerging countries have invested in setting up quality assurance programs with regular site audits.

Governments of emerging countries recognize the importance of data quality and GCP compliance as the foundation for a thriving clinical development industry. Hence, they have encouraged their regulatory authorities and national medical research councils to establish initiatives to ensure that international standards for clinical data quality are attained.

However, the most important validation of clinical data quality is its acceptability by international regulatory authorities. Both the FDA and EMA have accepted emerging world clinical data from pivotal Phase II and Phase III trials for over a decade. Reassuringly, FDA audits in emerging world countries have recorded findings similar to those in North America and Western Europe. There have been 14 FDA inspections in India since 2005 with no "Official Action Indicated" verdicts. Thus, from the international regulatory perspective the quality of data from emerging countries is not a concern.

Sponsors could have concerns that clinical data obtained from emerging countries may not be universally applicable because of the perceived differences in diets, lifestyle, attitudes to pain, concomitant usage of alternative medicines, and genetic differences.

A sound understanding of the characteristics of patient populations in different emerging countries is essential to overcome these concerns. For example, it is worth considering whether the widespread use of traditional Chinese medicine will affect the interpretation of the data from China.

16.3.3. Ethical Standards

There are concerns that the ethical standards in emerging countries may be different from those in North America and Western Europe; all the more so because in these countries there are clinical trial subject populations with unmet medical need, reverence for physicians who remain authority figures, and linguistic and cultural barriers for subjects to understand fully the implications of trial participation. The GCP guidelines for many emerging countries are based on the Declaration of Helsinki, World Health Organization (WHO), and International Conference on Harmonisation (ICH)-GCP. Notwithstanding this, ICH-GCP compliance in such an environment presents a challenge and there remain concerns about exploitative trial designs, weak informed consent, and inadequate post-trial access to study medication.

Stakeholders in the clinical research sector including government, industry, and academia have embraced the critical importance of compliance with the international ethical code and implemented numerous initiatives to do so. These include training, guidelines, legislation, and inspection processes. Informed consent given without any coercion is the cornerstone of GCP-compliant clinical research and arguably the inability to comply with this requirement should disqualify a country from participating in global clinical trials. Hence, meticulous attention to this tenet of clinical research is crucial, especially given the nascent clinical trial environment in emerging countries with large disparities in access to healthcare.

Access to biomedical innovation and free healthcare is an important motivator for clinical trial participation in all parts of the world. A survey of potential clinical trial participants in the USA revealed that about 50 percent of clinical trial participants claim that their primary motivation to participate in a clinical trial is to access free medication [3]. Thus, it is not surprising that within emerging countries with substantial inequity of healthcare resources access to free medication is an important motivator for clinical trial participation. This author submits that provided consent is free and informed, this remains wholly ethical and compliant with the principles of ICH-GCP.

16.3.4. Protection of Intellectual Property Rights

In the past, the legal system in countries such as India, Russia, Brazil, and China did not recognize intellectual property rights in accordance with international law. This has enabled their domestic generic pharmaceutical industry to flourish. From January 2005, as a full member of the World Trade Organization (WTO), India has been committed to recognizing and enforcing product patents in all fields of technology including pharmaceuticals.

Furthermore, as the economies of these countries have developed, their governments have realized the importance of sound intellectual property legislation to protect home-grown innovation. Hence, concerns that domestic generic pharmaceutical companies will be able to take inappropriate advantage from the conduct of innovator drug trials are not valid [4].

16.3.5. Low Commercial Potential

Although certain emerging world countries such as Brazil and China represent the fastest growing pharmaceutical markets in the world, the USA, Western Europe, and Japan continue to be the largest in revenue terms. The commercial potential of a country is important for country selection in a global clinical trial. Therefore, in terms of current market size, emerging world countries are commercially less attractive. However, they do represent future high-growth markets and offer the opportunity to shorten the clinical development timelines with earlier product launch in more lucrative markets.

16.4. RECOMMENDATIONS

Recognizing the opportunities and challenges listed above is essential for effectively designing and conducting clinical trials in emerging countries. Both trial design and conduct should be carried out so as to maximize the advantages and minimize the risks associated with emerging countries. Although the specifics of the trial will influence protocol design and trial planning, issues related to emerging country clinical trial conduct discussed below are universally applicable.

16.4.1. Regulatory Strategy

To fulfill the international regulatory requirements and meet the commercial imperatives, a proportion of the subjects in most global clinical trials should be enrolled from North America and/or Western Europe. The remainder could be recruited from one or more emerging countries to gain the advantages of expeditious and cost-effective clinical development that they so clearly offer. Given the above caveat, the best mix of countries should be selected on the basis of the product characteristics, country-specific regulatory approvals processes, and protocol-specific considerations.

The proportion of subjects to be recruited from emerging countries is dependent on the factors cited above as well as the trial objectives, commercial strategy for the product, and the scientific aspects of trial design. Unless there were compelling reasons, it would be unwise to recruit more than 80 percent of the total number of clinical trial participants from emerging countries. This is because the marketing authorization for the largest global markets is controlled by the FDA and the EMA. Further regulatory authorities in other countries continue to look to these more experienced regulatory agencies when making decisions about both marketing authorizations and clinical trial approvals for their own countries. Although neither the FDA nor the EMA has categorically stated its views on the proportion of evaluable subjects that may be enrolled from emerging countries, prevalent industry wisdom is that including a fair representation of subjects from the USA and Western Europe is essential so as to ensure a subject population that includes representation from their largest commercial markets. Given population flows around the world, it is possible to obtain a satisfactory racial mix by conducting trials only in the USA and Western Europe; however, the reverse may not be the case if the trial were to be conducted entirely in an emerging country. Ensuring that a proportion of subjects is white Caucasians is recommended.

Furthermore, it is currently perceived that some of the world's most experienced investigators and sites are in the USA and Western Europe. Hence, there is a concern, perhaps unfounded, that a pivotal clinical trial submission to the FDA or EMA with no developed world participation might signal inferior data quality or less than the best available expertise. It is likely that this perception will change, however, ensuring that representation of developed country sites and investigators does contribute to the data's credibility. In addition, the regulators of emerging countries take reassurance from the fact that the very same protocol has been approved by a more experienced regulator and is recruiting subjects in the USA and/or Western Europe. Given the sensitivity about the exploitation of emerging country populations, their regulators and ethics committees are suspicious of Phase II–III clinical trial applications that seek to recruit exclusively in their emerging country. Hence, unless there is a compelling scientific case to restrict recruitment to emerging countries, this is not recommended.

Given the similarities in the advantages of including emerging countries within Phase II–III clinical trials, arguably, a single well-selected emerging country could deliver the objectives of patient access, speedy recruitment, and cost savings. The risks of increasing the number of emerging countries in a global trial are loss of focus and requirement for increased resources. The counter-balancing advantages are a mitigation of the risks, by "placing eggs in more than one basket". If there are concerns that one of the selected emerging countries might experience delays with regulatory or ethics approval or there might be issues with enrollment of suitable subjects, then there is a case for increasing the number of emerging countries with the expectation that one of the other selected emerging countries could "step up" and take over the patient recruitment allocation of its emerging country counterpart. Thus, by including more than one emerging country there is a contingency solution built in from the outset. Including an emerging country as a rescue should other countries experience difficulties is also a viable approach; however, this would involve delays when regulatory approvals are awaited. If a thorough feasibility assessment is performed that addresses medical, clinical, and regulatory issues, the risks of the selected country not performing are negligible, and therefore a strong case could be made for selecting, albeit carefully, only one emerging country.

Regulatory approval processes, documents, and timelines vary among emerging countries. It is essential to understand these for the selected emerging countries and to anticipate potential hurdles. Up-to-date knowledge and meticulous compliance with the study start-up regulations are essential to mitigate the risk of delayed study start. At the time of study design it is advisable to factor in the requisite study start-up timelines with the intention of compensating for these with swift patient enrollment. Ensuring that all study documentation and start-up procedures are completed in accordance with local regulatory requirements with necessary language translations is vital to minimize unexpected delays. For certain indications or product types it may be prudent to avoid certain emerging countries in anticipation of prolonged regulatory deliberations. It is also worth understanding the process, documentation, and timelines for ethics approval. Advice should be sought on potential ethics committee objections with thought given to ways of pre-empting these.

16.4.2. Ethical Integrity

It is critical to try to ensure that all emerging country trials are conducted with ethical integrity and subject protection to the highest international standards. Sponsors and CROs with experience of clinical trial conduct in emerging countries know that careful investigator and site selection is essential to ensure the ethical integrity and scientific validity of the data. Investigators and site personnel play a vital role in ensuring that all clinical trial participants are provided with trial-related information in their own language with adequate provision for consultation with family and often community members. It is often required to have translations of patient-facing materials available in up to a dozen different local languages. In addition, sites provide counseling staff to discuss questions that participants and their families might have. Indeed, more experienced sponsors take the view that if there is any doubt over the investigators' ability to comply with a robust informed consent process it is best to exclude that site. Hence, the best in class investigators fully recognize the importance of informed consent, more so keeping in mind socioeconomic deprivation of their patients, and make every effort to comply. They do so not only for ethical reasons but also to ensure that clinical research in their country continues to flourish, recognizing that unethical practices will be the death knell of this nascent sector.

It must be acknowledged that the sociocultural approach to informed consent is different in certain countries. For example, it is not unusual for family members to give consent along with or on behalf of the subject. Access to subsidized healthcare is a strong motivator for patients to participate in clinical trials, and respect for physicians and their role as authority figures may make it difficult for subjects to refuse clinical trial participation. It is important for the sponsor and CRO to be aware of these factors and give particular attention to the informed consent process, endeavoring to ensure that subjects fully understand that they would be participating in a clinical experiment and appreciate the various consequences.

As yet, relatively few investigators, sites, or ethics committees within emerging countries insist that the sponsor provides post-trial access to their product. As the clinical trial environment in these countries matures, it is inevitable that investigators will demand post-trial product access for their patients, and sponsors should begin to plan for this. Most sponsors would not consider conducting a clinical trial in a country in which they have no intention of ever marketing a particular product. However, given the relatively lower commercial importance of emerging countries, product launch might be delayed, and so the timelag between trial completion and product launch could be considerable. Furthermore, there remains the thorny issue of subjects being unable to afford the medication. Post-trial and indeed post-launch access on a named patient basis will become an increasingly important consideration.

16.4.3. Feasibility Assessment

Protocol feasibility assessment from the medical, operational, and regulatory perspective is the first, crucial step for the successful clinical trial conduct in an emerging country.

Drawing on local experience, knowledge, and relationships, it is advisable to obtain an accurate feasibility assessment as early in the trial planning phase as possible.

It is important to ascertain the feasibility of the trial protocol within the emerging country's healthcare environment. This must include consideration of the standard of care for that particular disease state. Although increasingly there is a uniformity in healthcare standards across the world with a universal recognition for what is deemed to be best practice, in the underresourced healthcare environments of the emerging world there remain substantial differences in standard of care. This is perhaps most heightened in oncology, where there remains a discrepancy in the number and aggressiveness of the lines of treatment provided. Understanding these differences in standard of care not only prevents inappropriate protocol design but also could provide an opportunity to conduct certain trial designs that may no longer be possible in the USA or Western Europe.

There are relatively large numbers of treatment-naïve patients in emerging countries as a result of underresourced healthcare systems and differing clinical practices. Hence, it is possible to enroll steroid-naïve asthmatics or antiretroviral HIV patients in countries such as India or South Africa. This could be valuable for proof-of-concept clinical trials. In addition, differences in clinical practice may permit the inclusion of a placebo-controlled arm. For example, ethics committees in countries such as India may permit a placebo control arm in Alzheimer's disease trials because they could take the view that there is no available effective treatment. Understanding and taking advantage of such differences in healthcare standards could deliver much value to certain clinical trials.

Owing to ethnicity, diet, or cultural practices, disease profiles may be different from those seen in the Western world. Furthermore, differing standards of healthcare could result in the diagnosis of certain conditions being made at a later disease stage.

Seasonal differences between the countries in the northern and southern hemisphere enable year-round patient recruitment for disorders with a seasonal variation in incidence. Appreciating differences in disease profiles and accommodating for these is an important component of feasibility assessment.

A number of factors such as diet, alcohol intake, coprevalent disease, use of supplementary non-allopathic medications, and pain threshold modify disease behavior and responses to pharmaceutical intervention. The impact of these factors, if any, needs to be ascertained when selecting an emerging country. It is possible to identify and compensate for the effect of any such factors by minor modifications in the protocol such as additional baseline investigations or alterations to the inclusion–exclusion criteria. Clearly, there are substantial benefits to doing so early in the study planning process.

Quality of life questionnaires to be used in emerging countries need to be modified to accommodate country- and culture-specific differences. Attitudes to pain can be markedly different in subjects from emerging countries compared to their Western counterparts. This modification is above and beyond accurate translations into regional languages.

It is also essential to ensure access to prescribed diagnostic and therapeutic interventions with respect to the requirements of the trial protocol. It may be necessary to import comparator treatments or additional diagnostic kits, and ascertaining this at an early stage is good practice.

Certain trial protocols require long-term follow-up, but the socioeconomic situation in emerging countries may not be conducive to this. In such situations either special arrangements to ensure effective subject follow-up must be made or a decision taken not to use that particular emerging country. Similarly, if successful trial conduct requires access to comprehensive patient registries or databases and quality medical records, a number of emerging countries would need to be disqualified.

Racial differences have a limited impact on pharmacokinetics. The most significant effect is related to body mass index. However, it is important to explore whether the racial mix of the population of the considered emerging country would have a marked pharmacokinetic influence. In certain situations it may be appropriate to consider the human leukocyte antigen (HLA) subtypes prevalent in the population and, if necessary, include HLA subtyping or other genetic tests as part of the baseline trial investigations. Ideally, the study population should have a racial mix close to the population in which the product is finally expected to be used. It is worth avoiding conducting the entire study in a single racial group by trying to achieve a mixed population. However, the vast majority of marketed drugs have been developed in the Caucasian population, often 70 kilogram white males, with select bridging studies. The opportunity for heterogeneous study populations presented by emerging countries is a valuable one, provided the racial implications, if any, on pharmacokinetics and treatment outcomes are well understood.

Protocol feasibility assessment must address the above-mentioned medical and scientific issues and not focus only on estimating subject enrollment rates. If there are issues related to protocol feasibility, at an early planning stage an informed decision must be made about whether to modify the protocol so as to make it conducive for the selected emerging country. If such modifications could jeopardize the overall trial objectives, it would be best to abandon plans to conduct the trial in that emerging country. Such a timely decision could result in substantial time and resource savings.

16.4.4. Patient Recruitment Amongst Plenty

In spite of the fact that emerging countries have large numbers of potential clinical trial subjects, it is important to develop a customized patient recruitment strategy. Patient enrollment tools well suited for emerging countries include health camps where patients are offered free health check-ups and given general health education. During these visits they are also informed about ongoing trials in neighboring centers and, if they meet inclusion criteria, they are offered the opportunity to participate in the trial. Such an enrollment strategy is particularly effective for indications such as hypertension, obesity, lipid disorders, and undiagnosed diabetes mellitus. Patient referral from community clinics, general practitioners, and other specialists in the region is an important patient enrollment method and it is worth deploying local resources to affect this. Depending on local practices it may be necessary to arrange a reward system for referrals. Review of the institution's databases and medical records would be another useful way of identifying potential subjects. If advertising is used, it is essential that the medium and message are customized for the emerging country environment. Simply transferring advertising campaigns developed for the USA or Western Europe, perhaps with translations, is rarely effective. Compelling study design that carefully considers the participant's perspective remains one of the best ways to attract study participants, as is the case everywhere in the world.

Retention of subjects until study completion is important to ensure the study's statistical validity. In view of the less developed socioeconomic setting in emerging countries, close attention to facilitating patient follow-up and study continuation is necessary. Strong patient–physician relationships encourage high levels of patient retention as patients tend to be in awe of the investigator and hence desirous of complying with their instructions. However, the challenges of their daily lives may make it difficult for them to adhere to visit schedules. Site personnel who are able to establish rapport with participants can effectively motivate them to comply with visit and follow-up schedules. It would also be worthwhile to consider providing a budget to the site for supporting subject retention. These discretionary funds are used for transport and compensation for daily loss of wages for both the subject and their carers. Study design plays an equally important role in retention. Studies with fewer blood draws and less frequent site visits often have better subject retention. Designing trial logistics that have been customized for the local setting can also be valuable.

16.4.5. Selection of Vendors

The options facing a sponsor seeking to utilize emerging countries for their clinical trial would be to engage their own clinical operations staff in that country or to engage a CRO. Large pharmaceutical companies are likely to have clinical operations resources in the key emerging countries of the world. In the absence of this, the sponsor could engage a global CRO, offering the advantages of a one-stop shop service and standard global processes. Increasingly, sponsors seek to engage a specialist regional CRO for the emerging country because such specialists offer deep local knowledge and relationships, usually operate to a high quality standard, and deliver more of the emerging country-derived cost savings to the sponsor compared to their global generalist counterparts. The specialist CRO would work alongside other CROs that operate in other countries. Irrespective of which option is selected, it is essential to ensure that there is access to the local knowledge, expertise, and relationships essential for successful emerging country clinical trial conduct. Notably, relationships with high-performing sites, investigators, and the regulator's office are important.

Apart from clinical operations resources, it would be necessary to select local vendors to provide ancillary services such as central laboratory, imaging, and electronic data capture (EDC). The evolving clinical trial landscape in these countries has resulted in numerous high-quality vendors, making it possible to select vendors best suited for the trial requirements. There is also a number of specialist service providers providing clinical trial supply management, translation, centralized electrocardiogram (ECG) monitoring, and interactive voice response system (IVRS) services.

16.4.6. Smart Application of Technology

In recent years, there has been an exponential growth in technology tools, systems, and processes that claim to support drug development. The suite of offerings is now mature; hence the savvy user should be able to select technologies and craft solutions able effectively to meet operational challenges. This means progression from slavish adoption of technology to crafting solutions to serve real needs and achieve critical endpoints. Given the robust information technology (IT) infrastructure and widespread IT skills within most emerging countries, along with freedom from legacy biases, there is an opportunity for the smart application of technology to overcome some of the data quality issues that may exist.

EDC should become standard practice with a halt to legacy paper-based systems. This, combined with the deployment of clinical trial management systems and electronic document management, would help to address concerns that may exist regarding the loss of control sponsors located in the West may experience when conducting trials in distant countries across time zones. Such application of technology would enable real-time access to clinical trial data, providing the opportunities for remote, centralized, monitoring, timely detection of issues with prompt escalation and early trend and signal detection.

Furthermore, given the widespread availability of mobile telephony and wireless networks within emerging countries, there are numerous opportunities to deploy this technology for patient-reported outcomes and to ensure subject compliance. The hurdles at the moment are related to local language translations and subsequent validation.

It would be remiss not to highlight the compelling case for conducting data management in India for global studies, i.e. not only studies conducted in India. Access to a large resource pool with IT and business process skills in a low-cost labor environment provides the basis for cost-effective data management services. Excellent English language skills, a large number of IT-literate biomedical graduates, and a motivated and flexible work ethic contribute to the attractiveness of this proposition [5].

16.4.7. Compensating for Relative Inexperience

The amount of global pharmaceutical research and development dollars spent in the emerging regions of Latin America, Asia Pacific, and Central–Eastern Europe has multiplied many times over the past decade [6].

However, the total expenditure in these three regions continues to be less than a few percent of the global spend, implying that although the emerging countries are enjoying a period of brisk growth, their global contribution to clinical development continues to be a small one. Furthermore, although international clinical trial activity is growing at a substantial rate within emerging countries, this growth is occurring from a very low denominator base. Therefore, when designing and executing clinical trials within emerging countries, it is critical to recognize the relative inexperience and limited clinical trial infrastructure that may exist in these countries. It must also be noted that present international clinical trial activity within emerging countries involves a tiny fraction of the subject, investigator, and site pool available in these countries, and there exists considerable scope for continued growth. Recognizing this, investment in developing emerging country clinical trial capabilities has much medium- to long-term value. Investing in building investigator relationships and clinical trial site capabilities remains essential for both sponsors and CROs who seek to continue to conduct clinical trials in emerging countries.

16.4.8. Building Investigator Relationships

In view of the relatively nascent clinical trial environment in emerging countries, investigators play a critical role in that they not only enable access to trial participants but also are custodians of the data's scientific and ethical integrity. It is ultimately the investigator and site staff who ensure that subjects are recruited, the protocol is adhered to, and data are collated in a GCP-compliant manner and in keeping with the project expectations. In a country with limited healthcare resources the sanctity of the informed consent process remains a concern for sponsors. As investigators play a critical role in ensuring that informed consent is correctly obtained, attention to their selection and management is even more important in these countries than is the case in the more experienced clinical trial locales of the world.

Investigator selection is the first and perhaps most important step towards building a pool of high-quality investigators. Emerging countries offer a pool of physicians from which clinical trial investigators may be recruited. However, the vastness of this potential pool and the variation in their training and suitability for clinical research do pose a challenge. Although many physicians are keen to participate in global clinical trials motivated by the prestige, financial gains, and opportunity to offer better medical care to their patients, only a few have the requisite mindset and capabilities. Until recently, the experience that many of these physicians have had of clinical trial participation was restricted to domestic postmarketing studies usually conducted with limited resources and rigor. In view of the variability of available physician talent and their limited clinical research experience, it is imperative to circumvent the large numbers of enthusiastic ill-suited investigators and identify those with the necessary mindset, postgraduate qualifications, and patient access required for international clinical development.

As emerging countries contribute increasing amounts of pivotal clinical data for product registration studies, scientific and medical questions related to disease behavior, diagnostic tools, and genetic and environmental influence will become more important. Standard of care, availability and validity of rating scales, influence of HLA type on treatment outcome, impact of nutritional status, and sociocultural attitudes to management of terminal illness are examples of questions whose relevance will increase as the contribution of emerging countries to global clinical development accelerates. Determining the answers to these questions will be necessary for the correct interpretation of clinical trial data, appropriate protocol design, and correct subject selection. Experienced investigators within emerging countries are a valuable resource to provide guidance on these scientific and medical issues, and seeking their input on trial design and data interpretation will become increasingly important. Wise sponsors will be those who view emerging country investigators not just as reliable patient recruiters but as scientific and medical experts, and accordingly build long-term working relationships with them [7].

16.4.9. Developing Clinical Trial Sites

Clinical sites with a high potential to contribute subjects are those with heavy clinical workloads. Given centralized healthcare delivery in emerging countries, general hospitals located in urban areas that treat large numbers of patients have the potential to be highly productive clinical trial sites. However, these sites have to meet the huge demands of healthcare delivery with limited resources. In such a busy clinical environment dedicated resources to support clinical research are usually limited. As a result, sites often experience resource constraints and are hindered from delivering subject enrollment and the clinical data quality of which they could be capable. For such sites to deliver their clinical trial potential they must be supported by resources and staff clearly designated for clinical research.

High-potential sites need to be carefully selected, trained and supported. The necessary investment in capital, time, and resources curtails the number of sites that may be developed and requires discipline in rejecting sites unable to meet standards. Furthermore, sponsors and CROs need to make themselves attractive to their preferred sites. This approach would result in a move away from tactical feasibility assessments and initiation of large numbers of sites in numerous countries to a strategic and long-term relationship with select sites.

Contingency planning will minimize the risk of missed timelines and budget overspends. Within each of the selected countries, both developed and emerging, there should be contingency sites in case of unforeseen issues with the selected sites. While budgeting for emerging country utilization, managers often overlook the less obvious costs of conducting trials in emerging countries. Although there is much potential for cost savings, it is important to factor in costs associated with international travel, additional training, and extra audit visits which are often essential to provide emerging country sites with additional support.

The objective of site selection is to engage with only select investigators and clinical trial sites able and willing to meet international performance standards. Typically, these would be general and specialist hospitals with busy clinical workloads and a good standard of diagnostic, therapeutic, and infrastructure facilities. These institutions would need to be staffed with motivated physicians interested in and willing to be committed to clinical research; ideally, each site should have at least one lead investigator. The hospital would have a busy clinical workload with access to large numbers of patients, both those attending the hospital and referrals from satellite clinics. Ideally, the hospital should have its own ethics committee or institutional review board that is correctly constituted and meets regularly. If this was not the case, then easy access to a central ethics committee would be essential. Experience of participating in international clinical trials would be desirable; however, given the need to develop clinical trial sites within emerging countries, a lack of experience should not exclude a potential site if other criteria were met.

Although there is still relatively limited competitor trial activity in a number of emerging countries, this is rapidly changing at "first tier" sites that have developed experience and a track record. These sites are becoming increasingly busy with participation in numerous trials. To ensure that there will be sufficient capacity within emerging countries to accommodate the increasing numbers of clinical trials being earmarked for them, it is essential for "second tier" clinical sites to develop capabilities. Often these sites are located within smaller cities, still large by most standards, with populations of around two million, with healthcare facilities that also serve populations from surrounding rural and semirural areas. Initiatives to train potential investigators based in these places, allocation of funds to build infrastructure and resources at these sites, and commitment to growing the network of investigative sites will contribute to the development of necessary site capabilities and future capacity.

Site support should include training, equipment and, importantly, staff whose role is to facilitate the conduct of all ongoing clinical trials at the site. In the emerging country setting, the division between site management and study monitoring could be artificial and wasteful. Better productivity, data quality, GCP compliance, and subject protection can be achieved by coordinated team working, supported by proactive quality assurance measures. Support at the site level by dedicated, trained resources employed either by the site or by a site management organization is recommended.

Sponsor engagement, ideally in the form of site visits, either at the stage of investigator selection or early in the study cycle, positively influences site performance and recruitment. Demonstration of a sponsor's interest in the environment in which subjects are recruited can substantially influence study performance. If geographical challenges make site visits difficult, investigator interaction via electronic, telephone, and written communication is essential.

There is value in ongoing site development, which should include training. Ongoing clinical research and GCP training for investigators and site staff would meet the needs of new joiners and provide refresher courses to existing staff. Trial- and protocol-specific training should be delivered before the study start. Additional site development activities are the setting up of processes, systems, and tools to support effective clinical trial conduct; providing advice on investing in additional equipment, if required, such as freezers, computers, and fax machines, and preparing and supporting the site through site audits and participation in the quality assurance cycle.

Investment in developing clinical trials sites in the emerging clinical trial environment provides for a long-term relationship with ensuing benefits. It is possible to make the site even more conducive to clinical research by facilitating the compilation of patient databases for the site and building relationships with satellite clinics able to refer patients. By fostering strong investigator relationships at the site, it would be possible to collate the investigator's recommendations and advice on study design and accurately assess protocol feasibility and estimation of subject enrollment. Ease of addressing contractual issues and negotiation of study budgets would be the business benefits of a mutually beneficial long-term relationship.

An important source of clinical operations wastage is non-productive clinical trial sites. Apart from the waste of resources spent on initiating and monitoring these sites, non-productive sites have a detrimental effect on morale and arguably have poorer data quality because the investigator and site personnel have not developed any familiarity with the study protocol. Therefore, weeding out non-productive sites, by working predominantly, if not exclusively, with a relatively limited number of tried, tested, and trusted clinical trial sites would deliver value and prove to be worth the investment.

16.5. CONCLUSION

The inclusion of emerging countries in international clinical trials will continue to increase until a time in the not too distant future when the present separation of emerging and emerged countries will become unnecessary. Until then, attention must be paid to understanding how the medical, clinical, regulatory, and operational differences that exist within these emerging clinical trial countries may be effectively utilized to achieve the trial objectives. Addressing in advance potential hurdles that could arise from discrepancies between the product characteristics and protocol design and peculiarities of the emerging countries being considered will mitigate most of the risks associated with these regions. Therefore, access to local knowledge, relationships, and expertise will remain paramount.

REFERENCES

[1] Varawalla N. Unravelling the advantages of conducting clinical trials in the emerging world. European Pharmaceutical Contractor 2005:40–5.
[2] Bhatt A. Clinical trials in India: pangs of globalization. Editorial. Indian Journal of Pharmacology 2004;36:207–8.
[3] Center Watch Survey. What motivates participation in clinical research? 2004.
[4] Varawalla N. India's growing clinical research sector: opportunity for global companies. IDrugs 2007;10:391–4.
[5] Varawalla N. Conducting clinical trials in Asia. Applied Clinical Trials 2006:108–13.
[6] Tufts CSDD Analysis of FDA's Bioresearch Monitoring Information System File 2006.
[7] Varawalla N. Investigative sites unlock the door to success in India. Applied Clinical Trials 2007:48–54.

CHAPTER 17

Study Management

Jay Johnson*, Sebastian Antonelli, Jennifer Aquino[†], Sue Bailey***, Edwin Chia[‡], Mateusz Chrzaszcz[§], Gillian Corken***, Jogin Desai[†], Janos Filakovszky[|], Barbara Lilienfeld[¶], Monica Lizano[††], Vladimir Misik[|], Alan Ong[‡], Viola-Marie Raubenheimer[‡‡], Veronica Suarez**, J. Rick Turner*, Sorika van Niekerk***, Daniel Vazquez** and Johan Venter*****

*Quintiles, Durham, NC, USA
**Quintiles Latin America, Buenos Aires, Argentina
***Global Sales Operations, Quintiles Sub-Saharan Africa, Irene, South Africa
[†]Cenduit, Raleigh, NC, USA
[‡]Quintiles Asia, Singapore
[§]Quintiles Europe, Warsaw, Poland
[|]Quintiles Eastern Europe and Middle East, Vienna, Austria
[¶]International Corporate Events Network, Johannesburg, South Africa
[††]Quintiles Latin America, San José, Costa Rica
[‡‡]Quintiles Africa, Pretoria, South Africa

Contents

Global Clinical Trials
ISBN 978-0-12-381537-8, Doi:10.1016/B978-0-12-381537-8.10017-2

17.1. INTRODUCTION

This chapter's focus is study management of clinical trials conducted in developing regions of the world. Study management is a complex undertaking in any clinical trial, but considerably more so for large, multisite trials being conducted in developing countries. Simply having sites in more than one country (whether developed or developing) means that individual investigational sites may be many hundreds or thousands of miles apart. The contract research organization (CRO) selected to conduct the study has to have a footprint in the selected countries. Unless the sponsor has manufacturing plants in multiple countries, the drug product being tested will have to be shipped into at least one country, which brings travel and customs (import/export) considerations in to play. Biological samples collected from many investigational sites will have to be transported to the central laboratory chosen for this trial, which brings considerable logistical challenges. If the countries speak different languages, key documents will need to be translated into each appropriate language.

Cultural differences also need to be considered carefully. These can be much more subtle than language differences (there are often considerable cultural differences in countries that, ostensibly at least, "speak the same language"). Individuals in different cultures can tend to report the same degree of discomfort to a greater or lesser extent, leading to differences in self-reported adverse event occurrences. Some cultures may wish to "please the investigator" more than others, which influences how individuals from that culture interact with the investigator and his or her research staff. This phenomenon can be described in terms of Hofstede's power distance dimension [1], which is the degree to which individuals within a culture accept or challenge decisions by authority figures. Countries with large positive scores on Hofstede's power distance index (PDI) are more accepting of power distance: they are comfortable with a greater degree of social and class stratification and in accepting the decisions of authority figures; those with large negative PDI scores prefer small power distance, support minimizing social and class inequalities, and have a greater tendency to challenge authority figures. Countries with high PDI scores include Russia (+154), Arab countries, Bangladesh and China (all with +94), India (+90), and Brazil (+43); countries with low PDI scores include South Africa (−48), Hungary (−62), Jamaica (−67), Costa Rica (−113), and Israel (−214) [2].

Power distance may manifest itself in high-PDI cultures in terms of individuals being less likely to ask questions if they are not sure exactly what information contained in the informed consent form (ICF) actually means in practical terms for their participation (or otherwise) in the trial, and may therefore be more likely to agree to participate for a given level of uncertainty. All of these complexities in study management mean that advanced, detailed planning is critical, along with the preparation of worst case scenario contingency plans.

Throughout this chapter and the following chapter, examples are cited from several geographical regions: Asia Pacific, Africa, Central and Eastern Europe (CEE), Latin America, and the Middle East. Accordingly, before addressing specific project management issues, an overview of each geographical region is provided.

17.1.1. Asia Pacific

The Asia Pacific region has a very large population, containing the world's two most populous countries, China (about 1.3 billion) and India (about 1.2 million). Many languages are used, including some with strong regional dialects. There are some general similarities, and also some specific differences that require the sponsor's attention. With regard to similarities, all countries in the Asia Pacific region require an import license, and often an export license to ship samples or tissue. They all have regulatory or health authorities and site ethics committees, which evaluate and approve protocols, and provide ongoing supervision of the ethical conduct of the clinical trial as it is implemented.

Many physicians have gained clinical trials experience in Europe or the USA. In addition, the Internet facilitates access to global organizations, such as the International Conference on Harmonisation of Technical Requirements for Registration of Pharmaceuticals for Human Use (ICH) good clinical practice (GCP) (indicated as ICH-GCP from now on) and other appropriate documentation for physicians to stay current with training on therapeutic areas. Therefore, while the practice of medicine in some of the countries has traditionally been paternalistic, this has been changing over the last several years as more physicians participate in clinical research trials. Nonetheless, a significant challenge in many countries within the Asia Pacific region remains access to medical records. Most individuals own their respective records, carrying them back and forth to physicians.

Many countries means many different regulatory agencies. For example, China's regulatory agency is the State Food and Drug Association (SFDA). An extensive training program is underway for the SFDA inspectors and there will be increases in inspections in the years to come. The ICH-SFDA only allows sites that have been approved by the SFDA (trained on GCP) to conduct clinical trials, which can be a limiting factor. In addition, China has very long regulatory timelines.

In India, the Drug Controller General of India (DCGI) is responsible for oversight of any drugs in development. The Central Drugs Standard Control Organization (CDSCO) also regulates clinical trials. South Korea's agency, the Korean Food & Drug Administration (KFDA), provides oversight to clinical trials and also is responsible for qualifying sites to participate in clinical trials. This country is noteworthy for its very robust regulatory environment.

With regard to differences evident in the region's countries, healthcare infrastructure and resources vary significantly from country to country. Singapore and South Korea, for example, have more advanced hospital networks, including electronic medical records systems. Standards of care also vary. Therefore, in trials where the test drug is compared with an active control that is the standard of care for the disease or condition of clinical concern, potential variation needs to be addressed. Countries can also differ in their views concerning the appropriateness of specific study designs in certain cases. While major health centers, both public and private, are typically in densely populated urban areas and hence provide access to large patient populations, some of them have very limited infrastructures with regard to the implementation of clinical trials. Conducting appropriate feasibility studies, therefore, becomes particularly important in many instances.

From the CRO perspective, the training for CRO employees has been a (if not the) key factor. By providing training that meets international ICH-GCP standards, acquisition of data of consistent quality with those from other countries is achievable.

17.1.2. Africa, sub-Saharan Africa, and South Africa

Following Asia, Africa is the world's second largest and second most populated continent. The term sub-Saharan Africa describes the area of the African continent that lies south of the Sahara desert. An estimated 760 million people live in sub-Saharan Africa. This represents around 10 percent of the world's population, but the region has around two-thirds of all human immunodeficiency virus (HIV) infections, with an estimated 22–27 million people living with HIV infections. An estimated 2000 languages are spoken in Africa, but English is recognized as the language of business in many of its countries.

Table 17.1 Official languages in South Africa

Language	% of population's home language
Zulu	23.8
Xhosa	17.6
Afrikaans	13.3
Northern Sotho	9.4
Tswana	8.2
English	8.2
Sotho	7.9
Tsonga	4.4
Swati	2.7
Venda	2.3
Ndebele	1.6
Other	0.5

With the exception of the country of South Africa, the majority of the population in sub-Saharan Africa lives in rural areas, making communication and follow-up challenging when conducting clinical trials. While the increased use of cellular communication has made it somewhat easier to maintain contact with subjects, the lack of adequate infrastructure in the outermost areas (e.g. lack of street names and postal addresses) still causes difficulties. This further emphasizes the need for experienced staff to be used in conducting clinical trials in these regions.

The population of South Africa is around 50 million, and it has 11 official languages, listed in Table 17.1.

English is the joint fifth most common home language, but is understood in most urban areas and is the official language of business, government, and media. From a clinical trials perspective, informed consents and patient information leaflets are created in English and translated into at least two other languages, depending on regions. Protocols, investigator's brochures, and case report forms are not translated and are widely accepted in English.

17.1.3. Central and Eastern Europe

The geographical region called Central and Eastern Europe (CEE) has a combined population of more than 300 million people in 17 countries, 11 of which are members of the European Union (EU). Russia (with a population of 142 million), Ukraine (46 million), and Poland (38 million) account for about 70 percent of the regional population [3]. Like other developing regions containing multiple countries, it is difficult to make sweeping generalizations. However, three qualities of countries in this region are widely acknowledged: ready access to patients, attractive costs, and high quality. These factors have played a considerable role in the region's clinical trial success: it now delivers around 20 percent of the global participants in trials. This means that CEE now represents the largest number of global clinical trial initiations outside North America and Western Europe. The ready access to patients noted earlier is exemplified by the observation that recruitment rates per site are much higher in CEE than in any other global region, potentially up to 10 times faster than rates in Western countries. Companies usually position their headquarters for the CEE region in Vienna, Austria.

Classified as an emerging market region in the past decade, economists predict that growth across most of the region will slow substantially in the next 10 years and follow the steady market with reasonable growth pattern of Western Europe rather than the dynamic high-growth rate of emerging Asia. However, while countries such as the Czech Republic, Poland, and Hungary now adhere to a slower growth mature development model, other areas within the CEE, such as Russia, still exemplify the dynamic growth of an emerging market [4]. Thus, it is difficult to characterize the states as a homogeneous region. Overall, however, the region is still anticipated to experience significant growth in the future.

17.1.4. Latin America

Latin America comprises 21 countries and has a population of 570 million. The two majority languages are Spanish and Portuguese. As in the Asia Pacific region, the major health centers are located in high-density urban areas. Both public and private institutions have developed expertise to conduct large-scale clinical trials. Physicians have trained at high-level medical schools in Europe, the USA, and Latin America. The Internet provides investigators and other clinical trial professionals the opportunity to keep up to date with the global scientific environment.

Countries such as Argentina, Chile, and Brazil have excellent medical infrastructure and well-trained principal investigators, with standards and procedures in hospitals similar to those in the USA, a tradition of Western medicine, and well-established regulations for clinical trials. Some countries have a particularly large population of previously untreated patients since they have no tradition of clinical trials.

While Latin America has a long history in clinical research, starting in about 1950 in Argentina, Brazil, and Mexico, regulations were not initially in place. This has now changed considerably. For example, Brazil and

Argentina have aligned their regulations with 1996 ICH guidelines, placing these countries in the international clinical research community, and more recent adaptations in Chile, Colombia, Peru, and Venezuela have further homogenized the regulatory landscape across Latin American countries.

Brazil is the region's largest single component and a major pharmaceutical market. The South American region overall is particularly suitable for studies in oncology, cardiovascular, infectious diseases, metabolic disorders, pediatrics, and vaccines, because the incidence and prevalence of those conditions are similar to or greater than those in the USA or Western Europe. More than 80 percent of Latin Americans live in highly urbanized areas, meaning that patients are easily accessible. Many physicians are aware of clinical trials and are keen to participate. In addition, the region has reputable health science and academic institutions, which serve as tremendous resources for conducting clinical trials [3].

A main challenge in conducting clinical trials in Latin America is that the regulatory timelines tend to be longer than in other regions of the world. However, by using the period of time while waiting for regulatory input to help sites develop recruitment strategies, recruitment itself can be very speedy.

17.1.5. The Middle East

With an estimated population of 280 million, and one of the fastest growing in the world, the Middle East is an attractive place to run clinical trials. Countries in this geographical region include Israel, Qatar, Bahrain, Iran, Jordan, United Arab Emirates, Saudi Arabia, Lebanon, Kuwait, Oman, Turkey, Egypt, Iraq, Syria, and Yemen. A growing percentage of the region's population lives in cities, and urbanization is having a great impact on society and disease trends, with obesity, diabetes, and cardiovascular disease (CVD) being growing problems [5]. CVD, once thought to be confined primarily to industrialized countries, has emerged as a major health threat in many developing countries. CVD, especially coronary heart disease and stroke, now accounts for nearly 30 percent of deaths in low- and middle-income countries each year [6]. The prevalence of gastrointestinal disorders and cancer is also increasing. Populations in the Middle East are clinically naïve, making individuals with diseases and conditions of clinical concern potentially good candidates for clinical trials.

Within the Middle East, Israel, Turkey, Egypt, Jordan, and Saudi Arabia have the most well-established clinical trial regulatory agencies. Egypt, for example, has a well-established pharmaceutical background, and produces most of the pharmaceuticals used in the country. Israel is forward thinking and innovative, while new healthcare investment is notable in Saudi Arabia. Some countries, e.g. Jordan and Saudi Arabia, have established an agency similar to the US Food and Drug Administration (FDA), and are therefore at the forefront of regulators in the region by establishing procedures acceptable to Western agencies, and Egypt closely follows. Others countries do not have strong regulatory agencies in place. This heterogeneity must be emphasized if a sponsor has a tendency to think of this region in general terms.

Sponsors' interest in conducting and/or outsourcing trials in the Middle East is relatively recent but growing rapidly. Considerable expansion in Eastern Europe, Latin America, and South East Asia, along with growth in China, has overshadowed growth in this region, which currently accounts for less than 1 percent of global participants in trials. However, accelerated growth is forecast by many observers: in the next 10 years the Middle East may account for up to 5 percent of global participants in trials. This growth, if achieved, would result in a research and development (R&D) flow to the region from the biopharmaceutical industry in excess of US $1 billion [7].

17.2. PLANNING AND PROJECT MANAGEMENT

Once selected to conduct a specific clinical trial, a CRO has to create its own study team for that trial. This team will include the following individuals (the precise names used for these positions may differ somewhat from CRO to CRO, but the functions they perform will essentially be the same):

- investigational site selection specialists
- subject recruitment specialists
- physicians from the therapeutic area in which the drug falls: provide therapeutic expertise whenever necessary
- clinical trial monitors: visit the sites and check that the study protocol is being executed correctly
- medical monitors: physicians (often but not necessarily therapeutic area specialists) who are available to consult on any subject health (safety) concerns
- data managers: responsible for receiving, managing, and storing the data that will eventually be analyzed and interpreted
- project team leaders or managers: coordinate the efforts of all other individuals and specialty areas represented on the team.

In the present context, the term project is used for a new drug that is being developed.

As Turner observed, project management is a critical aspect of the clinical trial process whose importance in successful trial execution cannot be overstated [8]. Just a few components of project management include study budgets, timelines, marshaling and distributing resources, appraising metrics on a continual basis, and facilitating communication to all stakeholders. In addition, project managers must conduct their work against a backdrop of tremendous scientific, technical, and financial risks. Project managers need not be scientists or clinicians by profession, since their specialty is managing a small number of projects, but they often have scientific or medical backgrounds. As Spilker commented, managers are chosen for their administrative ability to move projects ahead rapidly, efficiently, and in the desired direction. Independence and interpersonal skills are key qualities for this specialty [9].

17.3. SELECTION, QUALIFICATION, AND MANAGEMENT OF CONTRACT RESEARCH ORGANIZATIONS

Selection, qualification, and management of CROs are discussed in turn.

17.3.1. Criteria for CRO Selection

Many components and characteristics of a clinical trial affect the choice of the most suitable CRO to conduct the trial. These include: the nature of the trial (design, sample size, etc.); the disease and population for which the investigational drug is intended, and the CRO's experience with that particular indication and therapeutic area; the resources of the CRO in the countries that are proposed to be included in the study; and competitive advantages that the CRO may be able to offer in complementary ways. The budgets proposed by the CROs bidding for the business of executing the study will certainly be a factor in the sponsor's choice of CRO, but it is unlikely to be the sole determinant. The sponsor may feel that the services offered by one CRO provide a better chance of successfully completing a study that has particular challenges, and this CRO may be selected even though other companies provided a lower budget in their proposal.

Key knowledge, understanding, and services that the CRO must be able to provide include:

- A feasibility study of the Sponsor's proposed Study Protocol. This will assess whether the trial as proposed has a high probability of being successfully completed.

- Knowledge of the regulatory landscape in each of the countries used, and an awareness of likely challenges and appropriate risk mitigation strategies and solutions. This enables such challenges to be proactively addressed and managed.
- The capability to provide appropriate training to principal investigators, coinvestigators, clinical research associates (CRAs), clinical study coordinators, etc. This training will need to focus on ICH-GCP and will need to be undertaken in multiple cultures and languages.
- Extending the previous point, the CRO needs to understand how best to implement international ICH-GCP requirements to acquire data that are acceptable to international regulatory agencies, how to fulfill local regulatory requirements and considerations, and how to do this in a culturally sensitive manner.
- Ideally, their ability to be a true partner or ally. This can be demonstrated, for example, by making suggestions concerning modifications to the proposed study protocol that increase the probabilities of the trial's successful execution at scientific, medical, timeline, and cost levels.

The logistics of operationally executing a clinical trial are enormous. Consider a therapeutic confirmatory trial involving 5000 subjects participating at a total of 100 investigational sites spread across several continents. Identifying potential subjects and then recruiting and retaining the required number is one challenge. Shipping the drug products for the clinical trial (the investigational and control drugs) to investigational sites located in various countries spread around the globe is another. Making sure that all necessary data are measured and recorded is a major task, as is managing and storing the data. And the list goes on. Of necessity, this chapter is far from exhaustive in its descriptions of operational aspects of running a clinical trial. Rather, its aim is to give you a feel for some of the challenges encountered, and the ways that organizations that conduct trials function at an operational level.

Many sponsors developing drug and biologicals outsource work to CROs. This means that they do not need as many permanent employees, and can hire specialized personnel to conduct their studies on a trial-by-trial basis. Relationships between sponsors and CROs can take many forms, and the nomenclature used by individual sponsors and CROs can vary. Therefore, you may hear different terms in your own work. However, the general picture painted in this chapter should provide a useful roadmap.

Historically, sponsors often hired CROs to carry out a trial exactly as specified in the study protocol that the sponsors had developed. They may also have hired several CROs, each fulfilling some of the overall responsibilities of running the trial. In such contexts, the term service provider was appropriate for the CRO. This trend is changing, and some sponsors and CROs are establishing relationships that are better described as partnerships or alliances, in which sponsors involve CROs early in their clinical development programs and welcome their medical and scientific expertise in the planning and protocol development stages. A single CRO who has provided good consulting at these stages may be given a large role in conducting the trial, or possibly become solely responsible for executing all aspects of the trial. There are mutual benefits. The sponsor gains additional expertise and economy of services, and the CROs obtain more business from a long-term relationship.

17.3.2. CRO Qualification

A CRO must meet a very large number of requirements in order to be eligible to fulfill the sponsor's requirements for some studies, such as large multisite trials being conducted in various developing regions of the world. Some of the most salient requirements are:

- It must have global standard operating procedures (SOPs). There must be documented training set up for these SOPs, and systems audits to make sure all the employees follow them.
- There must be an independent quality assurance (QA) department able to audit the systems and also the sites.
- The staff involved in the trials must have, or receive, training and experience in conducting and monitoring of the trials. This includes training in ICH-GCP, as discussed in the previous section.
- The CRO must have, or have access to, medical personnel who are able to follow medical concerns and issues that may arise before, during, and in follow-up periods after the conduct of a trial.
- The specific know-how of the countries to develop the trial, the therapeutic area, and the indication are essential.
- The CRO must be able to foresee potential issues in the regulatory submission, the recruitment of the patients, and the appropriate conduct of the trial, and have contingency plans.
- It must have a legal structure that takes care of all the legal aspects of the conduct of a trial.
- Most importantly, it must place patient safety as its top priority.

17.3.3. Outsourcing Strategy

A sponsor's outsourcing strategy (certainly for larger sponsors) typically consists of two phases. In the first step, sponsors will engage in extensive research of many CROs, and then qualification of a small group of CROs that have been carefully chosen according to certain performance criteria of particular importance to the sponsor. These CROs are called preferred providers. Once selected, these preferred providers sign confidentiality agreements with the sponsor. These agreements include a non-disclosure agreement (NDA, not to be confused with the abbreviation for a new drug application) and, thereafter, a master services agreement (MSA), which includes confidentiality provisions. The execution of these high-level confidentiality agreements means that, from that point forward, the sponsor can send any study-specific information in the knowledge that a new individual NDA need not be prepared and executed.

The second step is outsourcing a particular clinical trial. When preferred providers have been selected, the sponsor will look to them first for each individual piece of work. These providers have probably been selected since they are able to do the vast majority of the work the sponsor anticipates needing. However, if they are not able to perform a specific service or fulfill a special need, the sponsor will look to other CROs, or sometimes academic medical centers; for example, when a particular assay is needed and is not provided by the central laboratory services at any of the CRO preferred providers.

For each trial (or, on occasion, each subcomponent service), outsourcing specialists at the sponsor will prepare a request for proposal (RFP). This is discussed in the following section.

17.3.4. Writing Requests for Proposals

RFPs are usually sent to several CROs to enable the outsourcing specialists, and eventually the decision-making study team, to obtain a range of proposals that may differ to some extent in precise operational detail, and are likely to differ in cost. These CROs reply to the RFP by providing (proposing) a detailed plan describing how they would conduct the study, along with a line-by-line list of individual costs, leading to a grand total cost. Preferred providers will already have an MSA in place with the sponsor, and hence issues of confidentiality are already addressed in those cases. However, if the sponsor would like another CRO to provide a proposal (bid) to conduct the trial, a new NDA would need to be executed before the CRO could be sent the RFP and study protocol.

Outsourcing specialists may perform an initial inspection of the proposals submitted by various preferred providers, and then the proposals are forwarded to the study team that has been formed to be in charge of the trial. This team then selects perhaps two or three proposals that look promising, and invites representatives from these CROs to visit their premises for a Bid Defense. In this meeting, the study team asks the CRO representatives many questions, essentially interviewing them in regard to their ability to conduct the trial successfully, and at a price that is acceptable to the sponsor [8].

17.3.5. Managing the CRO/Allying with a CRO

The first part of the title of this section is interesting in the sense that it is perhaps more applicable to the "traditional model" that represented the interaction between a sponsor and a CRO, one in which CROs were hired by sponsors as service providers to provide a specific set of predetermined services. The CRO was not asked to provide medical, scientific, or operational input, but rather was asked to provide the price they would charge the sponsor for completing the requested tasks.

A new model is one in which the CRO functions as a full drug development partner, or ally [8]. In this model a sponsor and its CRO start working together much earlier in the drug development process, ideally as soon as the sponsor conceptualizes a clinical trial. While the sponsor will have expertise in many aspects of study design and operational execution "in house", a good CRO will also have expertise in all of these aspects and can therefore provide additional (and sometimes unique) beneficial input into the development of a study protocol. Optimizing the protocol before the trial commences can minimize the numbers of protocol amendments and protocol violations (on the part of both subjects and investigators).

In the allying with a CRO model, actual management of the CRO, in the traditional sense of the word management, becomes less onerous for the sponsor: the CRO is seen as a trusted partner that can execute the trial with minimum oversight. The benefits of this model, which has considerable advantages in all clinical trial settings, are perhaps even more powerful when conducting multisite trials in developing countries. Large CROs may have much more extensive footprints in such countries than a sponsor whose R&D and manufacturing facilities may be located in one or two (probably developed) countries. This allows the CRO to provide detailed advice to the sponsor on many operational aspects of conducting the trial, such as:

- site selection
- selection of the investigator(s) at each site
- subject recruitment and retention strategies
- facilitating transport of subjects to and from the site for each visit required by the trial protocol
- helping couriers to transport biological samples to central laboratories as needed
- facilitating electronic communication channels wherever possible
- facilitating medical (safety) oversight of all subjects participating in the study.

17.3.6. CRO Oversight in the Developing World

Although many companies use regional trials in developing regions to support products on a local basis, it is the ability to use locally generated data to support international regulatory submissions that is of greatest interest. Since ICH-GCP is currently less well known in many developing countries, running trials to the satisfaction of international authorities such as the FDA and the European Medicines Agency (EMA) requires huge training efforts for the local CRAs and principal investigators. CROs with global footprints can be very helpful in this regard, since they are knowledgeable about international standards and have been on the ground in developing regions for a number of years.

Via their chosen CROs, sponsors must also be prepared to make an investment in help call centers, and to ensure that bodies such as institutional review boards are appropriately operating. However, once training programs have been implemented and completed by all necessary parties, investigational sites in many developing regions have access to very large numbers of eligible patients, and can therefore quickly recruit subjects.

17.4. CENTRAL LABORATORY CONSIDERATIONS

The importance of laboratory data becomes extremely clear when one considers the fact that these data account for 70–90 percent of all data contained in marketing approval submissions to regulatory agencies, depending on the therapeutic area. They typically provide the primary scientific data for showing drug safety and efficacy. Appropriate selection of the laboratory or laboratories that will analyze these data is therefore a critical decision. In single-site studies, the degree of complexity is considerably less than in multinational, multisite studies, and the use of a central laboratory can be a very useful strategy. If a sponsor has selected CRO-preferred providers, it is likely that one or more of these providers will offer central laboratory services. Thus, selection of the central laboratory may well be tied to selection of the provider that will perform the other services (monitoring, project management, data management and analysis, etc.) discussed in Section 17.1.

Most therapeutic confirmatory trials are run at multiple investigative sites, since any one site would not be able to enroll the number of subjects required by the study's protocol. Central laboratories are analytical laboratories at which biological samples (blood, urine, and many others) collected at all of the investigative sites in a clinical trial are analyzed. This strategy has two important advantages compared with the alternative strategy of using local laboratories located in close proximity to each site:

- Assurance that the laboratory conducting the analyses of the samples is compliant with cGCP is much easier: only one laboratory has to be visited and audited.
- Statistical difficulties associated with analyzing data sets that have been pooled across laboratories are circumvented.

The ability to demonstrate data comparability for laboratory testing results from different centers is a crucial aspect with regard to efficacy, and a key reason for choosing a central laboratory over a local or diagnostic laboratory.

Laboratory selection factors therefore include turn-around time for laboratory results, importance of laboratory data comparability, laboratory quality, performance, and especially in developing countries, cost (including transportation, which is a big component in Asia), and constraining factors (such as inability to export samples easily out of China). Esoteric tests are increasingly being performed in Asia and these may require expertise in setting up and validating new biomarkers for the laboratory, which may limit selection to a few global central laboratories with such capability in Asia. Even when the esoteric tests are not novel, not all laboratories are equipped to support them for varied reasons such as lack of scientific and technical expertise, uneconomic volumes, or simply size and capacity of the laboratory. Central laboratories, especially the large global laboratories, will provide a standardized process for the set-up of the protocol in a laboratory information management system (LIMS) as well as provide a variety of support services including project management, call center support, and logistics.

From an operational execution perspective, the central laboratory chosen for a particular trial can be located many hundreds (or even thousands) of miles from some of the investigational sites, and even in another country or continent. Expedited shipping under very carefully controlled conditions is necessary to ensure that the samples arrive at the central laboratory quickly and safely. Many courier companies specialize in this transportation process.

Using central laboratories can also be considerably more expensive than would be the case if samples were sent to local laboratories close to each investigational site. Indeed, the transportation costs can outweigh the costs of the assays performed by considerable margins. However, obtaining optimum quality laboratory data is critical (as it is for all other data), and the necessary expenditure involved here is well worthwhile.

It is appropriate to acknowledge here that, while there are many advantages to using central laboratories, it is not uncommon to use local laboratories as well as a central laboratory for some studies. For example, when turn-around time is especially crucial (less than 24 hours) or where the requirements are primarily for safety purposes, local laboratories are more often used because of convenience, cost, and practicality.

17.4.1. Sample Shipment and Handling

For central laboratories, product types to be shipped include laboratory samples and laboratory-specific kits. Laboratory samples generally consist of body fluids and extracts (e.g. blood, urine, stools, swabs, DNA, RNA) as well as tissues (raw tissue, tissue blocks, tissue slides). Blood samples remain the most common samples shipped during clinical trials.

Consider shipping and handling requirements in Singapore. While movement of samples within the country is generally straightforward, shipments to other countries for clinical trials are shipped as UN 3373 (biological substance, category B).

While such shipments do not require specific training on the part of the shipper, frozen shipments where dry ice or liquid nitrogen is used do require special handling as these are classified as dangerous goods. Training for handlers includes International Air Transport Association (IATA) Dangerous Goods Regulations, and is carried out by a variety of training centers, logistics/courier companies, and central laboratories.

Shippers must ensure that shipments are prepared in such a manner that they present no hazard to persons or animals. Packaging for UN 3373 shipments generally conforms to Packaging Instruction (PI) 650 of the regulations. Examples include the following:

- **Inner packaging**: This consists of a watertight primary receptacle(s) and watertight secondary packaging. An absorbent material must be placed between these receptacles, and must be able to absorb the entire content of the primary receptacle(s).
- **Outer packaging**: This must be of adequate strength for its capacity, weight, and intended use. It must also be able to pass a "drop test", i.e. being dropped from a specified height and not sustaining damage to the inner contents.

While exceptional cases (e.g. shipment of whole organs) may require special packaging, the vast majority of diagnostic specimens must be packaged according to the following guidelines:

- **Substances shipped at ambient temperatures or higher**: Primary receptacle materials include glass, metal, and plastic. A positive means of ensuring a leakproof seal, such as heat seal, skirted stopper, or metal crimp seal, must be provided. If screw caps are used they must be reinforced with adhesive tape.
- **Substances shipped refrigerated or frozen** [wet ice, prefrozen packs, carbon dioxide (CO_2), solid (dry) ice, or other refrigerants] must be placed outside the secondary packaging(s) or alternatively in an overpack with one or more completed packagings. Interior support must be provided to secure the secondary packaging in the original position after the refrigerant has been dissipated. If ice is used the packing must be leakproof. If CO_2 is used as solid dry ice, the outer packaging must permit the release of CO_2 gas. The primary receptacle must maintain its containment integrity at the temperature of the refrigerant as well as the temperatures and pressure of air transport to which the receptacle could be subjected if refrigeration were to be lost.
- **Substances shipped in liquid nitrogen**: Plastic that is capable of withstanding very low temperatures must be used instead of glass for the primary packaging. Secondary packaging must also withstand very low temperatures and in most cases will need to be fitted over individual primary receptacles. Requirements for shipment of liquid nitrogen must also be observed. The primary receptacle must maintain its containment integrity at the temperature of the refrigerant as well as the temperatures and pressure of air transport to which the receptacle could be subjected if refrigeration were to be lost.

- **Lyophilized substances**: Primary receptacles must be either flame-sealed glass ampoules or rubber-stoppered glass vials with metal seals.

As additional examples of the attention to detail required by the sponsor and the central laboratory, some countries (e.g. Thailand) require material transfer agreements between the institution where samples are being sent and the laboratory where testing is done to be signed. China requires informed consent by patients for whole blood/tissue samples to be sent out of the country before shipping can occur.

17.4.2. Clearing Samples Through Customs

Table 17.2 provides information for South America concerning import, export, and transportation considerations.

For central laboratories, product types to be shipped include laboratory samples and laboratory-specific kits. Laboratory-specific kits refer to

Table 17.2 Import, export, and transportation considerations

Item	Considerations
Value and weight limitations	While there is no limitation, imported items must have a realistic declared value. If the value is below ZAR500, quick clearance can be arranged and no customs duty or value-added tax (VAT) will apply. Isolates/cultures are considered to be dangerous goods, as is anything shipped on dry ice. Infection shipments require an accompanying Dangerous Goods Declaration Form
Documents required	Imported drugs require an approval letter from the Medical Control Council, which serves as an import license. An AWB and Commercial Invoice are required for import and export. All biologicals require an Import/Export permit from the Health Department
Services	Premium couriers will export all kits and specimens, including infectious samples and specimens on dry ice. They will re-ice packages as necessary
Transit time	24—48 hours (import and export)
Couriers	Export: World Courier, Marken; Import: World Courier, UPS, DHL, FedEx; Local: TNT
Dried ice capabilities	Central service: Marken is the preferred service provider, and will supply dry ice upon request (with 24 hours' notice) Local service: site phones the local TNT depot to order dry ice. A TNT driver will then deliver the dry ice, and wait while the specimens are packed. TNT provides the contact telephone numbers for each country or a dry-ice instruction with the call-by cut-off information

clinical trial materials (CTMs) used for sample drawing and collection. Other more complex shipment types include medical devices, refrigerated consumables, and investigational drugs, which attract more stringent customs clearance.

As an example, Table 17.3 summarizes the permits required when shipping CTMs out of Singapore, and biological samples from various countries into Singapore.

For kit shipments, integrators such as FedEx and UPS, or express couriers such as Marken, DHL, and TNT, have the ability to do preclearance with a high percentage of shipments without formal inspection, including those countries that require import permits. Customs has the authority to pick randomly any shipment for inspection with no prior notice. When this happens, they have the authority to ask for any supporting documents to satisfy any concerns.

Table 17.3 Permits required to ship clinical trials material (CTM) out of Singapore, and biological samples into Singapore

		Export of:	
Country	**Import of CTM**	**Ambient (UN 3373 biological substance, category B)**	**Frozen (UN 3373 biological substance, category B)**
Australia	× *	×	×
China	×	√	√
Hong Kong	×	×	×
India	×	√	√
Indonesia	√	×	×
Japan	√	×	×
Korea	√	×	×
Malaysia	×	×	×
New Zealand	×	×	×
Pakistan	×	×	×
Philippines	×	√	√
Singapore	N/A	×	×
Sri Lanka	√	×	×
Taiwan	√	√	√
Thailand	√	×	×
Vietnam	√	×	×

An import permit for medical devices is required. Diagnostic kits with needles could be potentially classified under Medical Device.

N/A: not applicable.

* An Australian Quarantine and Inspection Service (AQIS) permit is required for the import of heparin tubes.

It should be noted that regulatory changes occur for countries over time. Therefore, sponsors are encouraged to reconfirm with central laboratories and courier companies before commencing a study involving sites in any of the countries listed in Table 17.3.

ACKNOWLEDGMENT

The authors gratefully thank Carolyn Moore for her administrative support.

REFERENCES

[1] Lustig MW, Koester J. Intercultural Competence: Interpersonal Communication across Cultures. 5th ed. Pearson Education; 2006.
[2] Hofstede G. Cultures and Organizations: Software of the Mind. London: McGraw-Hill; 1991.
[3] Fitzsimmons LD. Clinical convergence. R&D Directions, November/December, 2009:18–23.
[4] The Economist 2010.
[5] Kermani F. The Middle East: pharma's new frontier. Contract Pharma 2009:76–80. November/December.
[6] Institute of Medicine. Promoting cardiovascular health in the developing world: a critical challenge to achieve global health. Report Brief 2010. March 2010.
[7] Misik V. Role of the Middle East in global clinical trials. NextLevel Pharma: Advancing Clinical Research in Turkey, Middle East & North Africa. Istanbul: Turkey; 2010. September 2010:23–24.
[8] Turner JR. New Drug Development: An Introduction to Clinical Trials. 2nd ed. New York: Springer; 2010.
[9] Spilker B. Guide to Drug Development: A Comprehensive Review and Assessment, PA. Philadelphia: Wolters Kluwer/Lippincott Williams & Wilkins; 2009.

CHAPTER 18

Study Documents and Logistics

Jay Johnson*, Sebastian Antonelli, Jennifer Aquino[†], Sue Bailey***, Edwin Chia[‡], Gillian Corken***, Jogin Desai[†], Barbara Lilienfeld[§], Monica Lizano[|], Alan Ong[‡], Viola-Marie Raubenheimer[¶], Veronica Suarez**, J. Rick Turner*, Sorika van Niekerk***, Daniel Vazquez** and Johan Venter*****

*Quintiles, Durham, NC, USA
**Quintiles Latin America, Buenos Aires, Argentina
***Global Sales Operations, Quintiles Sub-Saharan Africa, Irene, South Africa
[†]Cenduit, Raleigh, NC, USA
[‡]Quintiles Asia, Singapore
[§]International Corporate Events Network, Johannesburg, South Africa
[|]Quintiles Latin America, San José, Costa Rica
[¶]Quintiles Africa, Pretoria, South Africa

Contents

Global Clinical Trials
ISBN 978-0-12-381537-8, Doi:10.1016/B978-0-12-381537-8.10018-4

18.1. INTRODUCTION

Following on from discussions in the previous chapter, this chapter focuses on key documents that are needed for the clinical trial to be conducted, and then on logistical considerations that need to be addressed to execute all aspects of the study in an optimum manner.

While clinical trials are designed to meet global International Conference on Harmonisation good clinical practice (ICH-GCP) standards in almost all countries around the world, local interpretations can vary and each country has its own procedures that need to be followed. The logistical challenges of running multinational trials are therefore considerable, and are likely to be more so in developing regions. It is therefore important to know the country-specific pathways that need to be followed during the start-up and maintenance phases, and the close-out of the study. It is important to have a balance of coordination and the right mixture of local/regional knowledge and global knowledge, and the establishment of good working relationships with local regulatory authorities is very important.

18.2. WRITING THE STUDY PROTOCOL

A study protocol includes precise accounts of the design, methodology, and statistical analysis considerations necessary to conduct the trial and analyze data collected. Frequently, most of the discussions concerning the statistical analyses to be performed are presented in an associated statistical analysis plan (SAP). If this approach is taken the study's SAP should be prepared at the same time as the protocol. Study design and statistical analysis are inextricably linked, and the design of the trial determines the appropriate mode of analysis. Therefore, writing the protocol and the SAP together is both feasible and good practice (although it must be acknowledged that, in reality, the SAP is often written at a later date).

The study protocol is "the most important document in clinical trials, since it ensures the quality and integrity of the clinical investigation in terms of its planning, execution, conduct, and the analysis of the data" [1]. The study protocol is a comprehensive plan of action that contains, for example, information concerning the goals of the study, details of subject recruitment, details of safety monitoring, and all aspects of design, methodology, and analysis. Input is therefore required, for example, from clinical scientists, medical safety officers, study managers, data managers, and statisticians. Consequently, while one clinical scientist or medical writer may take primary responsibility for its preparation, many members of the study team make important contributions.

Buncher and Tsay listed just some of the detailed requirements of a study protocol [2]:

- **Primary and secondary objectives**: These are stated as precisely as possible.
- **Measures of efficacy**: The criteria to be used to determine efficacy are provided.
- **Statistical analysis**: The precise analytical strategy needs to be detailed, here and/or in an associated SAP.
- **Diagnosis of the disease or condition**: When subjects are required to have the disease or condition for which the drug is intended, precise diagnostic criteria are provided.
- **Inclusion and exclusion criteria**: These provide detailed criteria for subject eligibility for participation in the trial.
- **Clinical and laboratory procedures**: Full details of the nature and timing of all procedures and tests are provided.
- **Drug treatment schedule**: Route of administration, dosage, and dosing regimen are detailed. This information is also provided for the control treatments.

18.2.1. Inclusion and Exclusion Criteria

A study's inclusion and exclusion criteria govern the subjects who may be admitted to the study, and they are listed in detail in the protocol. Criteria for inclusion in the study include items such as the following:

- reliable evidence of a diagnosis of the disease or condition of clinical concern
- being within a specified age range
- willingness to take measures to prevent pregnancies during the course of treatment (this can apply to male and female subjects).

Criteria for exclusion from the study include:

- taking certain medications for other reasons and which therefore cannot safely be stopped during the trial
- participation in another clinical trial within so many months prior to the commencement of this study
- liver or kidney disease.

Meeting all the inclusion criteria allows a subject to be considered as a study participant, but it is not sufficient. In addition, an individual must not meet any of the exclusion criteria. Inclusion and exclusion criteria precisely define the nature of the subject sample that participates in a clinical trial. Accordingly, they also define the study population to which statistical inferences may be made. That is, the inclusion and exclusion criteria define the study population to which the results of the trial can reasonably be generalized.

18.2.2. The Trial's Primary Objective

Clinical protocols should be clear, concise, unambiguous, and as short as possible while maintaining scientific integrity. However, this ideal is all too often not achieved [3]. One challenge is to "maintain a balance between brevity and completeness" [2]. All necessary procedural information must be included to allow the investigators to implement the protocol exactly as intended. However, as the length of the protocol increases, the chances of its being read throughout decreases. Clarity and conciseness are therefore beneficial characteristics of the protocol.

One aspect of conciseness is actually a study design feature. For many reasons, it is a very good idea to limit the number of primary objectives in a study, ideally to one but perhaps to two (many protocols contain far more objectives than are actually necessary to address the goals of the particular study). The designs used in most clinical trials are relatively simple. Clinical trials are certainly complex, but this complexity is not a direct function of the nature of the designs employed. Rather, it originates from other factors such as "ethics, biology, logistics, and execution" [4]. The logistics and execution can be particularly complex for trials where some or all of the investigational sites are located in developing countries.

18.3. WRITING THE INVESTIGATOR'S BROCHURE

Each drug has an investigator's brochure (IB) that is created for the use of investigators running clinical trials, and which is reviewed by institutional review boards (IRBs) and regulatory agencies [5]. The IB can be regarded as the repository of everything that is known about the drug, including non-clinical and clinical information, manufacturing details, and any adverse events that investigators should be particularly watchful for during clinical trials (i.e. adverse events of special interest). These brochures are often collated and then smoothed together by a medical writer using sections prepared by various experts within a biopharmaceutical company. Early information will focus on non-clinical investigations and the preparation of the drug compound used in that research. As the drug's clinical development program commences, information will be added on clinical pharmacology, including pharmacokinetics, pharmacodynamics, adverse events, and human responses to the drug. For this document, constant updating and revision is to be expected. Each sponsor will have a policy governing how often these updates should occur. Typically, this would be once a year as a basic rule, with the proviso that more immediate updates should occur if new information that needs to be communicated to investigators, IRBs, and regulatory agencies arises.

As a drug progresses through many preapproval clinical trials, more and more information becomes available. Once marketed, postmarketing trials may be conducted to gain additional information about the drug's safety and effectiveness. For example, relatively small studies may be conducted in certain populations that were not heavily represented in preapproval trials and, at the other end of the scale, very large therapeutic use trials may be conducted to study the drug's safety and effectiveness under "real-world" conditions. Large-scale comparative effectiveness trials may also be conducted, something that is likely to increase as comparative effective research becomes more routinely performed to provide third party payers with information on how best (often defined in terms of how cost-effectively) safe and effective pharmacotherapy can be implemented.

Whatever the research question being asked and answered by a trial, additional safety data continue to accrue. This is an extremely important facet of postmarketing studies since, even though great care has been paid to documenting a drug's safety in preapproval trials, those safety data are limited by the fact that, in a typical clinical development program, only some 5000–10,000 subjects will participate. While this provides a large enough database to identify adverse events that are relatively common, it is probabilistically very unlikely that rare adverse advents will be seen. As the Institute of Medicine has noted, "The [marketing] approval decision does not represent a singular moment of clarity about the risks and benefits associated with a drug – preapproval clinical trials do not obviate continuing formal evaluations after approval" [6]. Typically, a sponsor will declare a cut-off date for the collection of safety data to be reported in an annual update. This may be, perhaps, two to three months before the completed update is due. This allows time for these data to be analyzed and described in the updated IB (data collected in that two- to three-month period will be included in the following year's update).

Non-clinical data are very important, and can be very informative in clinical situations. For example, the best (and sometimes only) data available to investigators for certain potential occurrences come from animals: these include data on reproductive toxicology and genotoxicity. However, as more and more clinical data become available, the relative importance of non-clinical data is likely to decline in the eyes of investigators conducting clinical trials: they are most interested in all available clinical data, and particularly safety data collected to that point. To prevent the size of the IB continually increasing, some sponsors ask the IB writers commensurately to decrease the space allotted to non-clinical data. However, this is not always done, which means that investigators are sent larger and larger documents. While this may initially seem a good thing, since more information is available to the investigator, investigators should read the entire document before commencing the trial at their particular investigative site and, arguably, the larger the IB the less the chance that this will occur.

18.3.1. ICH Position on the Nature and Content of an Investigator's Brochure

ICH considers that the information presented in an IB should be concise, simple, objective, balanced, and non-promotional. It should enable a physician who may wish to participate in a specific clinical trial to understand all content, and to make an informed decision, in an unbiased manner, concerning the benefit–risk and the appropriateness of the proposed trial.

18.3.2. Typical Content of an Investigator's Brochure

While each sponsor can effectively structure an IB according to its own document standards and guidelines, the need to include certain key information tends to make them relatively similar. A hypothetical table of contents might be:

- Executive Summary
- Table of Contents
- Abbreviations and Descriptions Table
- Pharmacokinetic and Pharmacodynamic Information on Drug ABC123
- Rationale for Use of ABC123 for XYZ Indication
- Physicochemical Properties of ABC123 and Formulation Information
- Pharmaceutics
- List of Excipients Used in Addition to Active Pharmaceutical Product
- Manufacturing, Storage, and Handling
- Non-Clinical Studies Conducted to Date
- Clinical Studies Conducted to Date

- Guidance for the Investigator
- Dosing and Form of Administration
- Contraindications and Drug–Drug Interactions
- Warnings, Precautions, and Action for Instances of Overdose
- Adverse Events of Special Interest
- Appendices, References

As noted earlier, IBs can end up being lengthy documents. While investigators should read and become familiar with the entire document, it is realistic to assume that most of their attention is focused on the "Guidance for the Investigator" section, for pragmatic reasons.

18.3.3. One Comprehensive Brochure or Several Separate Ones?

If a drug is being considered for more than one indication, or prepared in various formulations, it is possible to create a set of individualized IBs, each one being tailored to a specific indication and/or formulation. If this strategy is followed, there would certainly become repetition in each version, but each version would also contain information uniquely informative to investigators conducting relevant trials. The investigators' "IB reading burden" would therefore be lessened.

While this is arguably a very good approach, the logistics are considerably more complex for the sponsor. Any new information that is relevant in every case (fundamental safety information, for example) needs to be inserted into every individualized IB. In addition, since various IBs could be updated at different times, version control becomes more important than ever. Accordingly, many sponsors maintain just one IB per drug.

18.4. INFORMED CONSENT AND THE INFORMED CONSENT FORM

The purpose of the clinical development program is to establish whether or not the drug is effective, and best way to do so is to conduct a randomized controlled trial using the drug and a comparator treatment. Often, the comparator is a placebo. One potential ethical concern with giving some subjects a placebo while others receive the active treatment is that those receiving the placebo are being done a "disservice" since they do not stand to gain any benefit from the test drug. However, there are well-established guidelines and procedures that enable investigators largely to overcome any ethical objections to the conduct of randomized trials. At the time of a trial's conduct, it is not known for certain whether the investigational drug is actually more effective than the placebo; that is, a state of equipoise exists. If it were known for sure that the test drug is more efficacious (and safe), it would indeed be unethical to deprive the subjects receiving placebo of therapeutic benefit since the trial would not provide any new information.

A cornerstone of conducting ethical trials is the process of informed consent. It is incumbent on sponsors and investigators to provide all potential subjects with an informed consent form (ICF) that explains the risks to the individual, the potential benefits at the public health level should the drug be found to be efficacious and safe, and full details of all procedures involved in the trial. Investigators should not simply be satisfied with the required signature from individuals consenting to participate in the trial: they should answer any and all questions each person has. Going one step further, some clinical trialists advocate that a "comprehension test" should be administered to individuals prior to their signing the form to ensure (to the greatest degree possible) that they truly understand the benefits and risks of their participation in the study. The informed consent process is undertaken to ensure that subject participation is undertaken in an informed and voluntary manner.

The US Food and Drug Administration (FDA) and other regulatory bodies elsewhere in the developed world have been encouraging more trials in developing regions (for diseases that are not of neglect). Such studies present challenges, opportunities, and barriers that are different from those in developed countries and regions. Consider the case of South Africa, and the example of recruiting pediatric subjects. There is a need to conduct studies in geographical regions in which the disease or condition of clinical interest is prevalent. The continent of Africa offers opportunities in vaccines, used either as preventive or curative measures. To satisfy regulators in the USA, there is an increased need for studies in black children. To meet these FDA requirements, sponsors must look outside the USA to recruit enough subjects. Historically, there have been few data on pediatric clinical pharmacology, hence the need to study drugs in a pediatric population to determine the most appropriate dosing levels from both safety and efficacy perspectives.

18.4.1. Recruitment of Pediatric Subjects

Recruitment is done primarily through doctors who work in clinics. Recruitment through word of mouth is also strong, for example with vaccines, because parents and community elders are keen to have all the children in their community group vaccinated.

18.4.2. Informed Consent

Adherence to the most stringent requirements is essential for ensuring that the informed consent process complies with the essential criteria for obtaining true informed consent. These include voluntary participation, the provision of complete and accurate information to the subject, and evidence of comprehension of this information by the subject.

Considerations to be taken into account when obtaining informed consent in African study populations include the language of consent, cultural sensitivity, sensitivity to potential illiteracy, and any specific considerations for pediatric subjects. These include the concept of community-based informed consent, which must be given by the local tribal elders as well as the child's parent(s).

When obtaining informed consent from African clinical trial subjects who speak an indigenous language, the following considerations are essential:

- ICFs are translated by certified/accredited translation agencies.
- When no translation agency is available for a specific language, a quality control procedure must be put in place to ensure the quality, completeness, and accuracy of the translation.
- If utilized, interpreters must be trained in the relevant medical terminology and be conversant with subtle aspects of the local language(s) usage by subjects (e.g. idioms, symbols, and metaphors).
- Comprehension of the information by potential subjects must be appropriately assessed, and individuals must not be enrolled in the study unless a predetermined level of comprehension is demonstrated.

The informed consent process needs to be sensitive to country- and/or tribal-specific beliefs and practices, examples of which include:

- respect for the elderly
- gender-related practices/beliefs
- disease-associated stigmatization
- autonomy of married female subjects
- role of tribal chiefs in the community
- awareness of superstitions possibly associated with the disease/therapeutic area under investigation and certain medical procedures (e.g. blood draws).

To address the issue of potential illiteracy, the following procedures can be helpful:

- If the illiterate person is competent to provide informed consent, it has to be orally obtained in his or her own language, the trial subject should make their mark on the ICF, and the person obtaining informed consent from them should sign the ICF.
- This process must be witnessed by a literate person who should also sign the ICF to attest to the fact that this process was followed.
- The witness should be independent of the study itself but may be from the site (such as another doctor or nurse) if a large medical center, and can also be someone who would otherwise qualify as a legally acceptable representative (LAR).
- Before obtaining informed consent from parents or legal guardians of pediatric subjects, regulatory requirements pertaining to legal guardianship, unmarried parents, cases where one parent is deceased, parents who are minors, parents who are illiterate, and the need for one or both parents to provide consent need to be ascertained and adhered to. In addition, a means to confirm the identity of legal guardians needs to be determined in advance.

18.5. CHEMISTRY, MANUFACTURING, AND CONTROLS

Chemistry, manufacturing, and controls (CMC) address issues related to drug identity, manufacturing control, and analysis. Drug manufacturing and processing procedures need to ensure that the compound is stable and can be consistently made to high standards.

The drug manufacturing process is much more complex than typically realized. As Monkhouse observed, "The difficulty of converting a laboratory concept into a consistent and well-characterized medical product that can be mass produced has been highly under-rated" [7]. In addition, while the approved drug must be manufactured on a commercial scale, smaller amounts are required throughout non-clinical and clinical development programs. Moreover, if there are any chemistry and manufacturing differences between the nature of the investigational drug proposed for clinical use and the drug product that was used in the animal toxicology trials that formed the basis for the sponsor's conclusion that it was safe to proceed to clinical studies, these need to be described in the investigational new drug (IND) application. If there are any such differences, the sponsor should discuss if and how these differences might affect the safety profile of the clinical drug product.

18.5.1. Manufacturing Drug Products for Non-Clinical Development

In early phases of non-clinical development, relatively small amounts of the drug product (test material) are needed, and manufacturing focuses on laboratory, small-scale synthesis of the drug substance. At this stage, the quantities needed are in the gram range. Studies do not need to be conducted to current good laboratory practice (cGLP) standards, and the drug compound does not need to be manufactured to current good manufacturing practice (cGMP) standards. However, both cGLP and cGMP standards must be met by later non-clinical studies.

18.5.2. Manufacturing Drug Products for Clinical Trials

Drug products include both the new investigative drug and the control materials, i.e. a placebo or an active comparator drug that will be administered to the control group. These materials need to be manufactured in a way that allows their double-blinded use in randomized, concurrently controlled, double-blinded clinical trials.

When an active drug is used as the comparator, an interesting challenge arises. It is conventional that the active comparator chosen is the gold standard of care at that time, and that, in the test treatment arm, the investigational drug is given as an "add-on" to the standard of care. However, this simple statement belies the complexity of defining the gold standard in each case [8]. Of many relevant issues, the pertinent question here is: Are there different gold standards in different countries? If so, which one should be chosen in multicenter trials being conducted in countries with differing views?

The manufacturing process of blinding makes the drug product and the comparator product appear the same. This involves two steps:

- making the test drug and the comparator drug as similar as possible in appearance (e.g. color and shape, if they are tablets) and other pertinent characteristics (e.g. taste and smell)
- packaging them in such a way that they cannot be distinguished by the package in which they are supplied to investigators.

While a relatively small amount of clinical drug products may be required for early-phase trials, later-phase trials can require considerable amounts, and therefore their manufacture is not a trivial undertaking. In addition, while the majority of blinded trials are conducted using solid dosage forms, blinding can also be needed for other drug forms, such as oral liquid formulations, injectable solutions, ointments, and metered dose inhalers. Since protein biologicals are typically administered by injection, manufacturing clinical drug products in this case can be a challenge.

The content of a package that contains clinical drug products cannot be deduced from the package itself. This presents special challenges. Once packaged, clinical drug products need to be distributed to all of the investigational sites participating in a trial. Given that many multicenter trials may now use sites in various countries, this adds several degrees of potential complexity to the process. If international shipping is required, each country's customs (import/export) authorities may need to be appraised of the drug products' entry into that country.

18.6. DRUG SUPPLY AND LABELING

Consider the example of South Africa. There, an investigational medicine must be properly labeled and the package must convey certain information adequately:

- the clinical trial to be conducted
- the medicine to be used
- the subject (identified in code form) to whom the medicine is to be administered
- the name and address of the premises where the clinical trial is to be carried out.

Labels are required in clearly legible, indelible letters in English and at least one other official language.

The Guidelines for Good Practice in the Conduct of Clinical Trials with Human Participants in South Africa (GCP) require that the sponsor is responsible for supplying the principal investigator with the investigational product(s). The sponsor should not supply an investigator/institution with the investigational product(s) until it has obtained all required documentation, e.g. approval from the appropriate ethics committee, the Medicines Control Council (MCC), and other applicable regulatory authorities.

The sponsor has many responsibilities, including:

- ensuring that written procedures include instructions that the investigator/institution should follow for handling and storage of investigational product(s) for the trial, and appropriate documentation; the procedures should address adequate and safe receipt, handling, storage, dispensing, retrieval of unused product from participants, and return of unused investigational product(s) to the sponsor (or alternative disposition if authorized by the sponsor and in compliance with the MCC-approved protocol and/or where available, applicable regulatory requirements)

- ensuring timely delivery of investigational product(s) to the principal investigator(s)
- maintaining records that document shipment, receipt, disposition, return, and destruction of the investigational product(s)
- maintaining a system for retrieving investigational products and documenting this retrieval (e.g. for deficient product recall, reclaim after trial completion, expired product reclaim)
- maintaining a system for the disposition of unused investigational product(s) and for the documentation of this disposition. Disposal must be done according to MCC regulation
- taking steps to ensure that the investigational product(s) is stable over the period of use
- maintaining sufficient quantities of the investigational product(s) used in the trials to reconfirm specifications, should this become necessary, and maintaining records of batch sample analyses and characteristics; to the extent stability permits, samples should be retained either until the analyses of the trial data are complete or as required by the applicable regulatory requirement(s), whichever represents the longer retention period.

The IP labeling requirements for countries in Africa have many similarities, but also some unique differences that require extensive knowledge of local regulations from country to country, and the ability to keep up with changes in regulations. For example, some countries, such as Namibia, require instructions on how to use the investigational product and/or route of administration, while others, such as Tanzania, do not.

18.6.1. Logistics of Clearing Customs

Consider the example of Latin America. Rules for importing investigational products and supplies to emerging markets are dependent on local regulations, which can vary greatly from country to country. The more mature the country is, the more likely it is that the regulatory agency will request an import license. Typically, this import license is requested to accompany the submission to conduct the trial. In many countries, the regulatory agency requests a declaration of the total amount of drug and supplies that will be imported and, with every shipment, they discount the approved amount. In cases where additional drug or supplies are required, a new import license must be requested. Additional procedures may be necessary for psychotropics or narcotics. Approval from the US Drug Enforcement Agency (DEA) and potentially additional permits are required in some countries.

The process of clearing customs can vary from country to country. A local expert needs to review the airway bills to make sure they comply with the local regulation. Such localized knowledge is the key to speeding up the customs release process.

18.6.2. Shipment and Cold Chain

Consider again the example of Latin America. In customs offices there are temperature-controlled facilities, such as cold rooms and ambient temperature rooms. The way drugs are imported depends on the type and specifications of the drugs. These include:

- ambient temperature drugs
- refrigerated drugs (vaccines, biologicals, etc.)
- frozen drugs (some vaccines).

The control employed during transportation also depends on the type of drug. For example, drugs that can be transported as ambient controlled drugs may not require monitoring during transportation. This will depend on the drug stability. Different strategies are applicable to refrigerated drugs. They may be shipped with different types of temperature monitoring devices (Temp-tails® is one such device) in order to track temperature deviations during shipping, and sometimes in validated coolers that control temperature for a specific length of time. In cases where the drug does not arrive in an appropriately controlled manner, the sponsor is made aware and the stability of the drug is reviewed; the monitoring of the temperature will allow the sponsor to compare these potential exclusions with stability data to make a decision to retain or release the drug.

It is important that sponsors establish and maintain a depot in these countries. This will ensure that the stability of the drug is appropriately addressed, and all GMPs are followed.

18.7. INTERACTIVE VOICE RESPONSE TECHNOLOGY

The software that enables the randomizing of subjects into clinical trials and dispensing medication to them in a blinded fashion is commonly known as an interactive voice response system (IVRS). With the increasing use of the web over the last few years, it is also referred to as an interactive web response system (IWRS) or interactive response technology (IRT). While some clinical trials still use paper-based systems, the majority of them now mandate the need for preconfigured, validated software to reduce human error, which could put the subjects and/or the trial itself at risk.

The process starts with the client identifying the specific requirements of the clinical trial. While some of the requirements are clear and universal (e.g. study design), many are very customer specific. Examples include how and where study drug (and control drug) should be dispensed and the study reports needed to monitor the trial. Once the specification process is completed, a telephone and/or web interface is built through a rigorous, quality-controlled process. A validated system is made available to the client and the investigator sites, with a typical timeline for this being 10–12 weeks. Users are sent secure envelopes which contain the unique identification information (ID) and password that are needed to access the system. In recent years, vendors have begun to use secure e-mail or web-based processes to distribute confidential user account information.

Once the sites receive the envelopes and the system is active, they start using the system for every transaction that they need to perform. A transaction may be in the form of randomizing the subject, dispensing them medication, or managing drug supplies (for the sponsor). With proactive system design, it is possible to reduce the drug wastage in a clinical trial to an absolute minimum by deftly managing the drug inventory.

Special attention needs to be paid to developing an IRT system that is primarily going to be utilized by sites in developing countries. Some of the unique problems that need to be addressed are:

- **Local language translation**: Most countries have a local language with which the investigator is comfortable. It is often necessary to translate all the phrases that are used by the system into the local language for the phone and the web. In a lot of countries, even the smallest of variations can lead to a very different meaning for the word, and even the way the word is pronounced can have different meanings. It is extremely important, therefore, to use a vendor who is very experienced in the local language and uses talented individuals who are experienced enough to notice the nuanced differences. The use of companies certified in medical translations for both the system and any supporting user documentation is also necessary
- **Web availability**: While Internet access is now fairly widely prevalent in developing countries, there still remain some pockets where it is not easily accessible. While more than 80 percent of users in the Western

world utilize and prefer the web over the telephone interface, the percentage may not be as high in the developing countries. Therefore, the telephone infrastructure that is set up needs to be robust and sophisticated to support the volumes.

- **Telecom infrastructure**: Unique attention needs to be paid to each country when deploying the telecom lines. The challenges include having rotary telephones, long lead times in obtaining toll free numbers, and sometimes the inability to obtain toll free numbers at all to dial out of country. The infrastructure also needs to account for possible line outages and hence needs to have built-in redundancy. There are some hospitals where all physicians do not have access to independent telephone lines, and hence the physical logistics might need to be worked out on a case-by-case basis. While mobile phone technology has alleviated some of these challenges, the need to provide free telephone access to sites is still an important consideration.
- **Patient scheduling**: In several countries, patients are brought to the hospital (investigational site) by social workers from quite far away. Should an issue prevent a successful visit when planned, a return visit might not be possible during the visit window. Such possibilities need to be kept in mind when designing the drug supply strategy of the trial. They also emphasize the criticality of system availability and accessibility, as described in previous points.
- **Drug supply management**: There are several countries where the time taken to ship the drug and clear it through customs is significant and needs to be taken into account when attempting just-in-time drug supplies. The increasing number of clinical trials for biologicals that have specific temperature controls that must be monitored, in addition to more stringent storage requirements, compounds this requirement. In addition to leveraging IRT technology, many sponsors choose to use local country drug depots or regional warehouses to facilitate this process. Finding the right balance of how much clinical supply to have available at sites versus in warehouses or depots requires experienced vendors working with appropriate technology to determine this based on initial assumptions and continuous refining and monitoring of these assumptions throughout the life of the trial.

Another feature that IRT technology provides is the ability to provide capping for situations where naïve subject populations have the potential to enroll quickly, while also utilizing the system to maintain potential stratifications or substudy attributes. Having immediate access to enrollment rates, subject visits, and subject terminations assists with the monitoring plans as well as ensuring that adequate clinical supplies are available.

Once all these factors have been taken into account and a system has been designed by an experienced IRT provider, the compliance and commitment levels in investigators using the system are typically no different than seen in the West. Currently, IRT technology is enabling the seamless execution of hundreds of trials in developing countries. Since a significant part of the overall subject population in large trials will be coming from these countries, sponsors and vendors will need to be able to adapt their processes and systems to country-specific challenges, and to create a solution that meets the requirements of the trial in that country. The quality and experience of sites and sponsors working in these emerging markets is improving every year, with more and more trials being successfully completed.

18.8. INVESTIGATORS' MEETING

While certain small trials can be conducted at one investigative site, it is much more common in large trials to see multicenter trials, and hence to have multiple principal investigators. Given that the expertise and previous experiences of these investigators will not be identical, the sponsor must work diligently to enable them all (to the greatest extent possible) to conduct every aspect of the trial in the standardized manner detailed in the study's protocol. Spilker provided some examples of areas in which principal investigators, and their study staff, may need training [5]:

- accurate implementation of inclusion/exclusion criteria such that (only) subjects who are appropriate for the trial are enrolled
- fully understanding (and accepting) the need for adherence to all study protocol procedures
- understanding cGCP
- completing case report forms (CRFs) accurately and completely, and making (and documenting) changes if and when necessary
- collecting, processing, storing, and shipping any biological samples (e.g. blood and urine) in a uniform manner
- diagnosing and rating the nature and severity of adverse events, and particularly adverse events of special interest, accurately, and in a uniform manner across all sites
- reporting these adverse events uniformly across all sites
- developing strategies for communication between sites and the sponsor [and the contract research organization (CRO) representing the sponsor].

Various strategies exist for implementing investigator training. One option is for all investigators to attend a training session, or investigator meeting, run by the CRO or possibly by a professional training organization. In trials for which the sites are in close proximity, it may be logistically possible to run just one meeting. However, particularly in the case of therapeutic confirmatory trials, the sites may be spread across a country, and even across various continents. In these cases, it may be advantageous to run several meetings. In this case, attention must be paid to standardizing the training course itself. Having the same trainers running multiple meetings is ideal, but when the investigators speak many different languages, this can be a challenge.

An alternative option is for the trainers to visit each site to conduct training. As in the previous strategy, more sites and greater geographical diversity make this potentially more challenging. At one time, it was considered somewhat of a perk for investigators to attend meetings, since they were often held at luxurious settings in exotic locations. This is much less the case now. In addition, some potential investigators would much rather have trainers come to visit them, since this is more time efficient for their busy schedules.

A third strategy is running web-based meetings. This can involve training modules that investigators and their staff must complete, and also interactive web-based sessions, potentially using video capabilities as well as just on-screen materials. Whichever method is chosen by the sponsor and/or their CRO, the more thorough the training, the more likely it is that investigators and their staff will implement the study protocol correctly.

18.8.1. Investigators' Meetings in Africa

The continent of Africa is associated for many individuals with images of many tribes and their differing traditional cultures. African tribe culture is distinguished by its great diversity of social patterns. For example, there are the anthropologist-termed hunter–gatherer tribes, as well as more technologically advanced pastoralist and horticulturalist tribes. In addition, African tribe culture is characterized by a great diversity of religions, ranging from animism to monotheistic religions such as Christianity, Judaism, and Islam. Furthermore, African tribe culture has a great diversity of visual arts and music.

Given this cultural diversity, planning investigator meetings (and many other meetings) need to take account of various factors, including:

- suitability of venue
- accommodation

- the welcome letter: this document provides attendees with detailed information on the meeting, e.g. times of sessions, conference room names and locations, and billing instructions
- language diversity
- dietary requirements
- travel
- religion
- meeting content
- communication
- information technology.

The majority of physicians are well traveled and used to attending meetings both locally and internationally. Some of them may have qualified or gained previous experience overseas. However, for health workers from rural areas, who often have very limited financial resources, particular care must be taken when choosing a venue. Access, distance to be traveled, and mode of transport must be considered. Appropriate accommodation must be provided (a five-star hotel, which may be attractive to some attendees, may be too intimidating for rural health workers). Single accommodation is normally booked for out-of-town delegates, unless a different request is made in advance (e.g. many married couples work together on trials). However, some of the rural health workers prefer to share accommodation with one or more of their colleagues.

The country of South Africa is multilingual. Besides the 11 officially recognized languages, many others (e.g. African, European, and Asian languages) are spoken as the country lies at the crossroads of southern Africa. English is generally understood across the country, being the language of business, politics, and the media, and the country's lingua franca. However, despite this, it ranks only joint fifth out of 11 as a home language.

The continent of Africa has four major language families and over 2000 existing languages. The largest language family in Africa is Niger-Congo, with over 1400 languages, illustrating the great diversity of African tribe culture. Certain parts of Africa speak French or Portuguese. When organizing meetings in certain parts of Africa, interpretation facilities becomes a necessity, which is a cost factor in meeting planning.

Delegate dietary requests typically fall into four categories:

- halaal
- vegetarian
- kosher
- no special dietary request.

Halaal and vegetarian requests are increasing. Most hotels are able to accommodate such requests. In certain rural areas, however, special dietary requests have to be flown in, again adding to the overall cost of the meeting.

With regard to transportation, doctors typically have some form of personal transport and are able to reach airports, hospitals, and venues by themselves. Out-of-town delegates sometimes have to arrive earlier and/or leave earlier owing to inconvenient flights. In the poorer rural areas, workers have to take taxis, walk, or make other arrangements. This can delay their arrival at the meetings, and these timing factors must be taken into account. Timing considerations are also necessary in the context of religion. Members of certain religions pray at certain times, and this must be taken into account when planning flights and commencement times of meeting. Religious holidays must also be considered.

Documentation intended to facilitate the on-time running of the meeting must be prepared in a thoughtful manner. Different cultures have differing perceptions of time, with some adhering more than others to the perception typical in Western cultures, and hence at Western investigational sites and meetings. Requests for acknowledgements and information from delegates and attendees should be sent out in plenty of time, and a reply date that is earlier than might otherwise be used requested. Postage can be unreliable, and so it is advisable to fax and/or e-mail correspondence and follow-up with a telephone call to ensure that the person concerned has received the information. "Personal reminder" telephone calls made before the meeting date are very important to prevent absenteeism.

At the meeting itself, advance planning must take into account the technological requirements of meeting sessions. Frequently, training takes place on personal computers, and web meetings run over the Internet are also common. Therefore, the meeting facility needs to be appropriately connected to the Internet, and computers provided for attendees who may not own one, or may find that transporting one they own from a distance is impractical.

18.9. ELECTRICAL GENERATORS AND POWER ISSUES

Any site is vulnerable to power outages, including those in Latin America. Latin American sites are very heterogeneous, with clinical trials being conducted in a variety of environments ranging from very large hospitals, private or public, to small offices where an independent investigator conducts the trial. Investigators who perform trials in hospitals always have back-up generators available as part of the site's infrastructure. However, this is often not the case for independent investigators at smaller sites. In these instances, other precautions must be considered, including:

- maintaining a supply of batteries for air conditioning and the refrigerator in which the drug is kept
- subcontracting a laboratory that does have back-up capabilities for storing blood samples when needed
- arrangements for handling emergencies in hospitals located near the independent investigator's office that has the required infrastructure.

When smaller sites become more proficient at conducting trials, and wish to continue to develop their revenue stream from this activity, they may decide to invest in a back-up generator.

18.10. TRANSLATIONS

In the planning of a clinical research trial, it is critical to take into account the logistics of the translations, and to ensure optimum quality of the translation itself. In Latin America, regulatory documents and clinical study documents must be submitted to the regulatory authorities in the legal language per country, and must be understandable by the site staff and subjects. The clinical team must determine the following on a country-by-country basis:

- the type of translation required: certified/qualified
- what documents require translation: considerations include determining which documents are required by regulators in the local region and in what language(s), when documents need to be submitted to them, when they need to be delivered to the site for use as recruitment tools, and when they will be needed for subject use
- which documents require back-translation
- which documents do not require translation.

A successful translation process requires all parties to be familiar with the timelines for securing translation, reviews, and back-translations ready on time. On average, the overall process can take from four to six weeks.

The level of translator depends on the nature of the document, which can require a certified or a qualified translator. One of these levels will always be required and, depending on the importance of the document, it may have one or two quality review steps. Even though most of the countries in Latin America use the Spanish language, there are notable variations in the words and expressions. Therefore, a native person from each country must make sure that the document's Spanish is appropriately tailored.

The other important language is Portuguese, as spoken in Brazil. Occasionally, it may be necessary to use some dialects, and to have documents submitted and approved in other languages depending on the target population of the study. For example, in some countries there may be large immigrant groups that contain many potential subjects. Therefore, providing documents in their native language may be very helpful.

ACKNOWLEDGMENT

The authors gratefully thank Carolyn Moore for her administrative support.

REFERENCES

[1] Chow S-C, Chang M. Adaptive Design Methods in Clinical Trials. Boca Raton, FL: Chapman & Hall/CRC; 2007.
[2] Buncher CR, Tsay J-Y, editors. Statistics in the Pharmaceutical Industry. 3rd ed. Boca Raton, FL: Chapman & Hall/CRC; 2006.
[3] Turner JR. New Drug Development: An Introduction to Clinical Trials. 2nd ed. New York: Springer; 2010.
[4] Piantadosi S. Clinical Trials: A Methodologic Perspective. 2nd ed. Hoboken, NJ: John Wiley & Sons; 2005.
[5] Spilker B. Guide to Drug Development: A Comprehensive Review and Assessment, PA. Philadelphia: Wolters Kluwer/Lippincott Williams & Wilkins; 2009.
[6] Institute of Medicine of the National Academies. The Future of Drug Safety: Promoting and Protecting the Health of the Public. Washington, DC: National Academies Press; 2007.
[7] Monkhouse DC. The clinical trials material professional: a changing role. In: Monkhouse DC, Carney CF, Clark JL, editors. Drug Products for Clinical Trials. 2nd ed. Boca Raton, FL: Taylor & Francis; 2006. p. 1–19.
[8] Brun P. Blinding of drug products. In: Monkhouse DC, Carney CF, Clark JL, editors. Drug Products for Clinical Trials. 2nd ed. Boca Raton, FL: Taylor & Francis; 2006. p. 149–72.

CHAPTER 19

Clinical Study Conduct and Monitoring

Edda Gomez-Panzani
Clinical Sciences – Neuroendocrinology, Ipsen US, Brisbane, CA, USA

Contents

Global Clinical Trials
ISBN 978-0-12-381537-8, Doi:10.1016/B978-0-12-381537-8.10019-6

19.1. REGULATIONS AND HARMONIZATION

The history of the evolution of regulations in the USA and Europe and the state of regulations around the globe have been discussed in previous chapters and will not be discussed individually here. However, with regard to harmonization, good clinical practice (GCP) was established in the USA in 1978 and has evolved constantly ever since. Around the same time GCP was instituted, the European Community increased its focus on harmonization of regulatory requirements for pharmaceuticals and talks on this topic began with the USA and Japan. During the World Health Organization (WHO) International Conference of Drug Regulatory Authorities (ICDRA) in Paris in 1989, an action plan began to develop. The following year regulatory agencies and industry representatives from Europe, Japan, and the USA began planning an International Conference on Harmonisation. The International Conference on Harmonsation of Technical Requirements for Registration of Pharmaceuticals for Human Use was the result of efforts mainly from the USA, Europe and Japan (plus Australia, Canada, the Nordic countries, and the WHO) and was adopted in 1997 in all three countries. In Japan it was adopted as a law and in Europe and the USA as a guideline (in the USA in the federal register).

As a result of this harmonization, and to ensure that only good quality, safe, and effective drugs are developed, GCP has most recently been redefined as the International Conference on Harmonisation (ICH) Harmonised Tripartite Guideline for Good Clinical Practice E6, and adopted as a Food and Drug Administration (FDA) guidance in May of 1997. All clinical trials involving human participants in the USA, regardless of the development phase, must be conducted following the guideline for GCP.

The harmonization of GCP across the three major developed countries/regions has been a major advance in drug development and has immensely aided the growth of global clinical trials. ICH continues to be developed further. Recently, for example, the format and content of submission of new drug approvals have been harmonized with the promulgation of the common technical dossier (CTD) format.

However, the regulations are not completely harmonized, since each region can add additional requirements. For example, the FDA often insists on more than the minimal database size specified by ICH, and the European Union (EU) has enacted requirements for patient privacy that are above and beyond requirements in the ICH guidelines. For the most part, most countries across the globe tend to follow ICH guidelines and very few have requirements that exceed the ICH requirements. So, as a general rule, following ICH guidelines will for the most part satisfy local regulations.

19.2. PRECLINICAL STUDY, AND CHEMISTRY, MANUFACTURING, AND CONTROL

Before studies in humans can begin, studies in animals (preclinical studies) must be completed. The purpose of these studies is to prove that the compound being studied is safe for use in humans. The chemistry of the drug must also be characterized.

The first step is to determine the chemical, physical, and biological characteristics of the compound. Data on the manufacturing of the compound as well as information to prove that the compound will remain stable (stability tests) throughout the time it will be used in the preclinical studies, are required prior to administering it to animals.

Once the compound has been characterized and deemed stable, preclinical studies can begin. The type of studies and their design can vary depending on the compound, but the purpose is to characterize the toxic effects of a compound on different organs and biological processes and their relationship to exposure. Simply stated, the majority of preclinical studies pertain to toxicology and safety of the molecule.

The minimum required preclinical studies include single and repeated dose toxicity studies, pharmacokinetic and pharmacodynamic studies, genotoxicity, carcinogenicity, and reproductive studies.

Repeated dose toxicity studies include multiple dose administrations done over a period similar to the indication and exposure expected during the clinical studies in humans (they rarely exceed 12 months). These studies also must be conducted in two mammalian species (one rodent and one non-rodent). Frequently, additional preclinical studies are performed while a drug is simultaneously undergoing studies in humans.

Carcinogenicity studies are necessary for any compound that is intended for use for six months (either continuous six months or for intermittent periods that add to a total six-month or more exposure). These studies entail daily administration of the compound, are usually conducted in rodents, and last approximately the life of the animal (approximately two years).

Reproductive toxicity studies assess the effects of the compound on fertility, reproduction, and fetal toxicity; they involve repeated administration of the compound to the animals before, during, and after a gestational period. These studies must be completed before submitting a New Drug Application (NDA) to the FDA.

All the studies must be conducted under the ICH good laboratory practices (GLPs) guidelines (the toxicology equivalent of the GCPs).

CMC and toxicology standards across the globe differ, but once again, in general the standards required by the FDA and European Medicines Agency (EMA) generally will satisfy local requirements. One notable exception, however, is China, which mandates that part of the manufacturing be performed in China.

19.3. PHASES OF CLINICAL DEVELOPMENT

Phases I–IV of clinical development are not strictly defined and are used mostly as milestones in the development process. Throughout this chapter reference is made to "investigational drug"; however, it should be kept in mind that most of the time the statements apply to an "investigational product", which could be a drug or a device.

Phase I is the first stage of the process and the time when the investigational drug is administered to humans for the first time. At this time the pharmacokinetic and pharmacodynamic parameters of the investigational drug will be assessed, as well as the safety of different doses. The focus of Phase I studies is subject safety, and therefore they are most often conducted on a relatively small number (fewer than 100) of healthy volunteers. In certain circumstances, Phase I studies could also involve patients suffering from the targeted disease and consequently allow for an initial efficacy reading.

Since it is the first time the investigational drug is administered to humans, these studies usually incorporate a dose-escalation/dose-ranging design where the participants are initially exposed to a very small dose of the investigational drug and closely monitored around the clock, sometimes for several days. Once it has been determined that the initial dose did not result in any serious or intolerable side-effects, a higher dose of the investigational drug can be tested. This pattern continues until the maximum tolerable dose in humans is determined. During the time the participants are at the research facility, laboratory samples will be collected at regular intervals to determine levels of the investigational drug as well as the effects of the different levels on different body systems; in addition, adverse events will be collected and the safety of the participants will be closely monitored. These studies are often conducted in settings specifically designed for this purpose [Phase I clinics or specialized contract research organizations (CROs)]. The results of Phase I will be used to help design the Phase II studies.

In general, many developing countries, including China and India, will not allow Phase I studies to be conducted on drugs discovered in another country or not owned by a local company. This is to prevent their citizens from being used as "guinea-pigs", particularly since, in many instances, the institutional review board (IRB) or ethics committee and the investigators may not be as experienced in detecting potentially dangerous or unethical studies. Therefore, the Phase I studies are usually conducted in a developed country. Australia is an excellent venue for Phase I studies because it allows certain Phase I studies to be conducted without regulatory review but has a very good ethics committee system, which can make the start-up period very short without endangering patients.

Phase II studies are conducted in patients with the targeted disease who have no other major underlying medical conditions; the inclusion and exclusion criteria for these studies are usually very strict. Two to four doses of the investigational drug, proven to be safe in the Phase I studies, are selected and compared to placebo in double-blind studies usually involving no more than 200 patients; these types of studies are also known as dose-ranging or dose–response studies. The main objective of this phase is to assess the efficacy of different doses of the compound; however, safety is also evaluated and a therapeutic window is identified. Pharmacokinetic data are frequently also collected to enable the correlation of blood levels of the investigational drug to their pharmacological effect. These studies mark a critical decision-making point in the drug development process; at this stage it will become evident whether or not the drug demonstrates efficacy for the proposed indication and whether the dose or doses that show efficacy has or have an acceptable safety profile.

Phase II studies are often conducted globally. The recruitment speed and availability of patients, along with reduced costs, can make global trials very attractive for Phase II.

If, based on the data collected from Phase II studies, the investigational drug has an acceptable safety profile for the indication being pursued, and if there is an adequate efficacy signal, Phase III studies can be initiated. It is important to note that the acceptability of the safety profile is strictly related to the indication; a safety profile that can be acceptable for a chemotherapeutic agent used in oncology would most probably not be acceptable for a contraceptive drug.

Phase III studies are also known as therapeutic confirmatory or pivotal clinical studies. At this stage, the doses with the best therapeutic window, selected from the Phase II studies, will be administered to a much larger number of patients. The purpose of these studies is to demonstrate the safety and efficacy of the investigational drug administered to the population with the disease for which the indication is being sought. Phase III studies must be well controlled and are almost always conducted by multiple clinical study sites (multicenter), sometimes in different countries, and in large numbers (hundreds to sometimes thousands) of patients. Most often, the regulatory agencies require positive results from two well-controlled studies for the registration of a drug; however, in the case of orphan diseases or in situations when the FDA believes that the data from one well-controlled study are enough to establish the safety and efficacy of the drug, the Agency may allow one well-controlled study to support the registration [1].

Once the Phase III studies have been completed and a final report has been written, all the information that has been collected throughout the clinical development path of the investigational drug will be summarized and submitted to the regulatory authorities to obtain authorization to market. In the USA, the information is submitted to the FDA as part of an NDA.

Global trials are very well suited for Phase III studies, for the reasons cited above for Phase II studies.

Occasionally, not all studies have been completed at the time of NDA submission; if studies are still ongoing these are usually designated as Phase

IIIb studies. The purpose of these studies is to provide additional safety information, to support the efficacy of a new indication, or to gather information on special populations (e.g. patients with renal failure, pediatric patients).

After regulatory approval has been obtained, additional studies may be conducted; these are designated as Phase IV or postmarketing studies.

Postmarketing studies can be research studies or observational studies; they are often open-label and conducted to gather specific information about an approved drug, device, procedure, or biological product. Conducting Phase IV studies could be mandated by the FDA or the result of the sponsor's decision. Sponsors usually conduct Phase IV studies to understand better real-life use of the product, to expand the labeling, to gather additional safety information, to assess the safety and efficacy in special populations, as marketing studies, or to pursue additional indications.

Examples of studies mandated by the FDA would include instances where the Agency authorized the NDA submission to be done with only one pivotal study on the condition that a second well-controlled study were conducted to confirm the results from the first study, or to provide additional long-term safety information.

19.4. CLINICAL STUDY CONDUCT

Several activities are involved in the planning and conduct of a clinical study. Depending on the phase of the study and the type of study, it is possible that not all activities are required, but for the most part a series of diverse and sometimes complex and time-consuming tasks, as outlined hereunder, must be completed. The tasks are presented in a sequential manner as they are usually performed in real life; some of these tasks can also be performed concurrently.

For the most part, the steps of conducting a global clinical trial are similar to conducting a non-global trial, but there are several key differences:

- Quality control is even more important in global trials because sites tend to be much more variable in quality.
- Site qualification is therefore very important.
- Training is also extremely important and the intensity of training must often be higher.
- In some instances, especially with trials involving neglected diseases, building site capacity (discussed elsewhere in the book) is often

necessary. This may involve activities ranging from building the physical clinic to teaching universal precautions.

- In some cases, stationing a clinical research associate (CRA) at the site may be necessary and cost effective if the volume is high enough.

19.5. CLINICAL STUDY TEAM

The clinical study team is composed of representatives of the areas that are involved in the conduct of the study:

- **Study clinician**: Physician with extensive clinical and research experience responsible for ensuring the scientific/medical rationale to conduct the study, for monitoring the study, and for providing medical/clinical input throughout the conduct of the study. Also responsible for assessing the integrity of the data and the safety of participants, and for writing and reviewing the pertinent sections of the final study report.
- **Clinical study manager**: Responsible for the day-to-day clinical operations of the study including planning the study timeline, budget, resources needed, leading the development of the case report forms (CRFs), informed consent forms (ICFs) and newsletters (if applicable), training and monitoring the clinical research associates, and reviewing and approving their monitoring reports, etc.
- **Clinical research associate/clinical study monitor (CRA/CSM)**: Oversees the progress and conduct of the clinical study to ensure the integrity of the data being collected and the safety of the participants, ensures the clinical study site complies with the protocol, monitors the study sites and documents findings on monitoring reports, reviews CRFs against source documents/medical records, verifies that the process of obtaining informed consent from all participants was done properly, in a timely manner (prior to any study-related procedures being performed), and documents and reviews adverse events to ensure they were properly collected, documented and reported, if applicable.
- **Data management representative**: Responsible for the development of the CRFs, ensuring that all the clinical study data are properly entered into the database, and generating tables, listings, and figures as described in the statistical analysis plan and the clinical study protocol.
- **Biostatistician**: In charge of determining the sample size, distribution (groups) and randomization ratio, developing the statistical analysis plan

(SAP), reviewing the CRFs to ensure all the data that will be used for analysis are being captured appropriately, and determining which tables, listings, and figures will be required. The biostatistician is also responsible for the analysis of the data and for summarizing this in the clinical study report.
- **Regulatory representative**: Ensures that all appropriate documents are submitted as part of the investigational new drug (IND) application, reviews the protocol, provides input regarding regulatory aspects of the study, and serves as a liaison between the FDA and the sponsor.

19.6. CLINICAL STUDY PROTOCOL

One of the earliest tasks in the conduct of a clinical study is the development of the clinical study protocol. To ensure optimal study design, the responsibility of developing a protocol lies with a multidisciplinary clinical study team; occasionally, it may also require input from experts in the field experienced in the specific indication and, optimally, in clinical research. This input may be obtained through advisory boards where all experts are brought together and the patient population, study endpoints, and protocol design are discussed and agreed to, or by individually approaching a selected group of experts for input.

The first step is the creation of a protocol summary (synopsis). This section of the protocol includes the title of the study, the study objectives, study endpoints, clinical study design, patient selection criteria (inclusion and exclusion), statistical considerations, and a brief description of safety data collection and reporting. Depending on the study, other sections (e.g. pharmacokinetics) may be included. The protocol summary serves as a guideline during protocol development.

The clinical study protocol is the bible to be followed in the conduct of any clinical study and its purpose is to describe the study in detail including, but not limited to:

- an overview of the investigational drug and the disease in which it is intended to be used
- study rationale
- duration of the study
- objectives
- study endpoints
- study design [length and phase of the study, type of study (e.g. double-blind, randomized, placebo controlled), size and patient population,

detailed description of the procedures/tests to be performed and the timepoints (study visits) for each]. It is extremely important to consider the burden all the processes and procedures will impose on the subject/patient: how will the number of visits and the time spent traveling to, from, and at the study center affect the subject/patient's daily life?
- study treatments, type of randomization and dosage regimens
- patient population (inclusion and exclusion criteria): when writing the protocol, it is critical to determine the exact criteria by which patients will be included or excluded from participation in the study, including age, gender, stage of the disease, whether they have been pretreated or not (and with what type of treatments), baseline markers of the disease (if applicable), time since the diagnosis, and restrictions on the use of concomitant medications/foods/therapies
- how the safety of the subjects/patients will be protected (adverse event definition, handling and reporting)
- statistical methods (statistical analysis plan)
- data quality control
- monitoring of the study
- reports, abstracts, and publications
- investigational drug storage, accountability, return and destruction
- investigator responsibilities
- study documentation, CRFs, and record-keeping
- appendices (if applicable).

The protocol design has been recognized as the single largest source of delay in study completion timelines. It is critical that, when designing a clinical study, a broad variety of factors beyond pure science and the protocol's compliance with good clinical practices are taken into account. These include, but are not limited to:
- The availability of alternative treatments, access to diagnostic tools/unique procedures (e.g. assessments may include PET/CT scans that are only available at the main hospital in the state's capital hundreds of miles away).
- Country-specific standards of care and regulatory requirements.
- The number of patients that could qualify for participation in the study (sometimes, in an effort to define strictly the patient population, the inclusion/exclusion criteria are so strict that most patients become screen failures).
- The number of procedures required: If a study requires a large number of time-consuming procedures, there will be a large financial impact on the study as well as an increased risk that some procedures are

performed out of the allowed time window or not done at all. The Tufts Center for the Study of Drug Development (an independent, academic, non-profit research group affiliated with Tufts University) conducted a study assessing the burden on clinical research sites and reported a 10.5 percent annual increase in study workload between 1999 and 2005 [2].

- The interest of the patients (what is in this study for them) and the investigators: Many placebo-controlled studies, although very well designed, fail to take into account the willingness of the patients and the investigators to participate in a study where the patients will not be getting treatment. The main goal of the investigator should be to provide optimal patient care and, therefore, when participating in a clinical study, they must ask themselves whether by attempting to enroll their patients in the study, they are offering the best standard of care to their patients. Are they putting their patients at risk?
- And last, but not least, operational factors that impact the study conduct such as the experience and ability of the clinical study staff/investigator to [3]:
  - obtain IRB and any other ethics committee approval
  - ensure timely study start-up and recruitment
  - follow protocol design/procedures
  - screen, enroll, and retain study participants
  - deliver within budget.

Researchers from the Tufts Center for the Study of Drug Development (Tufts University) conducted retrospective data analyses on 10,038 protocols to assess the impact of protocol design on study conduct and financial aspects. They found that the number of unique procedures and the frequency at which they had to be performed (most evident for studies Phase I and II) had escalated considerably in a period of seven years (from 1999 to 2005), increasing the workload for the clinical site. They also found that, as a result of the augmented procedural burden, the performance of the site conducting the study declined. The protocols conducted from 2003 to 2006 had longer patient recruitment times, longer data collection times, poorer patient randomization and completion dates, higher number of protocol amendments and adverse events, and more lengthy ICFs and CRFs than those conducted between 1999 and 2002 [3].

These findings substantiate the need to streamline clinical study protocol design. It is also important to approach the protocol design from a multidisciplinary perspective where every functional area can contribute to the design, and identify specific areas of concern early on during protocol development.

19.6.1. Protocol Amendments

The protocol is considered a dynamic document. Even when a clinical study protocol is considered final, some changes may be required during the course of the study. Any significant change that affects the safety of the subjects, the scientific quality of the data being collected, and/or the scope of the investigation would require a "protocol amendment". Any and all protocol amendments must be reviewed and approved by the IRB(s) and submitted to the FDA.

However, if a protocol requires only minor (administrative) changes (e.g. the drug will be supplied in a 5 ml vial instead of a 3 ml ampoule), an "administrative amendment" will suffice; this type of amendment is also submitted to the IRB; however, although they will acknowledge receipt, they do not need to approve it for the amendment to be implemented.

Every clinical research site participating in the study is expected to follow the protocol strictly. This cannot be stressed enough. Every year, the Center for Drug Evaluation and Research (CDER), a part of the FDA responsible for regulating over-the-counter, prescription, and generic drugs, as well as biological therapeutics, reports that failure to follow the investigational plan/clinical study protocol is the most frequent deficiency encountered during clinical site inspections [4].

Once the clinical study protocol has been finalized, and sometimes in parallel, the clinical study sites that will be conducting the study must be identified.

19.7. CASE REPORT FORMS

The CRFs constitute the key to data collection in any clinical study. These forms are study specific and are prepared based on the protocol; they contain fields for the collection of every piece of information that will be assessed during the clinical study. When designing the CRFs it is important to avoid collecting information that will not be used for further assessments as well as collecting the same information from two different sources (e.g. electrocardiogram interpretation by the physician and by the central reader); choose only one relevant source to avoid discrepancies and data errors.

There are two options for data capture in CRFs, electronic (eCRFs) or paper (CRFs). The decision to use one or the other depends on several different factors. There are some clear advantages to the use of electronic CRFs which include real-time access to data (which is extremely helpful for auditing site performance), the possibility of performing remote and efficient monitoring, and the speed at which queries can be generated and resolved. In addition, the set-up time, the cost, the data management capabilities of the sponsor, the elimination of storing paper-based CRFs, and whether the study will be collecting patient-generated data or not should be taken into account. Last, but not least, it is important to consider the acceptability of eCRFs by study coordinators.

19.8. DATA AND SAFETY MONITORING BOARD

A data and safety monitoring board (DSMB) is sometimes required for clinical studies. For early phase studies (Phases I and II), a DSMB may be needed if the studies are multicentric and/or blinded, if some of the assessments are considered of particularly high risk, or if they are conducted in vulnerable populations (children, elderly adults, etc.). A DSMB is common in Phase III studies, especially in the first Phase III study with the investigational drug.

The data safety monitoring (DSM) plan is developed as a guide for the monitoring of clinical studies to ensure that the subjects/patients are safe (the magnitude of the adverse event will be assessed) and to guarantee that the privacy of the participants is protected. The plan is designed to allow the members of the DSMB to take a look at the data (the frequency of this review will be specified in the plan) and readily identify any safety signals that could require the study to be stopped prior to completion. Occasionally, and usually upon request from the board members, they can also review efficacy data and that, too, could trigger the study's early termination.

19.9. INVESTIGATIONAL NEW DRUG APPLICATION AND THE FOOD AND DRUG ADMINISTRATION

Prior to initiating any clinical study with an investigational new drug, the sponsor must submit an IND application to the CDER. In the EU, an investigational medicinal product dossier (IMPD) must be submitted to the local national authority. Most other countries require a similar package to be submitted. In some countries, a local company or person must hold the IND equivalent. The holder of the IND or IND equivalent is usually legally responsible for conducting the study properly and safeguarding patient safety.

19.9.1. Local Versus Central Laboratory

As the study is being designed, it is important to decide whether a local or a central laboratory will be utilized. It is appropriate to use a local laboratory for single-center studies; however, for multicenter studies, a central laboratory is usually preferred.

The use of local laboratories for multicenter studies is discouraged mainly because the differences in their analytical methods and analytical quality, reference ranges and units, and even in staff training procedures can jeopardize the analysis of the data and therefore the study outcome.

By using a central laboratory, the methods and ranges are consistent, increasing the validity of the results. However, a careful selection of which central laboratory will be used is of importance.

When assessing and selecting central laboratories, the following criteria must be taken into account [5]:

- Are all the tests done in-house or are some of them outsourced?
- Laboratory accreditation certification, licensure, and registration: is the laboratory CLIA (Clinical Laboratory Improvement Amendments) and CAP (College of American Pathologists) certified? Any clinical laboratory that performs diagnostic, preventive, or treatment-based testing on human specimens must obtain a CLIA and a CAP certificate which must be renewed every two years. CLIA, CAP, and other laboratory certificates must be retained as essential documents with the clinical research records (or easily available and accessible if they need to be retrieved).
- Are validated methods used?
- How are the samples identified and tracked?
- What are their processes to ensure quality control?
- Are laboratory reports, data transfer, and progress reports customized?
- Will the sponsor have real-time access to sample status and results online?
- Will study-specific data software be available? Advantages include:
  - removes the need for double-data entry and avoids transcription errors
  - rapid access to laboratory results

- fast database clean-up
- lower study costs.

- Will there be a dedicated project manager assigned to the study?

Other aspects that need to be taken into account are the shipment of the samples (courier to be used, timing of shipping, etc.) and whether or not split samples (serum banking) will be available (as insurance).

19.9.2. Selection of Contract Research Organizations

In addition to utilizing the services of independent laboratories, the research and development (R&D) process is moving more and more to outsourcing. Many responsibilities and processes can be outsourced, but specifically critical to the conduct of a clinical study is the use of CROs. Having idle personnel during times in which only one or two studies are being conducted results in a major financial burden to the organization. By outsourcing the processes and activities that tend to ramp up just for a short period during the development process, the process can be accelerated and the development costs can be better controlled.

When deciding to outsource, there are different options:

- Outsource all aspects of the clinical trial and allow the CRO (or CROs; multiple CROs can be selected to perform different functions depending on their area of expertise) to assume all responsibility. This works particularly well for smaller organizations with a reduced in-house head count.
- Outsource only a few functions. In this case, the functions that are most often outsourced are data management and monitoring.
- Transfer the responsibility of one single area to a CRO.
- Contract temporary personnel responsible for specific tasks for the duration of the clinical study; this is usually CRAs/CSMs.

The selection of a CRO depends on the type and size of study, location of the study (single center versus multicenter and national versus international), indication, and experience with prior CROs. Aspects that must be taken into account when selecting a CRO include, but are not limited to:

- experience
  - phases of development
  - therapeutic areas
- integrity and credibility
  - consistently delivers the highest level of quality and performance
  - proven track record

  - consistently meets timelines and remains within budget
- compliance with regulations
- size (internal capabilities: human resources)
- location(s) (geographical capabilities)
- willingness of the CRO to accept penalty clauses for poor performance
- method for identifying and escalating issues
- process of identifying, recording, and reporting adverse events (and how and how often a trend analysis is performed)
- relationships with other vendors [interactive voice and web response systems (IVRS/IWRS) vendors, electronic patient recorded outcomes (ePRO) vendors, packaging and distribution services vendors, etc.] and contractual compliance
- experience with regulatory submissions
- central laboratory experience
- speed and ability to identify and manage clinical study sites
- financial stability
- proper quality assurance systems in place
- strong information technology (IT) systems and information management practices in place (must comply with CFR 21, Part 11)
- secure document storage (central file) locations.

19.9.3. Study Site Selection

It is critical to select motivated and well-qualified study sites and investigators to ensure the appropriate conduct of the study. There are different methods to approach study site selection. For first-in-human studies conducted in healthy volunteers, the most straightforward approach is usually to utilize the services of clinical research organizations that specialize in early phase research. They have very large databases of healthy volunteers and can ensure expedited recruitment.

When the indication being studied is one in which the company conducting the research already has vast experience, the investigators that have a proven track record of patient recruitment and retention as well as timely delivery of clean data are usually approached first. These investigators are also an invaluable source since they can further refer other colleagues with experience in the same area of research that could participate in the clinical studies. Other possible sources are investigators that have previously published in related fields, and clinical consultants expert in the field or indication.

As per FDA GCP regulations (21 CFR 312.53) and the ICH-GCP guideline, the principal investigators (PIs) "should be qualified by training and experience as appropriate experts to investigate the drug" and "to assume responsibility for the proper conduct of the trial and should meet all the qualifications specified by the applicable regulatory requirements". It is up to the sponsor to determine whether an investigator is qualified or not. In most cases, a physician functions as the PI; however, in the USA, another appropriately qualified individual (e.g. PhD) could take the PI role as long as a physician assumes all medical responsibility for the subjects enrolled in the study (this is not always the case for other countries such as Canada, where the PI must be a physician).

As soon as the initial sites have been identified, and before sharing any details regarding the investigational drug and the clinical study design, in order to ensure the confidentiality of the data, a confidential disclosure agreement (CDA) must be signed by the investigator and, on occasion, by the legal representative of the institution.

19.9.4. Confidential Disclosure Agreement

Once the initial sites have been identified and contacted, and their interest in learning more about the clinical study confirmed, a CDA must be set in place with each site. This is a legal document that will ensure that the investigator will not disclose proprietary information related to the company sponsoring the clinical study, the compound under investigation, and/or the study itself. The terms and conditions of the CDA have to be negotiated between the parties. Once both parties (sponsor and clinical investigator) agree to the terms and conditions, they must sign and date the CDA before the sponsor can share any proprietary information with the investigator.

Immediately after all parties have signed the CDA, the sponsor can proceed to share a brief description of the study (e.g. the protocol summary) with the potential investigators so that they can assess their interest in participating.

19.9.5. Feasibility Assessment

In addition to the study summary, study-specific feasibility questionnaires are sent to the sites to determine whether the site-specific capabilities match the clinical study objectives. These questionnaires collect information regarding the clinical site's (PI, subinvestigator, and study coordinators) experience with the indication being studied, the availability of patients meeting the main inclusion and exclusion criteria, access to resources, diagnostic and therapeutic tools as required by the protocol, and competing studies currently ongoing or to be initiated concurrently.

If, based on the answers to the feasibility questionnaire, the sponsor determines that the site appears to be qualified to conduct the study and the investigator is interested in participating, a prestudy/site qualification visit will be scheduled.

All site visits are conducted by a CRA/CSM. Sometimes, other individuals (as deemed necessary by the sponsor) such as the medical monitor may accompany the CRA/CSM to address specific topics or issues.

19.9.6. Prestudy/Site Qualification Visit

During the prestudy/site qualification visit the CRA/CSM will meet with the investigator and staff to assess the clinical study personnel and site capabilities. Points to be assessed include:

- **Interest of the investigator**: One of the major drivers for patient recruitment is the interest of the investigator in participating in a clinical study. Prior to the site qualification visit, the speed of response from the site and the completeness of the information provided can be a way to gage initial interest in the study. A site that had to be contacted five times before providing an initial reply is probably not interested in participating and will probably not do a good job.
- **Access to patients**: The experience of the PI in the therapeutic area in which the study is being conducted is of utmost importance. How many patients with the indication does the PI (and subinvestigator) have and see regularly? Are there centers/physicians that usually refer patients with this indication to them? Will the PI be conducting any other studies at the same time? Are the inclusion/exclusion criteria of any other concurrent studies similar enough that they will have to compete for the same patient population? What strategies does the site utilize for patient recruitment?
- **Staff and experience**: Have a subinvestigator and clinical study coordinator (CSC) been identified? Will the CSC be dedicated only to one study and, if not, how many studies is the CSC responsible for at any given time? The monitor must review the CV of the investigator, subinvestigator, clinical study coordinator, and any other member of the staff that has been identified to participate in the conduct of the study. Is the investigator considered an expert in the field? Have the investigators and coordinators conducted other studies in this indication? If so, how many and what phase of development were they in? Does the investigator have any experience with this type of active ingredient/method of action?

- **Facilities**: Will the PI be seeing the patients at a clinic, hospital, or medical center? Are all the facilities/resources immediately available and adequate for the proper conduct of the study? Are investigational drug and study binder storage areas adequate? There are some clinical study protocols that require access to specialized test/studies (e.g. a study on liver transplant that requires a transplant team available at all times) which in certain settings/locations, or even countries, may not be available. Is the area designated for drug storage appropriate? Some investigational drugs (e.g. controlled substances) may require special procedures be put in place for storage [double-locked cabinets that are firmly attached and secured (bolted to a permanent structure)], record-keeping, and disposal to comply with applicable laws and regulations.
- **Local/country-specific circumstances**: For international studies, this is an area that should not be overlooked. There may be some specialized studies that are not available, or sometimes things that would appear minor may be problematic, such as a medication that must be taken three times a day with food in areas where the majority of the inhabitants barely get one meal a day.
- **Other**: Can the investigator utilize a central IRB or must she or he use the local IRB (this can add considerable time to the process of obtaining approval)? Are there additional Ethics committees from which approval must be secured prior to initiating the study? Has the investigator, subinvestigator, and/or site ever been issued a Form FDA 483 (see Section 19.16)? Does the name of the investigator appear on the FDA debarment list?

The monitor will complete a prestudy/site qualification visit report that will be reviewed as part of the final site selection process. When the investigators and clinical sites have been selected, the sponsor will send out a letter to inform them that they have been selected for the study.

19.9.7. Budget

Although not directly part of the clinical conduct of a study, budgeting is a critical stepping-stone in the process of getting started. The institution or clinical site where the study will be conducted must assess the direct and indirect costs that will be incurred as a result of conducting the clinical study. Direct costs are those related to specific tests or procedures; indirect costs are incurred in support of activities that cannot readily be identified and that have no specific cost attached to them (e.g. the use of the facilities). The institution or clinical site then submits the budget to the sponsor; the budget usually includes not only the cost of the different procedures and supplies but also time estimates for key staff and laboratory tests. If the budget is within the amount the sponsor allocated for each site, the budget will be approved and a payment schedule agreed upon. If the budget submitted by the clinical site is outside the approved study budget, the budget will have to be negotiated; if an agreement is not reached, the sponsor may be obliged to remove this institution from the selected clinical study sites.

19.9.8. Clinical Trial Agreement

The clinical trial agreement (CTA) is the contract that defines the scope of the work and the expectations between the sponsor and the institution or clinical site that will conduct the study. If not already specified in the clinical study protocol, the sponsor and the institution or clinical site must agree on publication rights, patient injury, applicable laws, and any other terms that describe the scope of the study. In some instances (such as when the study is being conducted at a private clinic), the PI's signature may be sufficient; however, in others (for example when the study is being conducted at institutions or medical centers), the signature of the legal representative for the institution may be required to validate the agreement. Sometimes this is a straightforward process; however, other times, as is the case for any legal document, it could take some time for both parties to reach an agreement on the language; it is very important to factor time for these negotiations into the overall timeline of the study.

Once the CTA is in place or sometimes even in parallel, it is time for IRB or independent ethics committee (IEC) submissions. All the documents explaining the nature and design of the study, the risks and, if applicable, benefits to the subjects/patients, and any other document that could be used by the IRB/IEC to ensure that the research subjects will be protected during the conduct of the study should be submitted for review and approval.

19.10. PROTECTING THE RESEARCH SUBJECTS

Towards the end of the 1950s, Dr Henry K. Beecher, anesthetist-in-chief at the Massachusetts General Hospital, became aware of the lack of concern for the treatment of patients as research subjects. He believed that, if the patients were aware of the risks and consequences of their participation, they would never agree to be part of most clinical research. In 1966, he published "Ethics and clinical research" in the New England Journal of Medicine [6]. In this publication, he emphasizes two critical components of ethical research; one, the existence of an informed consent where a patient/subject agrees to participate only when he or she is perfectly aware of the risks and benefits of doing so, and two, the need to protect the safety of the participants. At that time, Beecher believed that this could be done solely by ensuring that the investigators were compassionate and had the best interest of the patients in mind. Although a step further than what he had envisioned, Beecher's thoughts prompted not only the need to obtain written informed consent from every research patient/subject but also changes to federal regulations and the establishment of IRBs.

19.10.1. Institutional Review Board

The IRB (called the IEC in the EU) is composed of a group of at least five members (and a few alternatives) that ensures that the rights and welfare of humans participating as subjects in a research study are maintained throughout the subjects' participation. To eliminate any diversity bias members must come from different disciplines [scientific (at least one physician) and non-scientific (at least one non-scientific or layperson)] and ethnic backgrounds and both genders must be represented. The IRB membership usually consists of physicians, nurses, pharmacists, statisticians, an ethicist, and a patient advocate. There are two main types of IRBs, a central IRB and a local IRB.

Central IRBs oversee multicenter trials and are at a location different from the clinical study sites. They can be established by public agencies [such as the National Institutes of Health (NIH)] or by private enterprises. They usually meet every week or two weeks and on occasion will agree to an unscheduled meeting if the lack of a prompt IRB approval could considerably delay the clinical study timelines.

Local IRBs are affiliated with the institution or study site that is conducting the clinical study (e.g. medical center, university) and are usually located at the same institution or study site. They usually meet less frequently than the central IRBs (once a month or even once every two months). This can result in lengthy approval times, especially if revisions are required after the first review and the documents have to be resubmitted for review at the next scheduled meeting.

It is very important to select an IRB properly. Some researchers have evaluated different IRBs and have found inconsistencies in both their review processes and their recommendations [1,7]; therefore, to prevent selection bias and to make the review process more efficient, it is advisable whenever possible to utilize one very experienced, single, qualified central IRB for the review of all documents. Occasionally, a local IRB will be preferred (e.g. for a single-center study) or required (some institutions, especially medical centers affiliated to academic institutions will not participate in any clinical research study that has not been approved by their own IRB). Either one is acceptable as long as such IRB has a clear understanding regarding the context of the study and the values of the community where the research is taking place.

When evaluating an IRB it helps to keep in mind the top IRB deficiencies identified by the CDER for the fiscal year 2009 [8] (Figure 19.1).

No research involving human subjects can commence without IRB approval except for research that involves the collection and analysis of existing information (e.g. records, biological and medical specimens) and where the subjects to which the information pertains cannot be identified.

Documents that require IRB approval include, but are not limited to, the clinical study protocol, study-specific informed consent, referral letters to be used during the study, investigator brochure (IB) and patient information brochure. The ICH-GCP guidelines describe the IB as "a

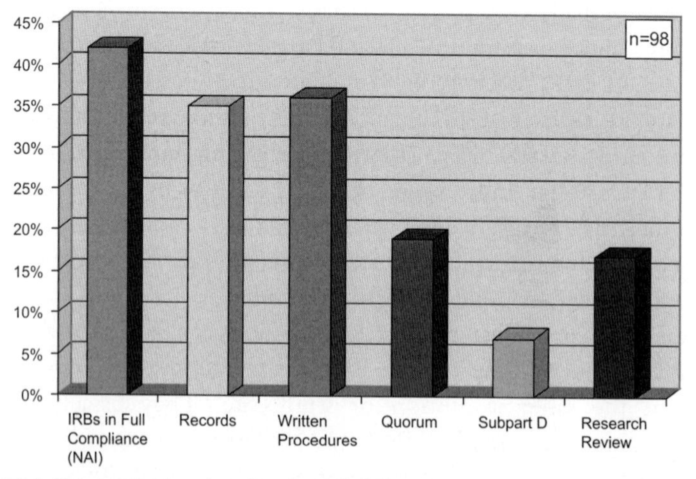

Figure 19.1 Top institutional review board deficiencies as per Center for Drug Evaluation and Research, fiscal year 2009. NAI: No Action Indicated *(With permission from Barnett International [8]).* (Please refer to color plate section)

compilation of the clinical and nonclinical data on the investigational product(s) that are relevant to the study of the product(s) in human subjects".

Approval should be obtained not only before the conduct of the clinical study, but also at any time any changes are made to the documents already approved by the IRB or if any new documents related to study-specific activities are generated.

The IRB will continue to review the study on an ongoing basis. The frequency with which the IRB chooses to receive updates is determined by the perceived risk to the subjects/patients; however, an annual report is expected throughout the duration of the study. Serious/life-threatening adverse events (SAEs) must also be reported to the IRB. During monitoring, ensuring that all SAE reports have been forwarded to the IRB is of utmost importance.

All IRBs have the authority to approve, require modifications (to secure approval), or disapprove research [9], and are subject to the rules and guidelines outlined by the US Department of Health and Human Services' Office for Human Research Protections (OHRP) and the FDA.

After completion of the study, a final report providing a summary of the final enrollment numbers, adverse event summaries and updates, as well as the conclusions of the study (as available) must be submitted to the IRB.

19.10.2. Other Committees

In some institutions, in addition to the IRB approval, the approval of other review committees is required. Examples of these committees are scientific review committees that focus on assessing the scientific merit of the study, feasibility committees that assess the feasibility of conducting the study from a cost, time, and effort perspective, protocol committees whose priority is study design, radiology committees looking at the radiology resources, and institutional biosafety committees that oversee all research involving biological materials and any other biohazards. As part of the site assessment, the sponsor should always determine which and how many committees will be involved in the review of a particular study and ensure that these reviews are appropriately reflected in the study timelines.

When conducting studies outside the USA, additional regulatory agencies and institutions may also be involved in the review process; in some countries this can be a considerably lengthy process.

19.10.3. Informed Consent Form

This is a document that ensures that any person participating as a subject in a clinical study (or the parent or tutor in the case of pediatric studies) fully understands the risks and benefits (if any) of participating in the study and that it is their decision to participate or not. The consent of the subject(s) must be documented by means of an ICF. The consent form explains the study and all procedures in detail and in a language that a layperson with at least a seventh grade level education can comprehend. Once the study participants have had time to review the consent form, discuss it with members of their family, and receive answers to any questions they may have, they can decide whether to participate in the clinical study or not. It is important that any prospective participant understands that if they choose not to participate there will be no repercussions or loss of benefits to which they may be otherwise entitled. If a participant agrees to participate, he or she will sign and date the consent form as acknowledgement that he or she understands all the information and is willing to participate in the study.

The ICF and the process of collecting the information is one of the areas that must be closely monitored not only because of the ethical implications but because, as per FDA inspections, deficiencies are common, especially in the following processes and sections:

- **Dates**: The monitors must pay attention to the date on which the consent form was signed and thoroughly review the source and the CRFs to ensure that no procedures were performed before the consent being signed.
- **Signatures and date:** Often the patient's (parent or tutor in case of a minor) or witness' signature (or thumb print in some instances) is missing.
- **Essential elements** of an informed consent are incorrect, such as:
  - the information is not appropriately communicated in a language that most lay people can understand
  - the consent does not state that the patient/subject is participating in a research study or why the study is being conducted

or missing, such as:

  - a statement identifying certain procedures as experimental or the investigational drug(s) as investigational [must clearly specify that the drug(s) have not yet been approved by the FDA]
  - the duration of the study

- the risks or benefits (if any) of participating
- a detailed list of potential side-effects and the likelihood of their occurring
- an explanation of alternative treatments available (if applicable) to treat the condition that qualified the patient/subject to participate in the clinical study
- detailed information regarding other clinical studies and treatment options including the option not to participate in the study
- a statement explaining that all efforts to maintain patient confidentiality will be undertaken but that, occasionally, some information may have to be shared with the sponsor, CRO, and/or regulatory authorities
- detailed explanation of the responsibility of the investigator and the sponsor if any injury were to occur to the patient/subject as a result of participating in the clinical study
- contact information for the IRB
- a section clearly explaining that the patient/subject's participation is completely voluntary and that if he or she were to choose not to participate there would be no loss of benefits to which the patient was otherwise entitled. Also, the patient/subject must be informed that he or she can withdraw from participating in the study at any time for any reason
- compliance with federal regulations 21 CFR part 5025.

No patient/subject should participate in a clinical study if he or she has not signed an ICF, except if the research involves subjects/patients in life-threatening situations where they may be unconscious and/or there is no time to find their authorized legal representative. For studies involving these patient populations, the IRB must approve an exception prior to the study start-up.

Special circumstances may arise such as the participation of patients/subjects with certain limitations such as illiteracy; this would not prevent a patient/subject from participating, but special provisions must be taken. The consent form will have to be read to them in presence of a witness and enough time should be given to the patient/subject to answer all questions and gain an understanding of what his or her participation will involve. The subject then may sign the form with a mark or fingerprint. This process must also be witnessed.

If a study involves special populations such as children, patients with mental disorders, elderly people, nursing home residents, and/or prisoners, special provisions will need to be in place. The requirements will not be discussed in detail in this chapter; however, when conducting research in these populations, they should be researched and considered.

When the study is to be conducted in multiple countries, it is imperative that the informed consent is translated to the corresponding language. Very often a reverse translation is also required to verify the accuracy of the translation.

19.11. STUDY START-UP

19.11.1. Investigator Meeting

Before initiating a clinical study, an investigator meeting is scheduled to ensure that the clinical study sites or institutions that will be responsible for conducting the study have an in-depth understanding of GCPs, the investigational drug, administration and accountability, the patient population being studied, the clinical study design, adverse event handling and reporting, appropriate specimen collection, handling and shipping, and all other study-related procedures. Study coordinators will also receive in-depth training on the appropriate manner in which to capture and record the information in the CRFs. Usually the PI and the study coordinator from each clinical site will attend this meeting and will be responsible for training the rest of their clinical staff as appropriate to ensure that all study activities are well understood and properly performed.

Unfortunately, investigator meetings will never achieve 100 percent site representation and therefore it is critical to ensure that appropriate training, such as that provided during the investigator meeting, is provided to those sites that were not able to attend. The initiation visit is the most appropriate time to do this.

19.11.2. Clinical Study Monitoring

The CRAs/CSMs should be qualified and experienced individuals responsible for the training of the study site as well as for managing the conduct of the study. The CRA/CSMs are the first line of defense. They are the first people the CSC at the site will contact if she or he has any questions and therefore they should be trained and highly familiar with the indication being studied (the monitor's qualifications must be documented), the drug or drugs that will be administered to the study participants, and the standard operating procedures for the qualification and selection of the investigator/clinical site, monitoring, adverse event reporting, and data management.

They should also be intimately familiar with the monitoring plan, describing the number and frequency of monitoring visits and the tasks that should be performed at each of the visits.

The number of CRA/CSMs will be mostly determined by the number of investigators and sites participating in the study and their location, as well as the complexity of the study. If a study is extremely complex and has a considerable number of CRFs that must be completed, it may require the CRA/CSM to remain at the site, even for a few patients, for quite a long time.

The obligation of the CRA/CSMs can be transferred to a CRO.

19.11.2.1. Monitoring Plan

The monitoring plan contains the written procedures for the monitoring of clinical studies as a guide for the monitors to follow as they review the data to ensure the quality of the study.

It must include a detailed plan regarding the number and frequency of monitoring visits, what will be monitored during the visits, and the duties and responsibilities of the monitor(s) at these visits. The determination of the extent of the monitoring is based on the objective, endpoints, design, complexity, study design (blinded versus open label) and number of subjects/patients. It should include a statement specifying the percentage of source data verification (SDV) that will be performed. SDV requires a comparison of the data collected in the source documents with the data collected in the CRFs to ensure that the transcription and data capture were accurate. Queries must be generated for each discrepancy.

When monitoring, attention to detail is critical to ensure that any errors or problems are detected early on and promptly corrected. Most common clinical site deficiencies are shown in Figures 19.2 and 19.3.

19.11.2.2. Site Initiation Visit

The site initiation visit is the last step prior to activating a clinical study site. This visit is a time for the sponsor's representative(s) to discuss and review, in detail, all aspects of the conduct of the clinical study, with an emphasis on evaluating the PI's and his or her staff's understanding of the protocol and their obligations. The initiation visit is scheduled several weeks before the date of the visit to ensure that all appropriate personnel can attend the meeting. Before the visit, the CRA/CSM should provide written confirmation of the date and scope of the planned visit to the PI and CSC; the letter will state who, if anyone, will accompany the CRA/CSM, which

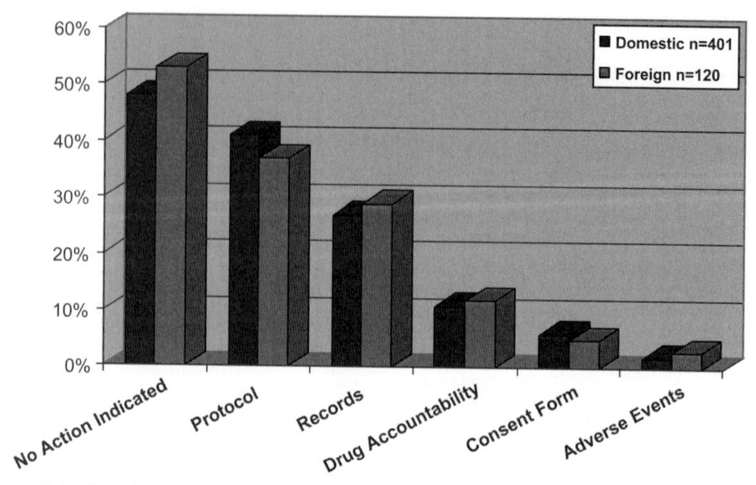

Figure 19.2 Top clinical investigator deficiencies as per Center for Drug Evaluation and Research Inspections, fiscal year 2009 *(With permission from Barnett International [8]).* (Please refer to color plate section)

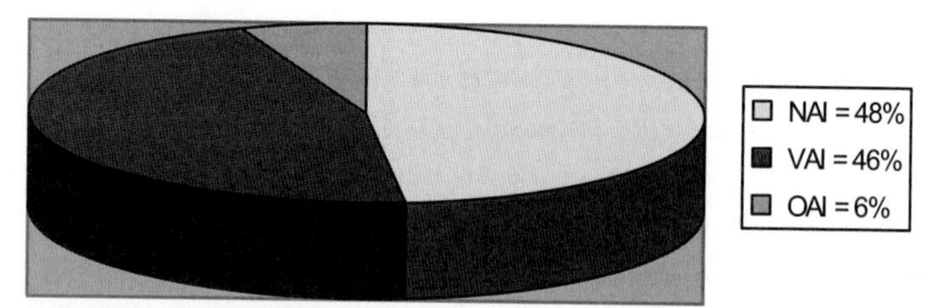

Figure 19.3 Clinical investigator inspections: final classification, fiscal year 2009. Total inspections with final classification = 417; NAI: No Action Indicated – sites in full compliance; VAI: Voluntary Action Indicated – sites found to have objectionable practices that do not represent major departures from regulations; OAI: Official Action Indicated – sites/investigators found to have objectionable practices or conditions that represent significant departures from regulations *(With permission from Barnett International [8]).* (Please refer to color plate section)

members of the staff are required to be present at the meeting, and what documents should be available for review. The objective of this meeting is to prepare the study staff to conduct the protocol as per GCP guidelines. At this time the CRA/CSM will ensure that all essential documents are at the site, that all approvals have been obtained, and that all logs and forms are in the study binder(s), will review the protocol and responsibilities of the staff, assess the facilities, and ensure they remain adequate, and will discuss the monitoring visits and any other study-specific requirements to ensure the staff and site are ready to initiate the study.

Essential documents (documents required as per ICH-GCP E6 are marked with an asterisk *) that must be on file at the site are as follows:

- CDA (signed and dated)
- finalized budget
- IB and an acknowledgment of receipt*
- the protocol and any protocol amendments approved at the time of the visit must be available and signed as an acknowledgment that they have been received*, reviewed, and understood; the IRB/IEC approval and date of approval* must also be on file
- the final ICF* (and any other written material to be provided to the subjects) and the IRB/IEC approval and date of approval; during the visit the CRA/CSM will discuss the informed consent with the study staff
- sample CRFs*, paper or electronic (eCRF): the information to be captured in the CRFs/eCRFs, proper method to make corrections on them and the query/clarification process will be reviewed; for eCRFs security/passwords and any other access information will also be discussed
- any letters/advertisements (television, radio, fliers) that will be used for subject recruitment must have IRB/IEC approval*
- any IRB/IEC correspondence as well as an IRB/IEC membership list and a statement that the IRB/IEC operates according to GCPs*
- sample source documents (if applicable)
- study manual
- diary cards and any other documents used during study conduct, if applicable
- curriculum vitae (CV) of the PI and subinvestigator(s), signed and dated*
- a CV on file for any other study site staff that will participate in the conduct of the clinical study (CRC, pharmacist, head of laboratory, etc.).
- financial disclosures* and licenses for the members of the clinical study staff, as applicable
- staff training records
- sponsor contact details sheet: the site staff should have contact information for the sponsor's and/or CRO's members of the study team, as appropriate
- site staff and site visit log (these are usually started at the time of the site initiation visit; ensure everybody who participated in the site initiation visit signs and dates the log)

- screening and recruitment log
- SAE forms and instructions for completing them
- CRF transmittal forms (if applicable)
- instructions for handling the investigational drug and trial-related materials*
- copy of drug label and acknowledgment of receipt*
- drug accountability log
- drug return form
- certificate of analysis for the drug*
- emergency code break instructions*
- drug administrator manual
- laboratory accreditation/certification*
- laboratory normal ranges*
- sample shipment log
- investigator/institutional final contract (must be signed and dated)*
- sponsor insurance/indemnity document*
- Form FDA 1572*: No investigator can participate in a clinical study until she or he provides a completed Form FDA 1572 to the Sponsor. The purposes of the 1572 are to provide the sponsor with all pertinent information to support the qualifications (education and experience) of the investigator and any member of the staff who will be involved with the conduct of the study (or significantly contribute to the data being collected), information about the physical location where the study will be conducted as well as the location of any other institution/group participating in the study (laboratory, additional clinics). This information will allow the sponsor to ascertain and document that the study staff is qualified and the facilities are adequate for the study to be conducted at that location. In addition, the Form 1572 serves to inform the investigator about her or his obligations and to obtain the investigator's signature signifying her or his commitment to follow the Code of Federal Regulations when conducting the study.
- Any clinical site/investigator conducting a study with a controlled substance must obtain a Drug Enforcement Administration (DEA) registration number. According to DEA regulation, some research activities may need to be approved separately from practice activities. All clinical staff who will have access to controlled substances must have a clean background check and no history of narcotic or recreational drug use. Controlled substances listed in Schedules I–V must be stored in a securely locked cabinet within a locked room.

- Health Insurance Portability and Accountability Act (HIPAA) Form (and Medical Records Release Form): In 1996, Congress passed the HIPAA to protect personal health information. By signing the HIPAA form, the subject/patient grants written permission to a specific entity (investigator, CRO, sponsor, and/or regulatory authority) to use his or her personal health information for the purpose of the clinical research study. If at any time it becomes necessary to use what is considered protected health information, prior to any party being allowed access to it, IRB approval is required.
- Standard operating procedures (SOPs): For studies conducted under an IND, all SOPs and manuals describing procedures performed during the conduct of the study must be retained.

During this visit, the CRA/CSM must go through all executional aspects of the study. Some of the issues that should be reviewed with the study staff include (but are not limited to):

- Study team roles and responsibilities: The CRA/CSM shall ensure that every member of the clinical study staff is adequately trained and has an adequate understanding of the protocol, the investigational drugs, and their study-related roles and responsibilities. The CRA/CSM should also emphasize the importance of the PI seeing the subjects/patients on a regular basis and not delegating *all* activities at every visit to the CRC or other clinical study staff; the time the PI spends with the subject/patient seems to have a significant impact on subject/patient retention [10].
- Local IRB/IEC requirements (if applicable):
    - submission dates: if available, a copy of the scheduled meeting dates
    - timelines: average time for approval of items requiring full committee review or expedite review
- Routine practice of patient referral
- Process of obtaining and documenting informed consent
- Study enrollment period and timelines
- Continued ability of the site and staff to conduct the study: Discuss new (competing) studies at the site as well as any staff, workload, process, budget, and/or facility changes since the prestudy/qualification visit
- Investigators' responsibilities regarding:
    - IRB/IEC requirements
    - audits and regulatory inspections
- Adverse event and SAE reporting procedures
- Investigational drug emergency unblinding procedures (if applicable)

- Use of rescue medication: How is the rescue medication going to be dispensed? Will a prescription be required? How will the usage be recorded?
- If using an electronic IVRS or IWS, discuss patient and clinical study staff instructions for system use; it is important to discuss also procedures in place for the staff to detect promptly any omissions/errors in data input and how to address them
- Drug administration, handling and accountability: If there is a need for an unblinded drug administrator, it is important that the CRA/CSM reviews the procedures with them
- Publications policy
- Whether the site/investigator has been audited by the FDA or any other regulatory agency since the prestudy/qualification visit
- Drug storage and drug security requirements will be reviewed. The CRA/CSM will also inquire as to whether a person has been identified as responsible for drug storage/dispensation and who that individual is
- Drug accountability, including shipment, dispensing, destruction and return procedures
- Treatment and study assessment schedules and procedures
- Emergency code-break procedures
- If a central laboratory will be used for the study, central laboratory requirements and sample transport procedures will be discussed
- Procedures for source data collection, data entry into the CRFs/eCRFs, diary cards, IVRS/IWRS, etc. The schedule and amount (percentage) of SDV as specified in the clinical monitoring plan will be presented; SDV can range up to 100 percent on-site verification.
- Process for query resolution and timing expectations
- Archiving and document retention requirements.

A critical task for the CRA/CSM at this visit is to set up the clinical study binder. The study binder must be set up before the first patient is enrolled at the clinical study site.

Finally, the CRA/CSM will discuss the expected frequency of the monitoring visits, the need to have direct access to the documentation and the importance of having the investigator and study staff available to answer questions at the time of the visits.

In addition, if the PI and/or the clinical study coordinator were unable to attend the investigator meeting, an overview of any other critical topics covered during the meeting that were not previously covered (e.g. an overview of the drug, data from previous studies) will be provided.

At the end of the visit, a record of the events, the topics covered, and a list of all the members of the staff who attended will be captured and filed.

The CRA/CSM is expected to complete a site initiation visit report which will be sent to the sponsor and shared with the CSC and the PI. Both the CSC and the PI will sign and date the report, which will be filed in the investigator site file.

19.12. CLINICAL STUDY BINDER

The clinical study binder (study binder) is a way of identifying, maintaining, and organizing all the documents essential for the conduct of a clinical study in one location. Having all the documents in one place minimizes the chance for errors or omissions during the study conduct and facilitates adherence to the GCP and protection of human subjects' standards.

A study binder must not contain any subject/patient-specific information; this type of information must be kept in a different binder (one for each subject/patient).

The documents to be collected in a study binder should suffice to demonstrate compliance with the regulations, the ethics committees, and the protocol.

The study binder, as well as all the study data, must be retained for a minimum of two years after the marketing application for the indication for which the drug is being investigated is approved. If no application is filed or if the application was not approved for the requested indication, the records must be retained for a minimum of two years after the investigation is discontinued and the FDA is notified (21 CFR 312.62).

19.12.1. Contents of the Study Binder

- List of contacts: A list of the contact details (e-mail, telephone, pager, mobile, fax number, mailing addresses and, if available, hours/dates of availability) for key people involved with the conduct of the study (and their back-ups if applicable):
  - main sponsor contacts
  - members of the clinical study staff
  - pharmacy contact
  - laboratory representative
  - contacts for other vendors
  - drug shipping and randomization representative
  - emergency contact details.

- CDA, CTA, budget and billing statements, and study subject/patient reimbursement records, if any
- Sponsor correspondence
- Telephone log
- Clinical study protocol:
  - A copy of the most recent version of the signed protocol. If there have been any amendments, they and all out-of-date versions of the protocol should also be filed.
- ICF:
  - A copy of the most recent version of the approved ICF and, as for the protocol, copies of all prior versions, if applicable. If the ICF has been translated to other languages, a copy of one ICF in each language will be included.
- IB
- Patient/subject documents:
  - Any documents that at any time during the study will be read by the patients/subjects must be filed. These include any patient questionnaires, diary cards, recruitment materials, etc. If the diaries/patient questionnaires are in electronic format, it is advisable to print them and place a hard copy of them in the file. Out-of-date versions of these documents should also be kept in the binder.
  - Screen/screen failure and enrollment logs (records of all the subjects/patients who are screened and enrolled into the study and those who upon screening did not meet the inclusion/exclusion criteria for enrollment and were therefore considered screen failures)
  - Subject visit schedule log
  - Signed ICFs (and copies of all approved versions).
- CRFs:
  - A blank copy of the most recent CRFs (and any out-of-date versions if applicable) as well as of any other data collection tools.
- Operations (procedures) manual:
  - This section will contain reference information regarding procedures to be performed during the study. These can include:
    - schedule of events
    - a flowchart illustrating the steps to follow when performing the study assessments (laboratory sample collection), and details on the processing, packaging, labeling, storage, and transport of samples
    - annotated forms

    - how to grade the severity of adverse events (if not detailed in the protocol)
    - instructions as to how to complete the CRFs
    - how to resolve data queries
  - equipment calibration records
  - shipping procedures.
- Ethics documentation:
  - human research ethics committee approval
  - IRB submission forms, outcome letters and correspondence
  - IRB approvals and renewals for:
    - protocol
    - amendments
    - informed consent
    - advertisements
    - payments to subjects/patients for participation.
- Regulatory documentation:
  - Form 1572
  - copy of the IND application
  - FDA correspondence
  - annual reports.
- Study staff signature and delegation log:
  - a quick reference detailing who is responsible for what throughout the study
  - CVs, licenses, financial disclosures and applicable certifications
  - delegation of responsibility log.
- Staff qualifications and training:
  - documentation supporting the education, training, and experience of the clinical study staff
  - any study-specific documents generated for training of the study staff.
- Laboratory documentation:
  - name of laboratory and contact information
  - certifications
  - list of test with normal values and ranges
  - sample collection, labeling, storing, and shipping procedures
  - biological specimen log
  - shipping records (if central laboratory is being utilized)
  - temperature logs
  - laboratory procedure manual.
- Data and safety monitoring:

- monitoring log
- monitoring visit summary
- monitoring reports
- correspondence
- if a DSMB has been established for the study, the charter and any correspondence between the site and the DSMB would need to be filed in this section.

- Drug accountability:
  - shipping records
  - drug accountability log
  - drug disposal records
  - temperature logs
  - documentation of investigational product returned to sponsor (with receipt) or drug disposal.
- Emergency unblinding log
- SAEs and any related correspondence
- IND safety reports and IRB acknowledgment letter
- Internal correspondence
- Agreements (e.g. publication agreements)
- Study reports:
  - study interim report (if applicable)
  - publications
  - final study report (and proof that it was sent to the IRB)
  - adverse event logs and reports
  - protocol deviation/violation log.
- Publications, presentations, and/or manuscripts arising from the study.

19.13. SUBJECT/PATIENT RECRUITMENT

Subject/patient identification and recruitment can become the most challenging aspect of a clinical study, especially for studies in Phase II, III, and IV. At this point, the sites should already have an estimate of the number of patients who could be eligible for participation in the study and a plan regarding how these patients will be reached and what will be the incentive for them to want to participate. The plan should be implemented as soon as the site is initiated. Recruitment goals should also be part of the plan, as well as alternative venues to pursue if the initial recruitment goals are not being met.

Recruitment for Phase I studies with healthy volunteers is somewhat different than recruitment for patients suffering from the disease or condition being studied. Recruiting healthy subjects is usually done via mass distribution media such as radio and television or, when working with an experienced Phase I unit, by contacting the volunteers who have signed up at the site.

- The first step in recruitment for a clinical study in patients with the disease/condition for which the investigational product is believed to be efficacious is to assess how many patients currently in the care of the PI have that disease or condition.
- The second step is for the PI to discuss this study with colleagues with nearby practices that may be interested in participating and ask them to refer the patients to the clinical study site.
- Approach other areas (e.g. emergency room) to discuss the protocol and ask them to refer patients to the PI.
- Contact support group organizations.
- Mass distribution (fliers, radio, television, etc.).

19.13.1. Screening

If upon review of medical records and discussion with the patient, the clinical study staff believes the patient could qualify for participation in the clinical study, the patient must undergo screening procedures. It is always preferable to begin with non-invasive procedures; these may disqualify a few of the prospective participants and avoid putting the patient through unnecessary procedures. Efficiency during the screening process improves enrollment and retention.

19.13.2. Subject/Patient Retention

Identifying and recruiting study participants is not enough; once they are taking part in the study it is important to encourage them to remain in the study until study completion [provided there are no major indications for the patient to withdraw prior to completing participation in the study (e.g. adverse event)] and to comply with all study procedures/visits. This can be particularly difficult especially in long-term studies or in studies where the patients' quality of life could be affected by the number or type of procedures they must undergo. To improve retention and compliance it is very important for the PI to play a very active role with the patients; they must be reachable and available and demonstrate interest in the patient's well-being. For the subject/patient the fact that he or she is participating in a clinical study does not change the fact that he or she is a patient and still expects the best possible medical care by his or her treating physician.

It is very important to educate the patient regarding their disease, the study drug, and the study itself; if patients understand the rationale behind the study and the procedures and assessments they are being asked to undertake, the chances of their continuing to participate in the study increase. Frequent reminders in the form of telephone calls, e-mail messages, and/or letters are also useful to keep the patient involved and aware of the study.

19.14. SITE MONITORING VISITS

The sponsor is responsible for monitoring the progress of all clinical investigations being conducted under its IND.

Throughout the period the study is being conducted, the CRA/CRMs will make regular monitoring visits (as described in the monitoring plan) to the sites to ensure:

- that the rights and well-being of the subjects/patients are being protected
- that the data collected are accurate, complete, and verifiable from source documents
- the conduct of the trial is in compliance with the currently approved protocol/amendments, with GCPs and with the applicable regulatory requirements
- the safety of the investigational drug being studied.

During monitoring visits, the CRA/CSMs must insist on:

- having access to the original records
- strict protocol adherence
- having the PI set aside some time during the visit to discuss visit findings.

Critical activities that the CRA/CSM must perform to ensure full compliance include:

- **Recruitment status/protocol adherence**:
    - Assess whether there are any recruitment issues (e.g. sites that have been active for a long time without enrolling subjects/patients).
    - Ensure that all the subjects/patients that have been enrolled since the previous visit are indeed eligible for the study.
    - Confirm that all informed consents were obtained and signed and that the HIPAA forms were signed by all patients (or their legal representative) prior to initiating any study-related activities.
    - Ensure that the study site performs subject compliance tracking on an ongoing basis and has a process in place to address lack of compliance.

  - Assess whether there have been any protocol violations since the last visit and if so, that they were properly documented and the appropriate action(s) taken (e.g. discontinue a patient's participation in the study).
  - Discuss the trial progress with the investigator and the staff.
  - Discuss any protocol amendments if available.
- **CRF review/source data verification**:
  - Review all CRFs to ensure they are being completed correctly and promptly, that if any corrections were made the erroneous data were crossed over with one single line, dated, and initialed by the person making the correction, and that they were signed. Perform source data verification.
  - Review the list of withdrawals and dropouts.
  - Confirm that the site is complying with the protocol.
  - Ensure that all source documents are adequate and available for review.
  - Ensure that all (or most) of the data queries have been reviewed and resolved.
- **Study medication/drug accountability**:
  - Inspect the investigational drug storage area to ensure that the storage conditions are still adequate. The investigational products should be stored in a locked cabinet or closet in order to restrict access to only designated members of the clinical site staff. When the study is conducted in a medical center/hospital, often the institution will require that the pharmacy stores and dispenses all medications. If the drug has special storage conditions (set temperature or humidity), the CRA/CSM will review the temperature and or humidity logs to ensure there have been no deviations outside the allowed windows for each of the conditions.
  - Review the shipment records to ensure they were properly documented.
  - Ensure that the drug accountability has been accurately performed and that there are enough supplies at the study site to continue the study.
  - Look at the expiry date to ensure that the investigational drug at the study site can still be utilized.
  - Make sure that the subjects/patients were given proper instructions on how properly to use, handle, and store the investigational drug and how to return unused supplies.

- Make certain that the investigational drug is being supplied only to the subjects/patients participating in the clinical study and at the dose specified in the protocol.
- Determine whether any investigational drugs have been returned for destruction or destroyed at the site and, if so, whether or not proper destruction procedures were performed.
- Review the pharmacy file to ensure it is still accurate.
- Review the code break/treatment and ensure it has not been tampered with.
- A drug accountability record should be in place to facilitate tracking of the investigational product. A drug accountability record has three parts:
  - initial inventory of supplies received
  - chronological record of supply dispensation and returns
  - final inventory of supplies returned to the sponsor or destroyed at the time of study close-out.

- **Adverse events/serious adverse events**:
  - Confirm that all adverse events are captured in the CRFs.
  - Ensure that adverse events are tracked and given proper and timely follow-up.
  - Ensure that if any SAEs have occurred since the last visit that they were all properly captured and documented. It is also critical to confirm that the SAEs were forwarded to the IRB and reported to regulatory authorities as appropriate.
- **Laboratory samples**
- **Biological samples**
- **Clinical study binder**:
  - The CRA/CSM must ensure that the study binder is organized in such a fashion that documents can easily be identified and that the filing process can be easily grasped by someone who is not directly working on or familiar with the study. Documents should be filed in a consistent manner, usually with the most recent documents filed on top of each section.
  - The study binder must be kept current and up to date.
  - If possible, ensure that the protocol unique identifier as well as the date is on all study documents.
  - Ensure that all changes and corrections have been made by crossing the error out with one single line, putting the initials of the person that made the change and adding the date of when it was made; then

the corrected information can be clearly written down. Never should white-out be used.
- Remove all irrelevant papers from the binder.
- Encourage the CSC to file all documents as soon as they are ready for filing.
- In the case of electronic records they must be backed up frequently. Occasionally, critical documents are printed and filed in the study binder.

- **Other duties**.

A few years ago, David A. Lepay, MD, PhD, Senior Advisor for the FDA, conducted an analysis of some notorious cases of monitoring failure. The most frequent issues were:

- Serious misconduct was not reported by sponsors.
- Seriously non-compliant investigators were used by multiple sponsors on multiple trials, impacting many applications by many sponsors.
- The majority of objectionable observations could have been detectable by adequate monitoring.

These findings highlight the importance of adequate monitoring and prompt reporting of any serious issues to the appropriate authorities.

19.15. END OF STUDY

Once the targeted number of enrolled patients has completed participation in the study or if the study is prematurely terminated, a few activities remain to be performed.

The CRA/CSM will visit each of the clinical study sites for a close-out/end-of-study visit. At the time of this visit, the CRA/CSM will:

- Meet with the PI and discuss his or her final responsibilities, which include the investigator's final report. The investigator shall prepare a final study report which he or she will provide to the sponsor shortly after completion of his or her participation in the investigation. A copy may also need to be provided to the regulatory agency (or agencies) as required by the applicable regulatory requirements, regardless of whether or not the trial is completed or prematurely terminated.
- Ensure that the follow-up of all adverse events and SAEs has been completed. The sponsor will continue to follow-up on all unresolved adverse events.
- Ensure that all data clarifications and queries have been resolved.

- Confirm that the subject/patient identification log has been completed and filed.
- Establish that all completed CRFs are accounted for and available and that unused CRFs have been returned or properly discarded.
- Ascertain that the patient's informed consent documents, including HIPAA forms, have been completed, and are present and filed.
- Confirm that the final cumulative query report has been signed by the investigator.
- Ensure that the full drug accountability has been performed and that all drug supplies reconcile with the records; also confirm that the entire study drug supply left behind has been returned to the sponsor or that the destruction records have been properly filed.
- Collect all code break/treatment allocation envelopes.
- Make sure that the pharmacy file is complete.
- Ensure that all local laboratory and biological samples have been collected and documented appropriately, that all laboratory supplies have been returned or discarded, and that the laboratory has been informed that the study has ended.
- Ascertain that all essential documents in the study binder were reviewed and found to be complete.
- If any equipment was provided to the site, ratify that it has been retrieved.
- Corroborate that all study payments have been made; if they have not, arrangements for final payments must be discussed with the investigator.
- Confirm that all essential study documents have been archived as per ICH-GCP requirements and remind the site of their retention responsibilities.
- Discuss transfer of responsibilities for all documentation in the event the investigator leaves the site and remind the site that, if documents were to be transferred, the sponsor must be immediately notified.
- Establish that the IRB has been notified that the study had ended and has received a copy of the final summary report.
- Ensure that all outstanding actions have been completed.

19.15.1. Clinical Study Report

At the end of the study, and regardless of whether the study was completed or prematurely terminated, the sponsor will write a clinical study report [11]. This report must integrate a detailed description of the study and how it was conducted, as well as all clinical (safety and efficacy) and statistical analyses of the data collected during the conduct of the study and a discussion and overall conclusions into a single report. Tables and figures (incorporated into the main text or at the end of the text) and the following appendices, are also part of the report:

- protocol
- sample CRFs
- investigator-related information
- investigational product information (including active control/ comparators)
- technical statistical documentation
- related publications
- patient data listings
- technical statistical details.

19.15.2. Drug Approval Process

When all the studies have been completed, the sponsor makes a formal application to the FDA to approve a new drug for use in the USA by submitting an NDA. All the available information including results and analysis from all animal and human studies, as well as a description of how the drug was manufactured, is submitted as part of the NDA. The information submitted should be enough for the FDA to be able to assess the purity and integrity of the drug, confirm that the manufacturing methods are acceptable, and determine whether the drug is safe and effective and whether its benefits outweigh its risks.

When the FDA's review is complete, the Agency will send a letter to the sponsor either granting approval or stating that the drug is "approvable" (if the sponsor provides additional information, makes some changes, or further explains certain areas) or not approvable (e.g. lack of efficacy or a disturbing safety signal).

Before 1992, the review times for applications were extremely long owing to personnel limitations at the FDA. To address this issue, in 1992, Congress passed the Prescription Drug User Fee Act (PDUFA), which authorizes the FDA to collect user fees for reviewing and processing applications for the approval of certain human drug and biological products from the companies that were developing them. These fees were meant to allow CDER to increase the number of people to review the applications and to reduce the review time without compromising quality. This act has undergone reviews and changes since; it is currently in its fourth iteration. With the establishment of the PDUFA, PDUFA dates came along. A PDUFA date is a target date by which the FDA commits to complete the review of a drug application; this date is usually 10 months after the sponsor submits the information to the FDA, except for cases in which a drug warrants priority review (e.g. drugs intended to treat serious diseases or conditions for which no treatment is currently available), in which case the FDA has six months to complete the review of the application. These dates are not set in stone; however, in some instances the Agency may complete the review sooner, and in others, later than the PDUFA date [12].

19.16. AUDITS AND INSPECTIONS

The types of audits or inspections a site participating in clinical study or a CRO can undergo are in-house audits performed or contracted by the sponsor and external audits, usually in the form of FDA inspections.

Internal audits are sometimes triggered by the high number of enrolled patients. Sponsors usually audit sites when they reach a predetermined number of patients to ensure the accuracy of the data. This is even more important in the case of registration studies when the results of the study will be a critical part of the NDA application.

The FDA can decide to perform an inspection to monitor and review different aspects of the conduct and reporting of any clinical study under an IND at any time. The Agency can inspect the clinical site(s), the CRO if one is being used and/or the sponsor. During these visits the inspectors will review the documents (and electronic records if applicable) collected during the conduct of the study to ensure that the research complies with the regulations. The visits can be preannounced (if so, the FDA usually gives three to 10 days' advance notice for routine audits) or the inspector may choose to show up without prior notice. There are two types of inspection, for routine surveillance or "for cause". The latter are the result of complaints submitted to the Agency. The majority of inspections are for routine surveillance; for the fiscal year 2008, CDER performed 449 inspections, of which 81 percent of those were routine and 19 percent were complaint related [8].

19.16.1. Clinical Site Inspection

If the inspection will be of one of the clinical study sites, and if the FDA gives advanced notice of this visit, often the sponsor will send a representative to the site to aid with the preparation of all study-related documents/records. It is important for the CRA/CSMs to ensure that all the study documents are saved and properly organized so they can easily be found during an FDA inspection.

The FDA will want to see the paper trail of the study, starting with the Form FDA 1572 and the investigator agreement. The documents and records that the FDA inspects can include the protocol and its amendments, if any, the investigator agreements and financial disclosures, organizational charts, correspondence, vendor agreements, investigational drug accountability records, monitoring plan and reports, CRA/CSM and clinical study staff qualification and training records, patient/subject records, adverse event reports, data listings, and SOPs.

The inspectors may also want to inspect the facilities, including the drug storage area.

The most common investigator deficiencies are: failure to keep proper study records, failure to follow the protocol, informed consent process not properly documented, failure to report and/or follow-up on adverse events (subject safety), and some instances of fraudulent research (false data).

If we look specifically at the final classification of clinical investigator deficiencies where official action was indicated (Figure 19.4), we can see that record-keeping and protocol deficiencies constitute more than 80 percent of the problem.

Other deficiencies often found were failure to inform the IRB of any significant change in a prompt manner, drug accountability, and necessary equipment not available.

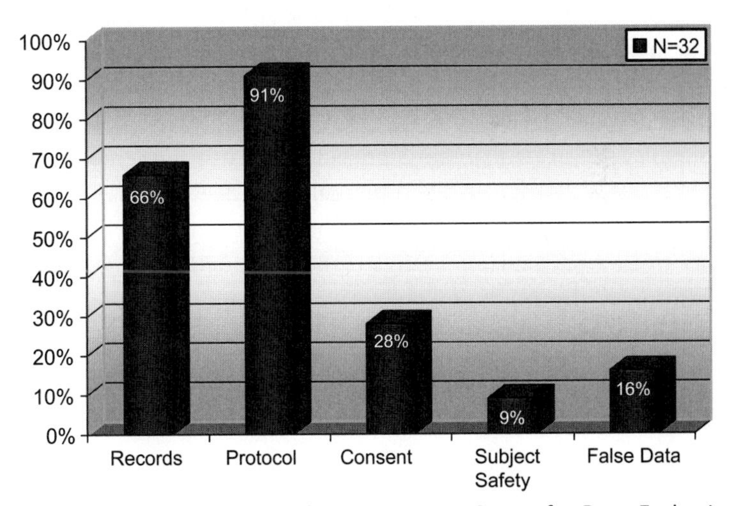

Figure 19.4 Clinical investigator deficiencies as per Center for Drug Evaluation and Research Inspections, fiscal year 2009: final classification of Official Action Indicated (OAI) *(With permission from Barnett International [8]).* (Please refer to color plate section)

At the end of the visit, the FDA inspector will provide a notice of inspection (Form FDA 482) and meet the assigned representative and provide a review of the findings. If the inspector does not provide a review, the clinical site's representative, CRO or sponsor can request it. If there were any significant deviations or violations, the FDA representative will issue a Form FDA 483. The FDA reviews the deviations and violations on the 483 and assigns one of three classifications: NAI (No Action Indicated), meaning they determined there were no objectionable conditions or findings, VAI (Voluntary Action Indicated), where they found objectionable conditions but not to the point that regulatory action was required, and OAI (Official Action Indicated), where regulatory action is recommended. If the site or sponsor receives a 483, the FDA recommends that an answer should be provided.

When selecting a research site for participation in a clinical study, the sponsor will always inquire whether the site has received a Form FDA 483 at any time; if the deviation(s) or violation(s) are significant, an investigator could be disqualified from participating in other studies.

If the FDA proves fraudulent research or other serious offenses, the investigator can have his or her involvement restricted, can be prohibited from participating in further clinical studies, or can even be debarred. Lists of restricted, disqualified, and debarred investigators are available from the FDA.

19.16.2. Sponsor/CRO Inspection

The Agency can also inspect the sponsor or the CRO. According to the FDA, the most common sponsor deficiencies are: inadequate monitoring, failure to secure investigator compliance, failure to submit progress reports, inadequate adverse event reporting and analysis, inadequate drug/device accountability, failure to obtain signed investigator agreement, failure to obtain FDA or IRB approval, and unqualified CRA/CSMs.

If serious problems are found they could result in the FDA demanding that data from a particular investigator be discarded. Sponsor deficiencies could also delay the NDA review and approval process, or even result in a non-approval of the NDA.

REFERENCES

[1] Zimmerman JF. The Belmont Report: an ethical framework for protecting research subjects. The Monitor, Publication of the Association of Clinical Research Professionals 1997.

[2] Gold JL, Dewa CS. Institutional review boards and multisite studies in health services research: is there a better way? Health Services Research 2005;40:291–307.

[3] Getz K. The heavy burden of protocol design – more complex and demanding protocols are hurting clinical trial performance and success. Applied Clinical Trials 2008 (online).
[4] Getz KA, Wenger J, Campo RA, Seguine ES, Kaitin KI. Assessing the impact of protocol design changes on clinical trial performance. American Journal of Therapeutics 2008;15:450–7.
[5] Code of Federal Regulations. Title 21 – Food and Drugs. Subpart B – Investigational New Drug Application 2009. 312.20 Requirement for an IND.
[6] Code of Federal Regulations. Title 21 – Food and Drugs. Subpart C – Administrative Actions 2009. 312.40 General requirements for use of an investigational new drug in a clinical investigation.
[7] Schulz H. The central laboratory – a partner for multinational clinical trials. Clinical Researcher 2001;1(5):1–14.
[8] Beecher HK. Ethics and clinical research. New England Journal of Medicine 1966;274:1354–60.
[9] US Food and Drug Administration. Center for Drug Evaluation and Research. Institutional Review Boards (IRBs) and Protection of Human Subjects in Clinical Trials, http://www.fda.gov/AboutFDA/CentersOffices/CDER /ucm164171.htm; 2008.
[10] Greene SM, Geiger AM. A review finds that multicenter studies face substantial challenges but strategies exist to achieve institutional review board approval. Journal of Clinical Epidemiology 2006;59:784–90.
[11] Mathiew MP. Section 21: GCP Compliance Statistics and Trends (Data on US FDA Inspections) in Good Clinical Practice: A Question & Answer Reference Guide. Barnett Educational Services 2009.
[12] Moench E. The partially involved principal investigator. Applied Clinical Trials 2002;1:30.
[13] Structure and Content of Clinical Study Reports. Guideline for Industry. ICHE3, www.fda.gov/downloads/Regulatory Information/Guidances/UCM129456.pdf; 1996.
[14] Dombrowski C. PDUFA old school: industry wants user fees to get back to basics. The Pink Sheet 2010. June 21, 28.

CHAPTER 20

Data Collection, Data Management, and Electronic Data Capture

Margaret Ann Snowden* and Yaw Asare-Aboagye**
*Biostatistics and Data Management, Aeras, Rockville, MD, USA
**Clinical Trial Services, United Therapeutics, Durnham, NC, USA

Contents

20.1. INTRODUCTION

Many of the best practices for data management in global clinical trials mirror those of local clinical trials. However, the growth of global clinical trials and the growth of electronic data capture (EDC) have been complementary, as one has enabled the expansion and advancement of the other. Globalization of clinical trials has meant that the ability to process data without traditional shuffling of paper has become even more important from a practical standpoint. This is particularly true as not only the clinical sites but also the data processing centers have become globalized, with many companies setting up data processing centers in India and other developing nations. Globalization of the studies has accelerated the adoption of EDC.

Global Clinical Trials
ISBN 978-0-12-381537-8, Doi:10.1016/B978-0-12-381537-8.10020-2

Similarly, the availability of EDC has made international trials less cumbersome, and has made some studies possible that previously may have presented logistical challenges. EDC is an excellent way to reduce data errors, cleaning time, and data processing costs, and the technological power of EDC systems continues to grow. EDC solutions are partnering with other systems to produce integrated options: EDC and interactive voice response systems (IVRS) systems, EDC and clinical trial management systems (CTMS), EDC and electronic health records (EHR). The sponsor or contract research organization (CRO) selecting an EDC system has a range of options; however, increases in options will lead to an increase in the amount of time required by the clinical trial sponsor to evaluate these options.

Most of this chapter will therefore focus on the evaluation of EDC systems.

There are a few important considerations unique to global clinical trials that may not be pertinent to local trials. The first is privacy. Privacy laws in different jurisdictions differ and in some, such as the European Union (EU), the restrictions are very stringent. As the clinical data are collected and analyzed, it is very important to respect the privacy laws of each jurisdiction. A good way to address privacy concerns is to adopt the strictest privacy law, which currently restricts the collection of identifiable information such as patient names, initials, and date of birth, and build a data collection policy around it.

Another important consideration is the difficulties with EDC in developing countries where Internet access is poor or where the electricity supply is not very reliable. Fortunately, Internet access in the areas where one is likely to run a clinical trial has improved considerably over the last few years. Several countries now have wireless carriers that are capable of providing 3G access, comparable to the USA and other major countries, so this gap is likely to be temporary. In areas where Internet access is poor, however, a workaround will need to be devised. One workaround is to revert to the traditional paper-based system. Another alternative is a fax-based system, or a pseudo-fax system where case report forms (CRFs) are scanned in and batch e-mailed to the data processing center.

In areas where the electricity supply is unreliable, serious consideration should be given to providing laptops and wireless modems to the clinical trial sites.

The use of EDC itself can pose several challenges to the global study team. It will affect several study processes including study start-up, investigator site set-up, and staff training. Many roles may have to be redefined to ensure successful adoption of the system and technical support should be readily available to the study team. The team should be prepared to manage changes at sites that are using EDC for the first time and sufficient time should be given to extensive testing of available technology (including broadband and wireless Internet access) and provisioning of computers to sites, where necessary.

20.2. THE ELECTRONIC DATA CAPTURE EVALUATION TEAM

The first step in EDC implementation is the planning. Timing of the initial discussions should occur as early in the development process as possible. For example, in a traditional approach, a company may have used paper CRFs for data collection in a series of Phase I studies and would like to consider an EDC system for larger scale data collection in Phases II and III. If a company is already emerging from Phase I into Phase II studies, it may be too late to implement a new system until a second Phase II. It is not possible to delay a decision to move to EDC until the start of Phase II. Starting the decision process for exploring an EDC option two years prior to the implementation of the first Phase II study allows sufficient time to review options and make initial determinations of cost.

20.2.1. Building the Team

Building the EDC evaluation team is the first step in the assessment process. The decision to explore the use of an EDC system should be performed by a multidisciplinary management-level and functional-level team with representation from clinical, biostatistics and data management, clinical operations, and information technology (IT), with additional members from quality assurance (QA), site project management, in-house project management, regulatory and finance departments, as well as from clinical sites (Table 20.1). Team members should be familiar with the clinical trials

Table 20.1 Electronic data capture evaluation team

Data managers*	Site monitors *
Database developers*	Clinical operations*
Biostatisticians*	Project management*
SAS programmers	IT support*
Quality assurance	Contracts and finance
Regulatory	Clinical site

* Core team functions.

process and the clinical development plan. A team leader (an experienced senior data management or a database developer manager) should be appointed to the project. The team leader will serve as the point of contact for all vendor interactions, facilitate minutes, and provide support for all aspects of the vendor process. The team leader should also provide input into selection of the core team to his or her management, and will ultimately present the evaluations to drive final decision-making and vendor selection.

When the team convenes for the first time, objectives and decision points around selection of an EDC option for the company should be outlined and discussed. A common understanding of objectives for selection of an EDC system should be discussed, and these discussions captured by the team leader for future reference.

20.2.2. Objectives

Objectives for pursuing an EDC from a multidisciplinary EDC evaluation team may include:

- access to data in real time for adaptive design trials
- cleaner data and faster impact on time to market
- avoiding shipping of paper CRFs from remote locations
- synergy between serious adverse event reporting and the clinical database
- reduced on-site monitoring costs.

While decision points for decisions may seem premature at this initial planning stage, these objectives can guide discussions based on the adage "begin with the end in mind". The decision points should address realistic assessment parameters. For example, a small biotech operating with paper CRFs for current Phase I trials must evaluate the expected number and size of upcoming trials and logistical concerns of extracting paper CRFs from remote locations for large clinical trials. Resultant decision points should address which technology solutions are practical. Depending on the company size and on the scope of development in clinical trials, many of the decision points may be more logistical and financial, rather than data technology based; if the decision to explore EDC technology is deemed appropriate, these decision points will expand to include technological aspects of the systems under consideration following vendor reviews.

The EDC evaluation team should also explore:

- A general review of the company's current data management vendor or software/processes. Information on current processes could be provided to the EDC evaluation team as a presentation by team members from data management. In addition to establishing a baseline understanding of company needs in data management, a review will help the team to respond easily to one of the earliest questions from senior management: "Why is our current data management system not going to work?", if an EDC option is ultimately chosen to be explored.

- Initial contact with EDC vendors to gather some initial information on pricing and technology options.
- Timelines, both in the context of the clinical development plan and in terms of expectations around data cleaning and availability. While this topic will be revisited at various stages in the process, early discussion will help to establish expectations and set realistic goals for the EDC evaluation team for next steps and planning.

The team leader should record all decisions and establish a location on a shared drive accessible to both data and clinical team members to maintain minutes, vendor specifications, draft contracts, and other documentation. The team leader should also identify a back-up team member, in case regular work scheduling impacts on the progress of the EDC evaluation.

20.2.3. Management Interaction

If the decision is taken actively to explore an EDC option for the company, the team leader should arrange to meet with the appropriate senior management representatives, assuming they are not part of the EDC evaluation team. Without at least initial awareness and support from the appropriate level of management, further exploration should not continue. The team leader should provide initial timelines and present the objectives from the first meeting of the EDC evaluation team. To manage expectations, the criteria for vendor selection should be specified in advance.

20.3. ELECTRONIC DATA CAPTURE VENDOR SELECTION

The team leader should decide on timelines and next steps to inform initial vendor selection. If the team leader is a member of the data management functional area, many options for initial vendor selection will exist, including previous vendor organizations utilized, professional contacts, and attendance at any number of regional or national data management conferences. A number of conferences for e-clinical trials also offer the opportunity to hear presentations and view demonstations in an efficient manner.

20.3.1. Strategies for First Round Selection

Once some initial vendors have been selected, the team leader should construct a first round vendor selection specification. This specification should consist of a high-level yet targeted series of questions. The specification should be provided to all first round vendors before the initial demonstration or sponsor presentation. Providing targeted questions will result in a more efficient and targeted vendor presentation. The team leader should request that the EDC vendor responds to the series of questions, ideally before the first demonstration. All responses should be compiled and disseminated to members of the EDC evaluation team. An initial specification for initial vendor selection should address the questions in Table 20.2.

During the first round vendor demonstration, most EDC vendors will offer an in-house option presentation and will request the positions of participating team members in advance. As an in-house presentation is a considerable time commitment for the EDC evaluation team, the evaluation team should consider other options such as web-based presentations, if possible. If the EDC vendor does come in house, the EDC vendor should be instructed by the team leader to limit the presentation to a set period and

Table 20.2 Initial specification for electronic data capture (EDC) Vendor X (first round selection)

Company basics, stability parameters (years in operation, major industry clients)
What sets Vendor X's EDC system apart from other EDC vendors – in general and in {X} venue (e.g. in developing countries)?
Percentage of Vendor X's non-US studies which are hosted by Vendor X?
Summary of countries in which Vendor X has worked
Description of hosting and security
Approximate number of size of studies used with Vendor X's system
Timelines from contract to first patient in, etc. Additional metrics around timelines for past projects would be helpful as well
Please describe the major stumbling blocks to success in these countries, as applicable, and necessary accommodations around site training and user support
Length of Vendor X's average client relationship?
What metrics does Vendor X maintain to evaluate customer satisfaction? Provide a summary if possible
What metrics does Vendor X maintain around time/cost savings? Provide a summary if possible
Average amount of downtime for Vendor X's clients in the past two years
For participation in our initial demo, can Vendor X provide a data management representative, in addition to a business development representative?

to adhere as closely as possible to the series of questions provided by the EDC evaluation team. Also, the EDC vendor should be requested to bring a non-business development team member such as a project manager or a data manager if possible.

20.3.2. Initial Assessment of Technology

Following the initial presentations by first round vendors, members of the EDC evaluation team with technical expertise in database design and build or in database administration should begin to assess vendor database configuration options. The assessment should include questions raised by the initial vendor selection, and general questions regarding functionality, accessibility for in-house database builds, standard operating procedures (SOPs), and training requirements for database build. A targeted discussion with individual vendors from the first round may be requested and should be limited to vendor technical experts as well as the developers from the sponsor EDC evaluation team.

For sponsors conducting clinical trials in a low-power or developing country, this initial assessment of technology is expanded to include reviews and questions concerning the operation of the EDC system in select countries, or under certain connectivity challenges. Connectivity metrics for systems in similar settings should be requested and carefully reviewed. Also, risk assessment for data loss should be addressed and carefully explored. Finally, the database administrators on the EDC evaluation team should consider whether the database design or page layout makes the system increasingly cumbersome without sufficient bandwidth at the intended site.

Follow-up by quality assurance and IT to address server functionality, security, SOPs, and options such as maintenance of user accounts for the EDC vendors will also help to narrow down the list of vendors under consideration. Based on individual company SOPs, a quality assurance review of the SOPs for the vendor may not take place until final vendor selection; however, initial discussions relative to the in-house capacity of the sponsor are important to narrow the initial vendor roster. For example, a sponsor without an adequately staffed IT group will not be able to select a vendor that requires 24-hour in-house server maintenance and monitoring.

20.3.3. Team Review Process

Once first round vendors have been vetted, the team leader should hold a meeting of EDC evaluation team members and solicit their feedback.

Applying a Likert scale or "grades" to each vendor may be useful to the EDC evaluation team members in narrowing down the field. Additional questions for individual EDC vendors can also be developed at this time, and addressed via e-mail if required. The team leader should record all evaluations and decisions, and compile a shortlist of EDC vendors who will enter the second round review. At this stage, the team leader should provide an update on status to management and to the overall clinical development team, as appropriate, and also check and update timelines for the review process if needed.

20.4. NARROWING THE FIELD: STRATEGIES FOR SECOND ROUND SELECTION

From the vendors initially selected, a shortlist of vendors for second round reviews will be selected (Table 20.3). The number of vendors on the shortlist will be a reflection of the size of the initial list, but three to four vendors may be the maximum number for which further in-depth evaluation is possible.

Once second round vendors have been reviewed, the team leader should convene the EDC evaluation team members and solicit their feedback. Lists of pros and cons may be considered for each vendor. The team leader should provide an update on status to management and confirm timelines.

20.5. FINAL REVIEW

Assuming the second round selection results in two remaining vendors, the team leader may choose one or more of the following elements to incorporate into final EDC vendor selection. Options for reviews include:

- **contracts review**: in-house meeting with sponsor finance and contract team members; review of all cost parameters (multisite license options, restrictions on numbers of end users, charges for data exports, archiving costs, project management costs, hosting fees, end-user license fees)
- **quality assurance review of SOPs and corrective action strategies for deviations**: in-house audit conducted by sponsor QA group to review SOPs
- **on-site testing** of technology
- **database build**: if timelines allow, a mock database build by both final round vendors will allow review of edit checks, screen design, and output

Table 20.3 Second round selection options

On-site testing of technology	Request that the vendor provide a UAT environment for hands-on use by EDC evaluation team members
Request interviews with current sponsors	Requesting references earlier in the process may provide additional insight and result in additional questions to be addressed in the second round review
Clinical site and study monitor feedback	For a sponsor working with established sites and study monitors, the second round selection may provide a good opportunity to observe a "typical" site interacting with the system. Can user reports be exported? How user-friendly is the interface?
Data export functionality	How readily can the sponsor export data, and in what format. Are exports CDISC compliant and user-friendly?
Costing profiles	Request initial cost estimates from the EDC vendor using existing clinical trials design and number of subjects. Compare current vendor cost, as applicable, to these estimates
Technology reviews	Set up a separate discussion for database developers to focus on database build efficiencies or issues
Technology roll-out and implementation plan	Does the vendor have an implementation plan? For a company considering a roll-out of new technology and new SOPs, such a plan would be a time-saving tool
Targeted demo	Request that the vendor perform an additional demonstration of only a certain component of their system. For example, loading external laboratory data into the clinical database, integration of an IVR flat file

UAT: user acceptance testing; EDC: electronic data capture; CDISC: Clinical Data Interchange Standards Consortium; SOP: standard operating procedure; IVR: interactive voice response.

- **interviews** with current customers
- **staff training strategies**: in-house meeting with site trainers to review their strategies for training; options for training of remote sites
- **user acceptance testing (UAT)**: additional hands-on work in a UAT environment.

Table 20.4 Electronic data capture (EDC) interview questionnaire

Date of reference:
Name of reference:
Title/Position in company:
Type of company:

(1) Background: So I can better understand the relative positions of sponsor's companies, please briefly describe the size/structure of your clinical data management group, overall focus (e.g. CNS, oncology), and any special considerations (e.g. specialize in orphan drug status, use of outside programming consultants, multinational groups, etc.).
   a. Could you please describe your department prior to initiating {EDC vendor} (e.g. were you using all in-house CDMS?)
   b. Are you using {EDC vendor} with a vendor hosted option?

(2) Implementation:
   a. When did you roll out {EDC vendor}for your group? (Month/Year)
   b. How long did it take to roll it out from signing of contract to live data in-house for your first study?
   c. Did you perform a pilot study with {EDC vendor}?
   d. How satisfied overall were you with the implementation experience on a scale of 1 to 5 (5 = outstanding)? What issues occurred during the implementation, if any?
   e. How would you rate {EDC vendor}'s performance on a scale of 1 to 5 with respect to project and timeline management during the implementation process?
   f. Did {EDC vendor} provide you with a consistent project management team during the entire process?

(3) Vendor selection:
   a. Please describe the two primary reasons you selected {EDC vendor}. On a scale of 1 to 5, please indicate to date how well your expectations have been met with regard to each reason.
   b. If rating less than 4, please briefly describe the pitfalls or issues which have stood in the way of attaining your primary goal(s).

(4) Internal or external measures of success:
   a. What measures of success/criteria do you use to evaluate the performance or value-added benefit of {EDC vendor}for your organization?
   b. Has performance met with your expectations?
   c. Do you feel {EDC vendor}is providing a good value for your investment based on these criteria?

(5) On a scale of 1 to 5, how satisfied is your staff with {EDC vendor}?

(6) Given the opportunity to change one thing about the {EDC vendor} product, what would it be? Final thoughts or lessons learned?

CNS: central nervous system; CDMS: clinical data management system; EDC: electronic data capture.

20.5.1. Key Strategies

A pilot study or on-site testing of technology may consist of a demonstration by the final round vendor candidate, conducted live at a current sponsor site, using a real database designed for the sponsor or a UAT database. This opportunity will test Internet connectivity issues, solicit real-time site feedback, and allow the site, the sponsor, and the EDC vendor a hands-on opportunity to explore a working example of both the EDC technology and their teams' interaction potential.

Interviews with current customers are a cost-effective way to assess the current use of the system, areas of weakness in the implementation of the system, and any growing pains that occurred. Table 20.4 provides a sample interview questionnaire.

20.5.2. Team Assessment and Final Selection

Following contracts and other final round review, the team leader should distribute an evaluation matrix to all members of the EDC evaluation team. A sample matrix is provided in Table 20.5. The matrix should rate the two final round vendor candidates side by side, and should include cumulative feedback as received in all vendor reviews.

Once all feedback is received, the team leader should generate summary statistics and release the findings to the EDC evaluation team. A formal presentation to management should follow. If possible, the selected EDC vendor should be asked to attend, particularly if a contracts review will follow.

Table 20.5 Electronic data capture (EDC) vendor evaluation matrix: final selection

	Rating/description	
Element for evaluation	**Vendor X1**	**Vendor X2**
Database builds		
Structure of database (relational, hierarchical)		
Programming language: commonalities with current data management system, other		
Adaptability of prepackaged options for builds, GUI, other		
Number of developers required to build a study (build by sponsor, build by vendor)		
Database build training		
Prepared training available to bring database build in-house		

(*Continued*)

Table 20.5 Electronic data capture (EDC) vendor evaluation matrix: final selection—cont'd

Element for evaluation	Rating/description	
	Vendor X1	Vendor X2
Average training time/impact on time to first in-house database build		
Ongoing support/mentoring option		
Supporting the sites		
How is site assessment accomplished in remote locations? Is there any burden on the site IT or existing infrastructure?		
Evaluate how robust the overall system is for the end user, e.g. flexibility in design options, level of automation on edits and configuration, other?		
Describe any hardware or other on-site requirements		
Can workflows and/or user roles be easily configured to match current work processes at individual sites?		
Comment on seamlessness of communications interface (if connection is lost, what happens on the user end?). Comment on the quality of the system-monitoring applications used. Can alerts be sent to sponsor as well as to the site?		
Comment on overall quality of eCRFs. User-friendly interface? Dynamic pages/visit possible?		
Change control process considerations; how long (if at all) is the site workflow impacted by changes made to a live database? Describe process post go-live including testing, validation, UAT, etc.		
Describe process for bringing a new site on board mid-study		
Describe a mid-study change to a live database		
Connectivity		
Real-time evidence of connectivity speed at sites		
Network latency. Options for web performance acceleration, optimization, and impact on latency?		

Table 20.5 Electronic data capture (EDC) vendor evaluation matrix: final selection—cont'd

Element for evaluation	Rating/description	
	Vendor X1	Vendor X2
Describe inbound and outbound communication		
Resolution of branching mechanism		
Burden on the site – is the connectivity transparent from their perspective?		
Review of bandwidth requirements		
Scalability		
Number of kilobytes per eCRF (e.g. evaluate in terms of x kilobytes per page). Request all technical specifications available		
Screen refresh times, request any end-user statistics maintained by vendor		
Downtime statistics (other studies), if available		
Recommendations on first and second connectivity options (and back-up)		
Database elements (front and back end)		
Quality of edit checks, ease of use (front end), and ease of updates as required		
User-friendly self-designing of custom edits (sponsor designs)		
Overall quality of online reports, other reporting tools		
User-friendly self-designing of custom reports (sponsor designs)		
Regulatory		
21 CFR Part 11 compliance (review certificates), confirm 128-bit encryption SSL and other encryption algorithms		
User passwords maintenance		
Audit trails		
Password options (e.g. biometric fingerprints to save the site time in logging in)		
SOPs available for adoption/purchase?		
Training		
Interview dedicated training staff member		

(Continued)

Table 20.5 Electronic data capture (EDC) vendor evaluation matrix: final selection—cont'd

Element for evaluation	Rating/description	
	Vendor X1	Vendor X2
Review of a training plan or other training materials		
Approach to training data management (not DB build training), typical timelines for various components		
Recommended approach to training monitors (using their own PCs)		
Approach to training study sites (nurses, etc., who are not tech savvy)		
Online training manuals		
Flexibility, degree of customization possible if required, follow-up training		
Outside the box solutions for trainings held outside the USA		
Technical support		
Language support		
Locations (time zones), who staffs their various help centers, number of staff supporting Africa?		
Tracking of support calls, how relayed to sponsor		
Interaction/accessibility, ease of communication – please base this on pilot study		
Data uploads, coding, and archiving		
Describe batch uploads		
Lab data, loading of lab normals		
Interface with IVRS vendors		
Accept XML (E2B formatted) imports from safety systems?		
Ease of performing medical coding for AEs, Conmeds; is a separate dictionary license required by the sponsor to code?		
Adoption of sponsor coding conventions (e.g. can it support multiple dictionaries?)		
Define process for archiving a study		
Data export procedures		
Ease of use of export procedures, options for file type export (SAS, .csv)		

Table 20.5 Electronic data capture (EDC) vendor evaluation matrix: final selection—cont'd

Element for evaluation	Rating/description	
	Vendor X1	Vendor X2
Is there a programmed template available for SAS exports? Is it fully validated?		
Is adaptation of CDISC standards addressed by their tools?		
Servers		
Location of servers for optimizing speed		
Evaluation of redundancies (with sponsor IT)		
Vendor maintenance schedules		
Security of hosting environment		
Validation (assume vendor-provided except UAT)		
Describe validation process and documentation (general and system-specific)		
Additional timeline for validation		
UAT test scripts preparation (who will write these?), UAT implementation, timelines. Number of typical UAT cycles?		
Process for validation of change made to a live database		
Pretraining for UAT (discuss general process flow for the UAT, perhaps in conjunction with training components)		
License fees		
Address under contracts review		
Corporate (general)		
Date of initial commercial release of software		
Number of clients similar to sponsor, e.g. small biotechs? Number with sites outside the USA?		
Largest studies (number of patients, number of pages) to date?		
Commitment of senior management to project		
Review of case studies, references (relevant in company size/scope)		
Stability, financials		
Study management and timelines		

(*Continued*)

Table 20.5 Electronic data capture (EDC) vendor evaluation matrix: final selection—cont'd

	Rating/description	
Element for evaluation	**Vendor X1**	**Vendor X2**
Review timelines for large and small studies. Comment on demonstrated quality of planning/timelines and other recommendations. Was any outside the box thinking demonstrated in discussions?		
Review of personnel on core team supporting each study		
Adequacy of staffing and availability of PMs, e.g. number of studies the typical PM is responsible for		
Staff average time with company, attrition rates		
Staff proximity to sites/familiarity with non-US site issues		
Is the PM tech or clinical trial savvy, or just a timeline monitor?		
Speed of returned e-mail/phone call request for information		
Discussion on changes to timelines and impact on budget, etc.		

GUI: graphical user interface; eCRF: electronic case report form; UAT: user acceptance testing; CFR: Code of Federal Regulations; SSL: secure sockets layer; SOP: standard operating procedure; DB: database; IVRS: interactive voice response system; AE: adverse event; CDISC: Clinical Data Interchange Standards Consortium; IT: information technology; PM: project manager.

INDEX

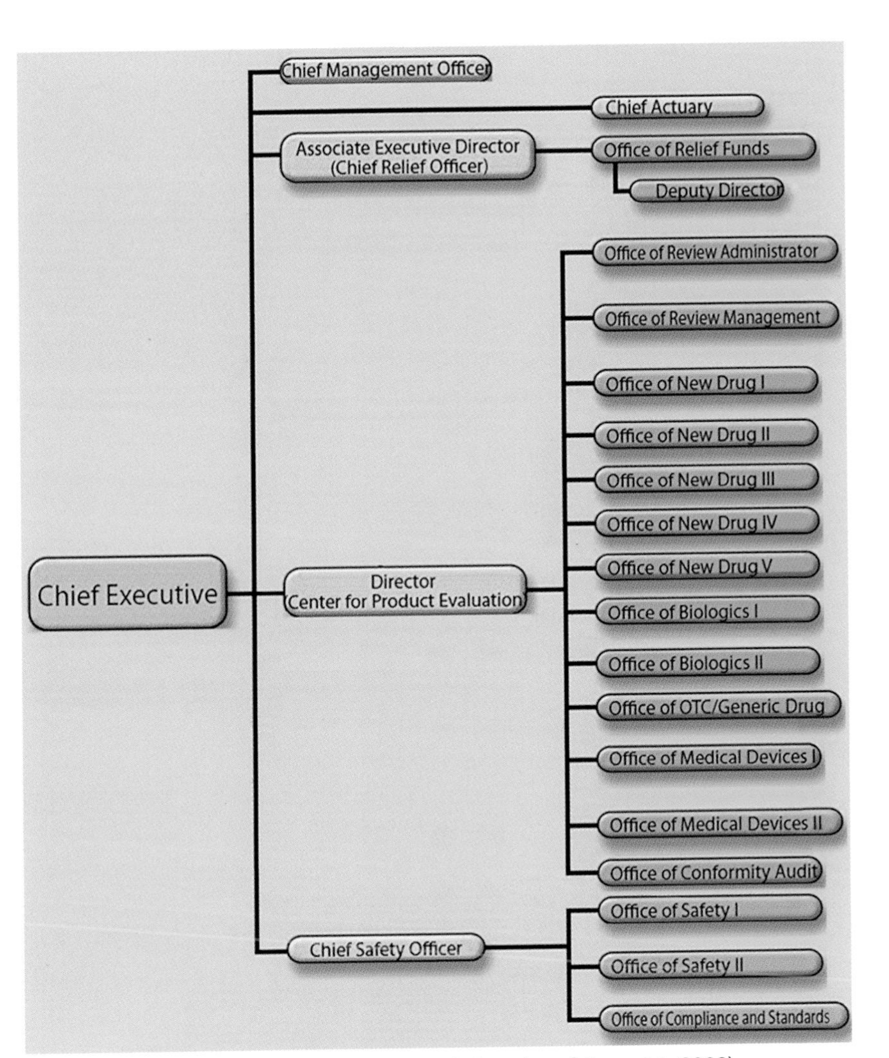

Figure 5.1 PMDA organizational chart (as of August 1, 2009).

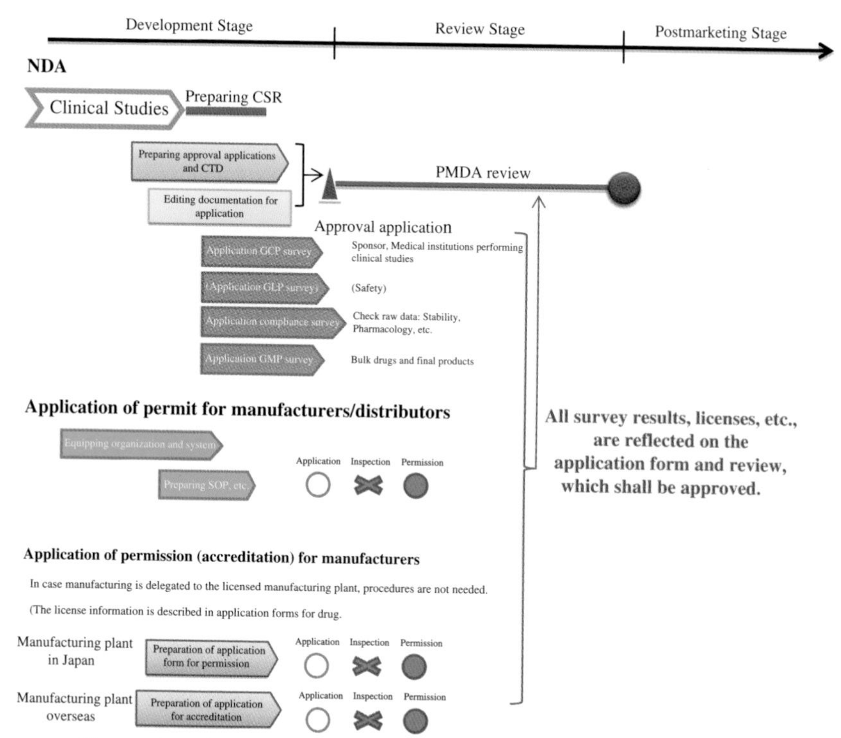

Figure 5.4 Pharmaceutical procedures in initial approval applications.

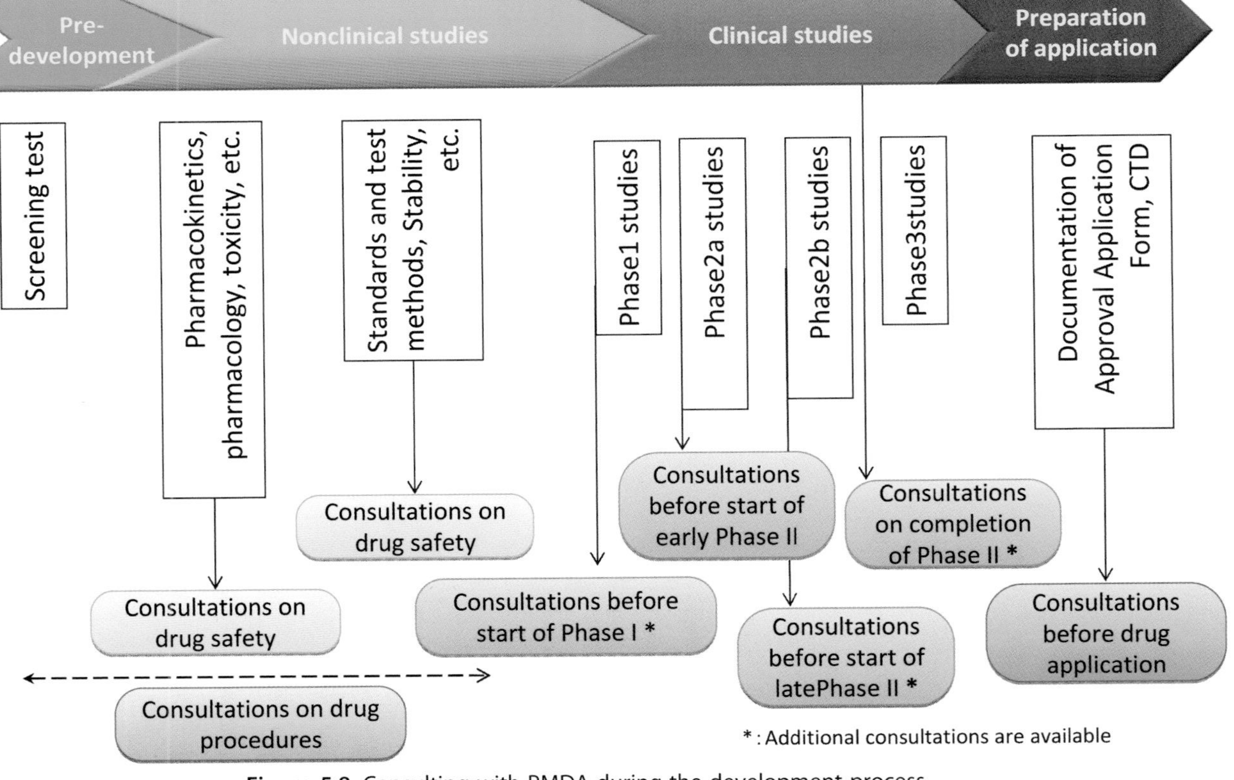

Figure 5.8 Consulting with PMDA during the development process.

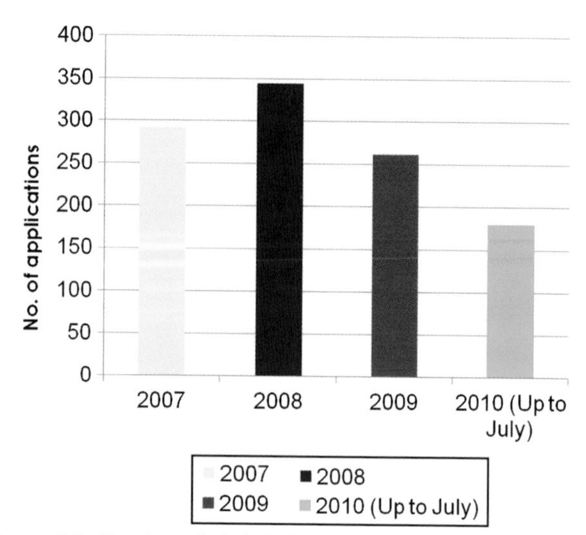

Figure 6.1 Number of global clinical trial applications received.

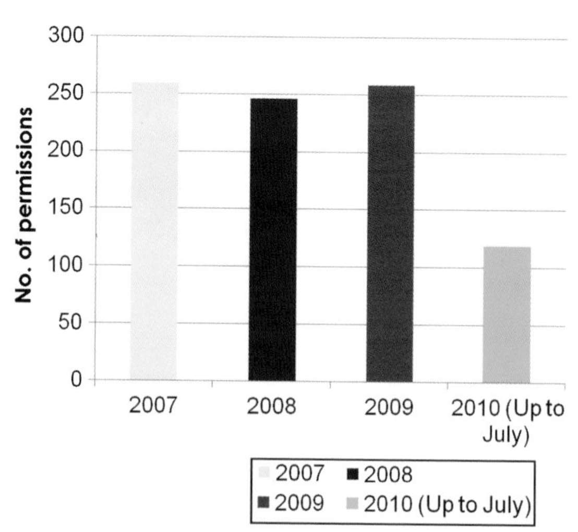

Figure 6.2 Number of global clinical trials granted permission.

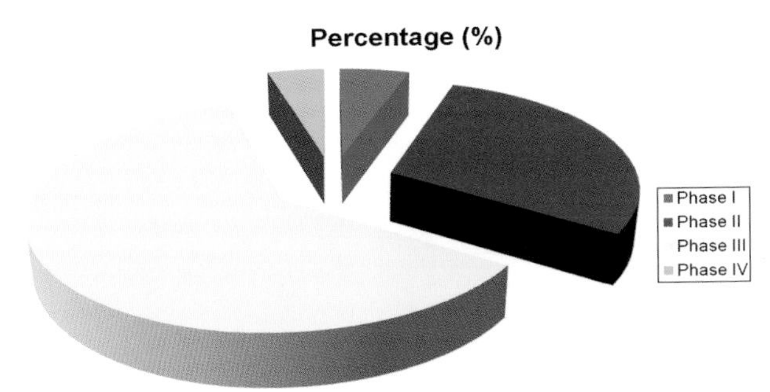

Figure 6.3 Phase-wise breakdown of clinical trials conducted in India.

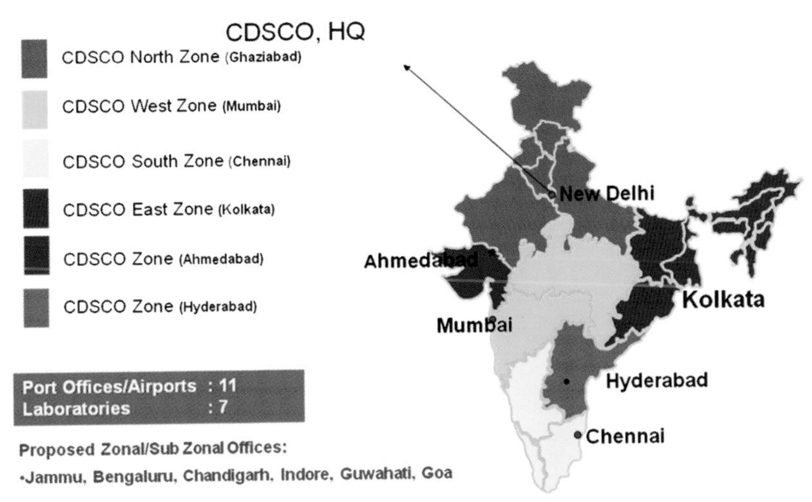

Figure 6.5 Geographical locations of Central Drugs Standard Control Organization (CDSCO) offices in India.

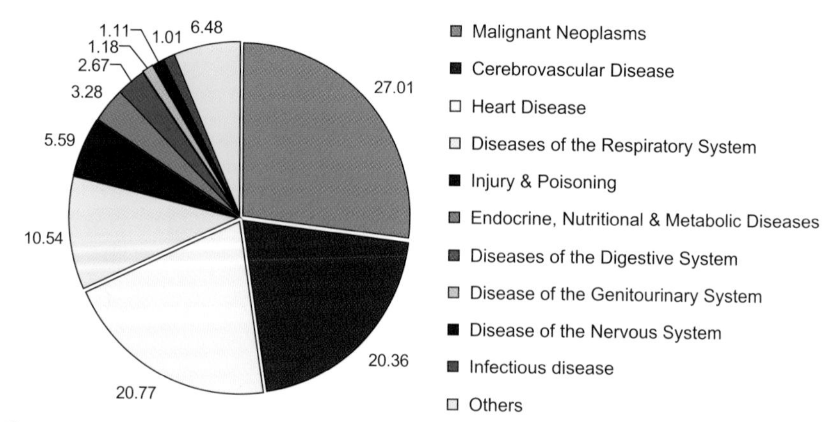

Figure 9.1 Percentages of the 10 main diseases causing death in China's cities in 2009.

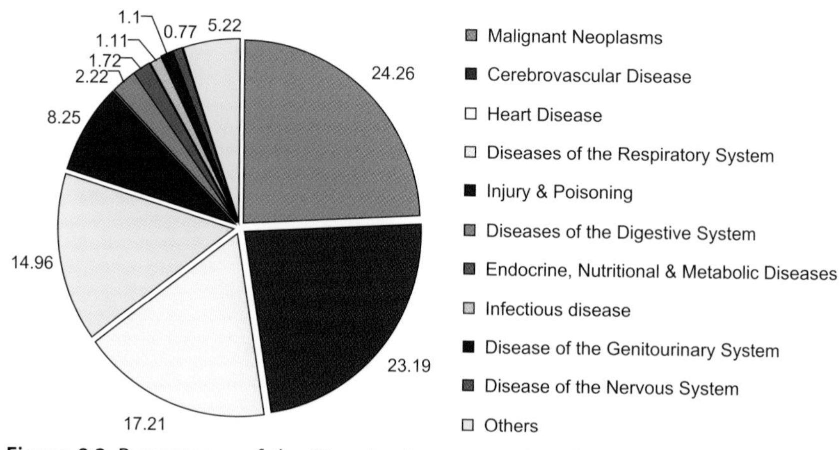

Figure 9.2 Percentages of the 10 main diseases causing death in China's counties in 2009.

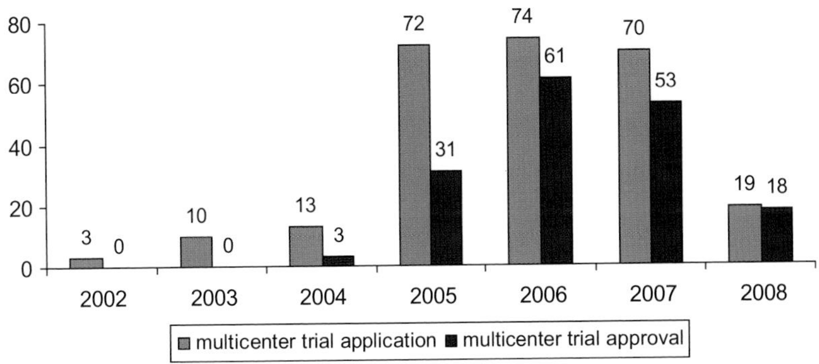

Figure 9.3 Number of multicenter trials in China from 2002 to the first half of 2008 *(Source: adapted from China Prescription Drug, 2009).*

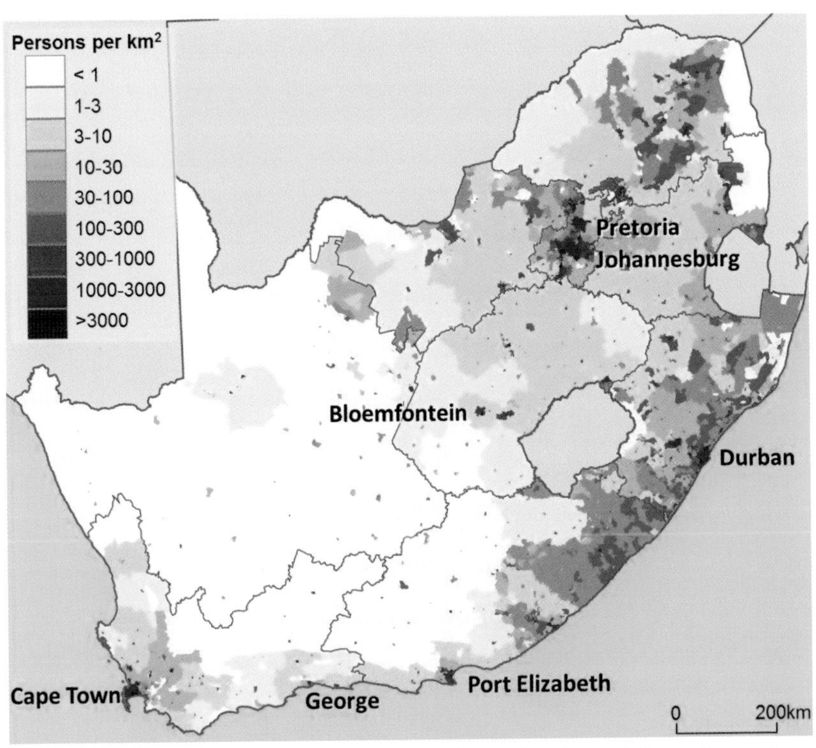

Figure 13.1 Map of South Africa showing population density and the major centers of clinical trial activity *(Source: Adapted from Statistics South Africa).*

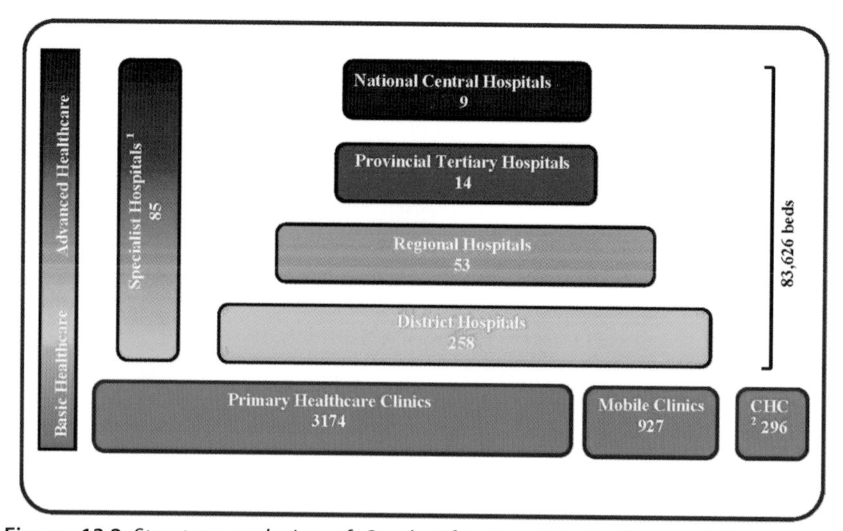

Figure 13.2 Structure and size of South Africa's public healthcare infrastructure. Specialist hospitals include TB, psychiatric/mental health units, and other specialties; CHC: community healthcare clinic. *(Source: Adapted from District Health Barometer, 2008/09).*

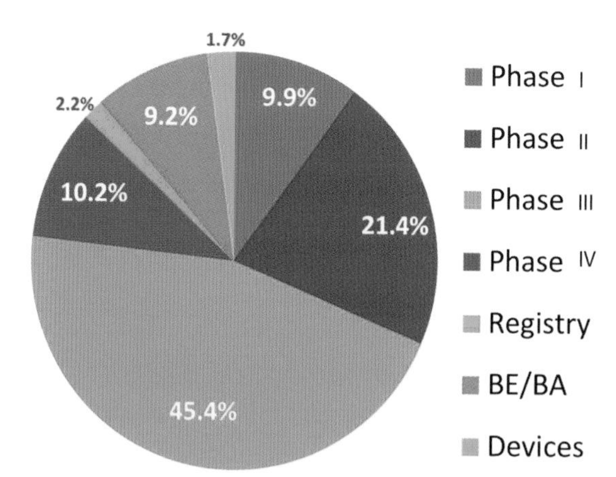

Figure 13.5 Clinical trials registered on the SANCTR between July 2005 and October 2010: breakdown by phase/type *(Source: Data extracted from SANCTR).*

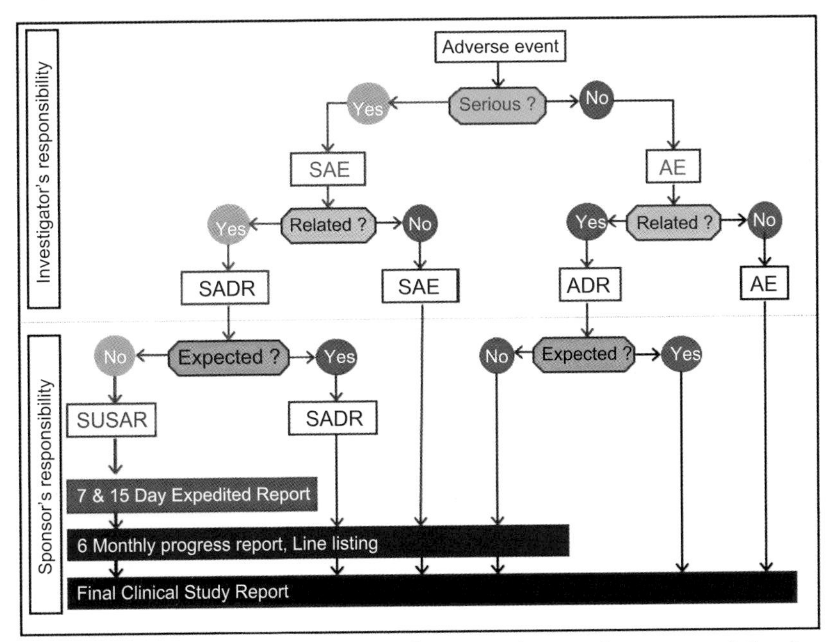

Figure 13.6 Algorithm showing MCC safety reporting requirements and timelines. BE/BA: *(Source: Adapted from MCC regulations).*

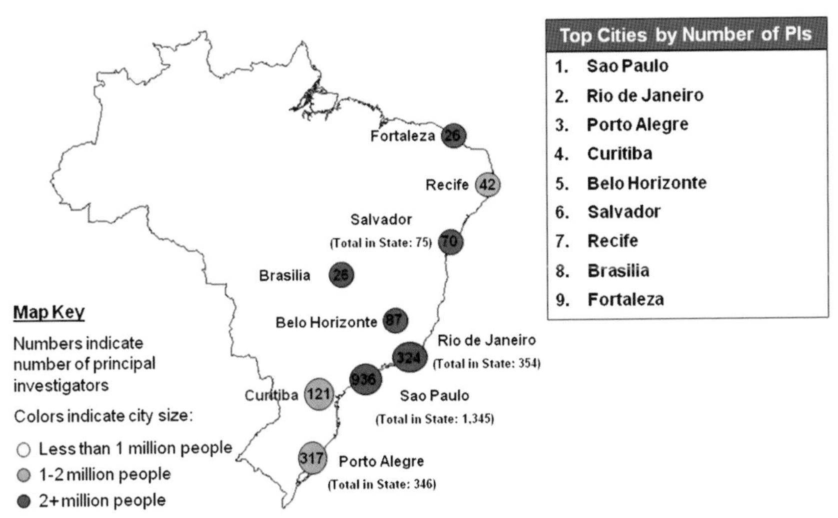

Figure 14.1 Principal investigators in Brazil.

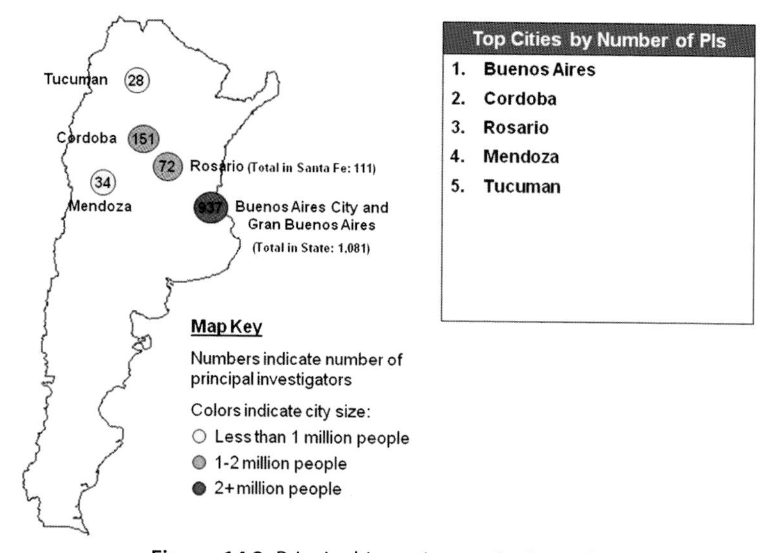

Figure 14.2 Principal investigators in Argentina.

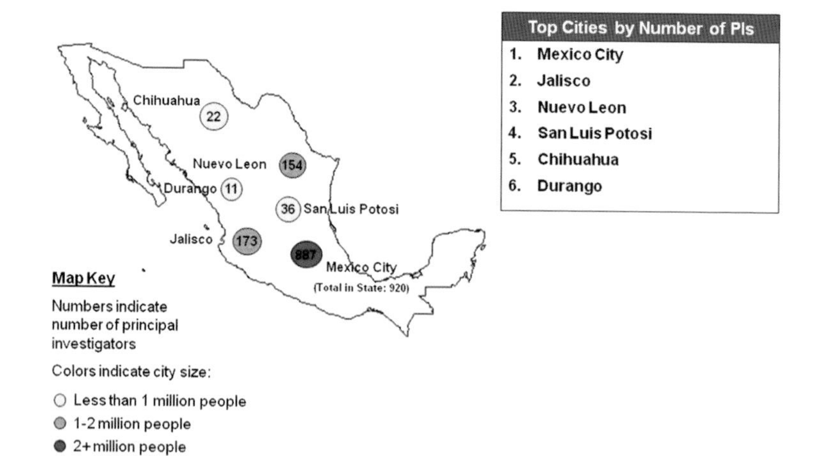

Figure 14.3 Principal investigators in Mexico.

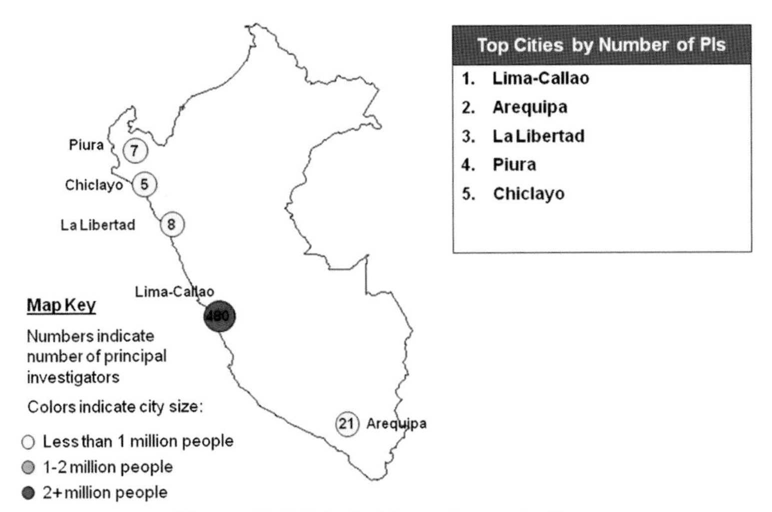

Figure 14.4 Principal investigators in Peru.

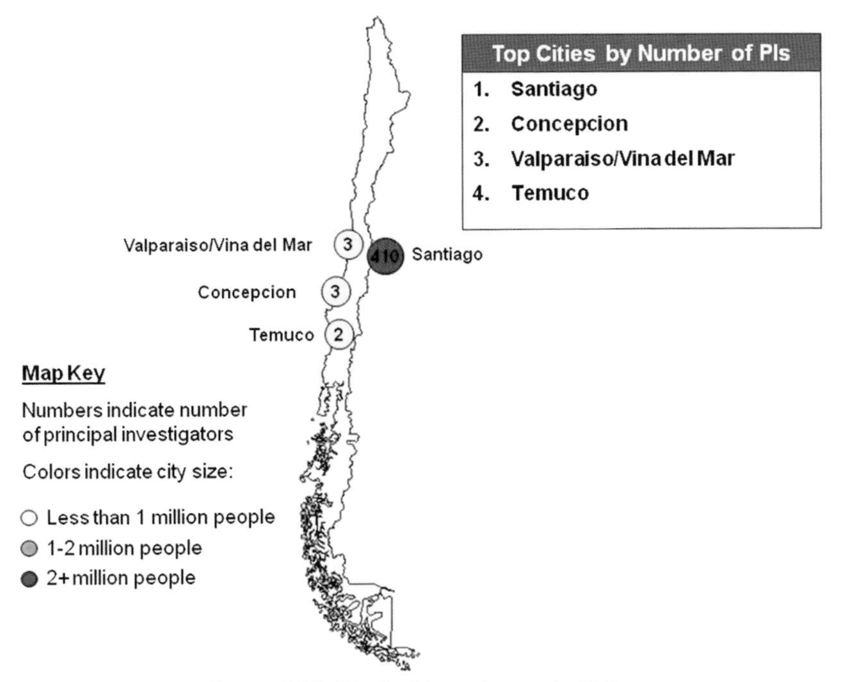

Figure 14.5 Principal investigators in Chile.

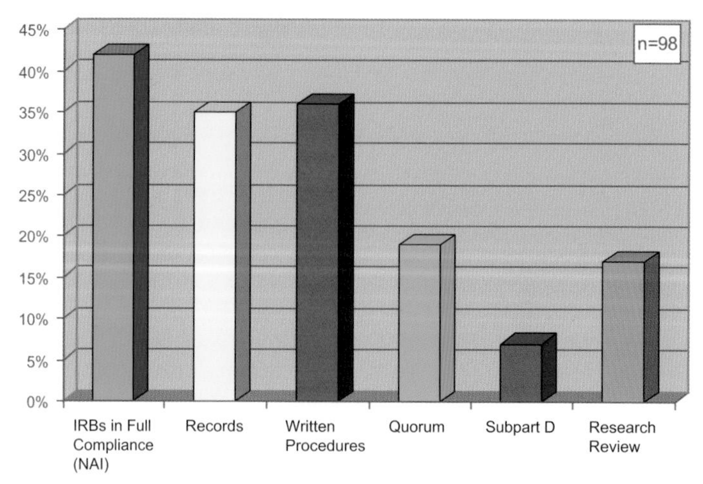

Figure 19.1 Top institutional review board deficiencies as per Center for Drug Evaluation and Research, fiscal year 2009. NAI: No Action Indicated *(With permission from Barnett International).*

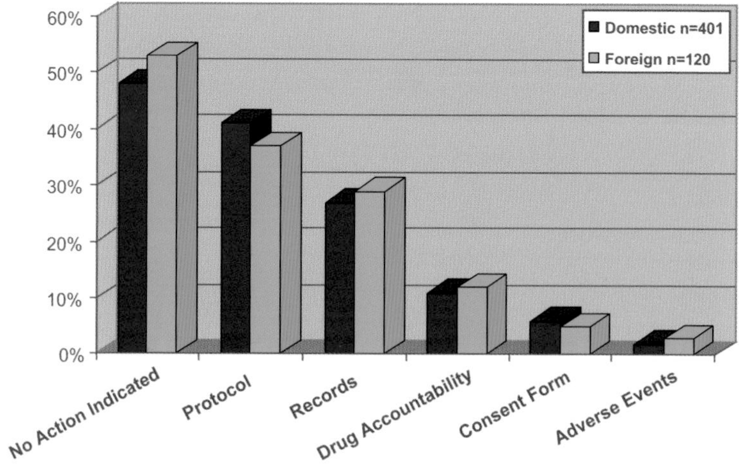

Figure 19.2 Top clinical investigator deficiencies as per Center for Drug Evaluation and Research Inspections, fiscal year 2009 *(With permission from Barnett International).*

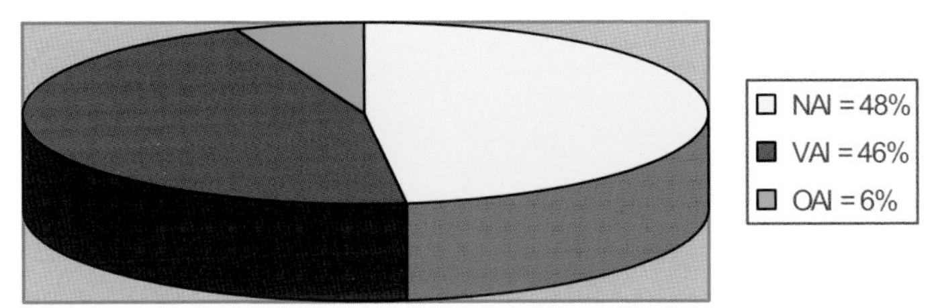

Figure 19.3 Clinical investigator inspections: final classification, fiscal year 2009. Total inspections with final classification = 417; NAI: No Action Indicated – sites in full compliance; VAI: Voluntary Action Indicated – sites found to have objectionable practices that do not represent major departures from regulations; OAI: Official Action Indicated – sites/investigators found to have objectionable practices or conditions that represent significant departures from regulations *(With permission from Barnett International)*.

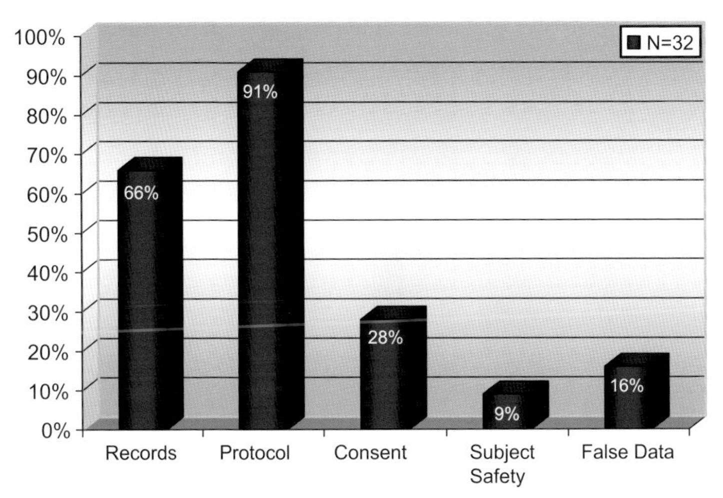

Figure 19.4 Clinical investigator deficiencies as per Center for Drug Evaluation and Research Inspections, fiscal year 2009: final classification of Official Action Indicated (OAI) *(With permission from Barnett International)*.